# BARRON'S

# NEW JERSEY

GRADE 8

# MATH TEST

**Judith T. Brendel, M.Ed.**
**CCSS Mathematics Specialist in Curriculum and Assessment, Foundation for Educational Administration (FEA) and The Madison Institute (TMI)**

## About the Author

Judith T. Brendel is former District Supervisor of Mathematics and Art at the Pascack Valley Regional High School District in New Jersey, and chaired their K–8 Math Articulation Committee. She has taught art and math in all K–12 grade levels, plus graduate math- and technology-education courses at Fairleigh Dickinson University in N.J., Mercy College in N.Y., and at the Rutgers New Math/Science Teacher Institute. Judith is also a member of the Board of Directors of FEA and former member of the Board of Directors of NJPSA (Principal Supervisors Association) and former board member of AMTNJ (Association of Math Teachers of New Jersey.) She is a frequent presenter at national and state conferences and is an evaluator of math textbooks for middle and high school students. Formerly, Judith was a member of the New Jersey Coalition of Concerned Math Educators, recommending changes to the Pre-K to 12 Core Curriculum Content Math Standards and, more recently, to the new Common Core State Standards.



*All inquiries should be addressed to:*

Barron's Educational Series, Inc.
250 Wireless Boulevard
Hauppauge, New York 11788
**www.barronseduc.com**

ISBN: 978-1-4380-0558-4

PCN 2014950959

Printed in the United States of America

9 8 7 6 5 4 3 2 1

# Contents


**Introduction** 1

- About the New Jersey Grade 8 Math Test 1
- Summative Assessment Components 1
- About This Book 1
- Types of Questions on the PARCC 2
- Material from Grade 6 and Grade 7 11
- Score and Test Results 11
- Helpful Resources 11
- Calculators 12
- Accommodations and Modifications 12

**Chapter 1**
**The Number System, 8.NS** 13

- What Do PARCC 8 Number System Questions Look Like? 13
- Our Number System (Number Sets) 16
- Rational and Irrational Numbers 18
- Numerical Operations 20
- Multiplying and Dividing Positive and Negative Numbers 20
- Exponents, Square Roots, and Cube Roots (New and Important!) 23
- Estimating Square Roots 32
- Scientific Notation 33
- Computating with Scientific Notation 36
- Adding and Subtracting Numbers in Scientific Notation Form 38
- Comparing Data in Real-Life Situations 39
- Algebraic Order of Operations (A Review) 41
- Ratio and Proportion (from Grade 7.RP and 7.G1) 44
- Chapter 1 Test: The Number System 53

**Chapter 2**
**Geometry, 8.G** 59

- What Do PARCC 8 Geometry Questions Look Like? 59
- Points, Lines, Angles, and Planes 62
- Relationships of Lines 66
- Planes 67

- Polygons: Area of Two-Dimensional Shapes 79
- Area of Irregular Shapes (Some from Grade 6.G.1) 85
- Volume of Three-Dimensional Forms 94
- Perimeter 98
- Triangles and Other Polygons 104
- Right Triangles 111
- Coordinate Geometry 116
- Area and Perimeter on the Coordinate Plane 117
- Congruency 121
- Transforming Shapes 124
- Chapter 2 Test: Geometry 142

**Chapter 3**
**Expressions, Equations, and Functions, 8.EE and 8.F 151**

- What Do PARCC 8 Expression, Equation, and Function Questions Look Like? 151
- Monomials, Terms, and Expressions (From Grade 7.EE.4) 155
- Simplify and Evaluate Expressions (From Grade 7.EE.1) 156
- Equations 157
- Functions and Relations 160
- Slope and Rate of Change 168
- Systems of Equations 177
- Combining Algebra and Geometry 181
- Writing Expressions and Equations 186
- Real-Life Applications 187
- Chapter 3 Test: Expressions, Equations, and Functions 192

**Chapter 4**
**Probability and Statistics, 8.SP 197**

- What Do PARCC 8 Probability and Statistics Questions Look Like? 197
- Facts You Should Know 200
- Probability and Statistics 201
- Experimental Probability 203
- Finding Probabilities of Outcomes in a Sample Space 206
- Finding Probabilities of Events 207
- Data Collection and Analysis (A Review from Grade 6.SP.5C) 208
- Scatter Plots (Scattergrams) 210
- Circle Graphs 214
- Chapter 4 Test: Probability and Statistics 222

**Chapters 1–4 Solutions** **229**

**Practice Test Performance-Based Assessment Session I (No Calculator Permitted.)** **279**

**Practice Test Performance-Based Assessment Session II (Calculator Permitted.)** **288**

- Performance-Based Assessment Answers Session I 303
- Performance-Based Assessment Answers Session II 309

**Practice Test End-of-Year Assessment Session I (No Calculator Permitted.)** **319**

**Practice Test End-of-Year Assessment Session II (Calculator Permitted.)** **329**

- End-of-Year Assessment Answers Session I 341
- End-of-Year Assessment Answers Session II 345

**APPENDIX**

- A: Grade 8 Common Core State Standards 349
- B: Mathematical Practices 355
- C: Additional Online Resources 357
- D: PARCC Grade 8 Mathematics Assessment Reference Sheet 359

# Introduction

## About the New Jersey Grade 8 Math Test

This mathematics test is actually a series of assessments for all students. PARCC assessments are designed to indicate the progress students are making in mastering the knowledge and skills specified in the Common Core State Standards (CCSS) and is an indicator for identifying 8th grade students who may need instructional intervention in two content areas: language arts literacy and/or mathematics. The tests will help teachers, parents, and students know how students are progressing toward being prepared for college and/or a career after high school. The PARCC design includes four components–two required summative and two optional non-summative assessments.

## Summative Assessment Components

- ***Performance-Based Assessment*** *(PBA)* is administered after approximately 75% of the school year (March/April). The PBA will focus on applying skills, concepts, and understandings to solve multi-step problems requiring abstract reasoning, precision, perseverance, and strategic use of tools.
- ***End-of-Year Assessment*** *(EOY)* is administered after approximately 90% of the school year (May/June). The EOY will call on students to demonstrate further conceptual understanding of the Major Content and Additional and Supporting Content of the grade/course (as outlined in the PARCC Model Content Frameworks), and demonstrate mathematical fluency, applicable to the grade. The grades 3–8 assessments will include a range of item types, including innovative constructed response, extended performance tasks, and selected response (all of which will be computer-based).

## About This Book

This book is designed for use by students in school, at home, or for use in tutorial or remedial sessions. It is designed to review the math skills and applications that students should master before their PARCC 8 exam. This book also is a resource for those 9th grade students who are in a high school remedial program because they scored less than proficient on their PARCC 8 math exam or as a resource for Pre-Algebra, Basic Algebra, or Resource Room math students. It also is recommended for private and parochial school 7th and 8th graders who plan to attend a public high school.

Each chapter includes the different types of questions used on the PARCC 8: (MC) multiple-choice, (NMC) the new type of multiple choice, (CR) constructed response, and (ECR or PBT) extended response or performance-based tasks. Each chapter includes an introduction and review followed by practice examples. At the end of each chapter there are sections with additional constructed response questions, extended performance tasks, and a chapter test.

At the end of the book you will find a full-length test modeling a PARCC PBA (Performance-Based Assessment) and another for the PARCC EOY (End-of-Year Assessment). These sessions give students experience with the variety of question types and the content that will be on their online PARCC assessments.

## Types of Questions on the PARCC

### Short Constructed-Response Questions (SCR)

Short constructed response questions are questions where the student must write (or type) the answer without having answer choices. There will be a space provided for students to give their responses.

### Performance-Based Questions

These questions are quite different. They require students to write, chart, or graph their responses to questions and often require a written explanation to demonstrate understanding. Each performance-based question may be worth 2, 3, 4, 5, or even 6 points.

### Multiple-Choice Questions

There will still be traditional multiple-choice questions with one correct answer chosen from lettered options.

New multiple-choice type questions may take a variety of formats. They also may have more than one correct answer.

> (In this book you can write your explanation on paper. When taking the online PARCC assessment you will be able to draft your answer on scrap paper, but then you must type your answer on the computer. The scrap paper is not part of the test.)

For more on what the online examples or types of questions look like, see the following pages.

# WHAT DO ONLINE PARCC 8 QUESTIONS LOOK LIKE?

(1) This is a ***fill-in type question with only one correct answer***.

In this particular question you are asked to solve an equation (on scrap paper) and to enter **only** your solution on the computer screen. You will be able to click and drag the numbers and symbols you need into the open space provided on the screen. You usually can also type the numbers and symbols from your keyboard.

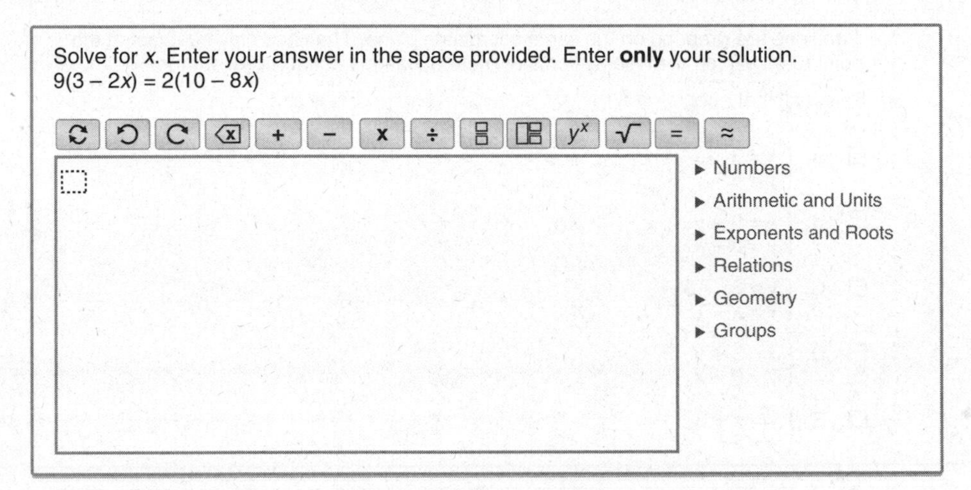


(2) The next question type is a typical ***multiple-choice question with one correct solution***. You do your work on scrap paper (no calculator is permitted here), and then click in one of the small circles to select your answer (A, B, C, or D). When there is one correct answer, your bubble-in symbols will be circles.

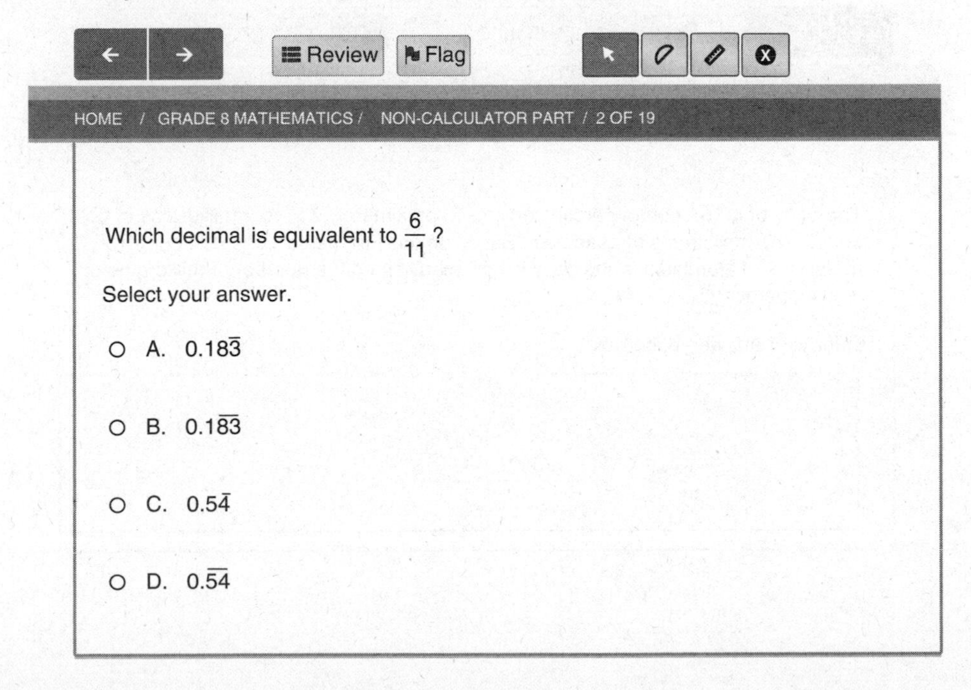

(3) This is a "new" type of multiple-choice question. It has ***more than one correct answer***. You are expected to ***select all that apply***; select all that are correct. When there is more than one correct answer, your bubble-in symbols will be small squares.

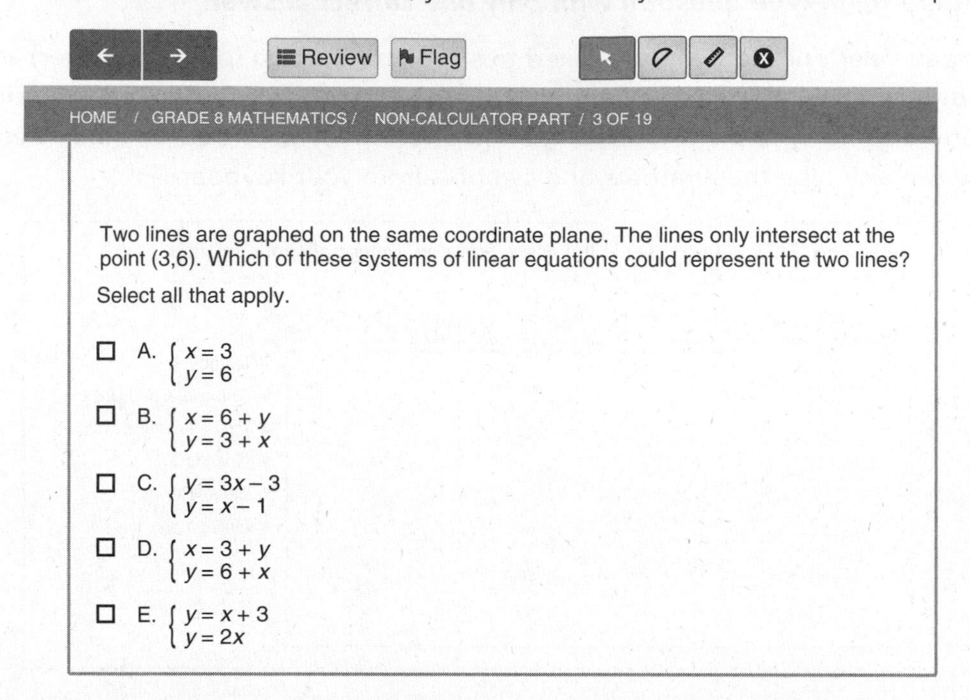


(4) This kind of question is a ***word problem*** with ***one correct answer***.

Here, you do your work on scrap paper and then just ***type your answer*** in the box on the screen.

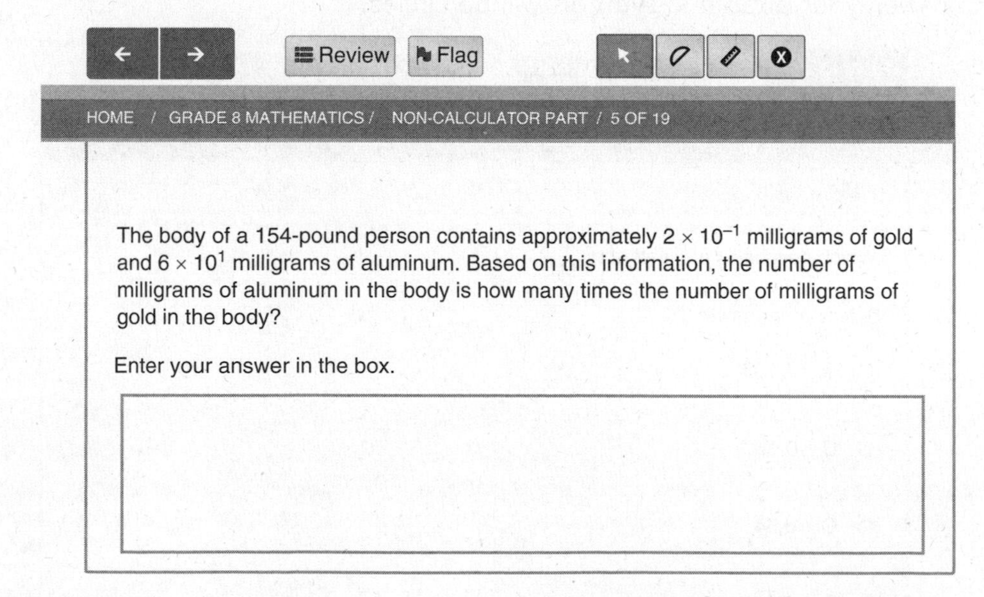

(5) For some questions you need to ***check off multiple answers in a table***. In this example you are given six different sections of a function; you should select an answer for each section and ***click*** the appropriate boxes.

The graph shows $y$ as a function of $x$.

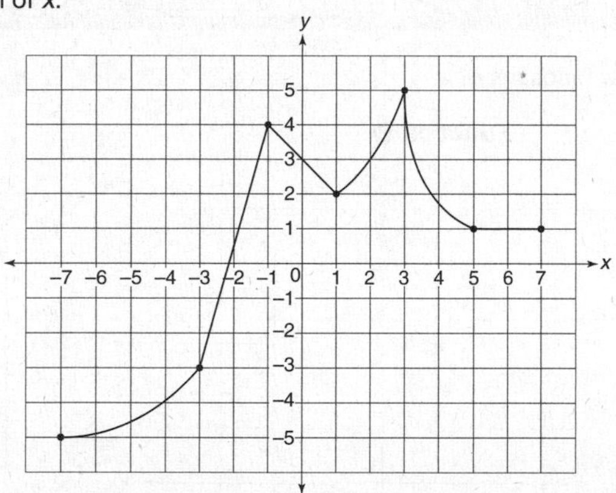


For each interval in the table, indicate whether the function is increasing, decreasing, or neither increasing nor decreasing over the interval.

| Interval | Increasing | Decreasing | Neither Increasing nor Decreasing |
|---|---|---|---|
| $-7 < x < -3$ | | | |
| $-3 < x < -1$ | | | |
| $-1 < x < 1$ | | | |
| $I < x < 3$ | | | |
| $3 < x < 5$ | | | |
| $5 < x < 7$ | | | |

(6) Sometimes you will be asked to **compare two sets of data given in different formats**. Here you are comparing a function W shown in a graph, and a function Z shown as a table of values.

HOME / GRADE 8 MATHEMATICS / CALCULATOR PART / 5 OF 13

Functions W and Z are both linear functions of *x*.

**Function W**

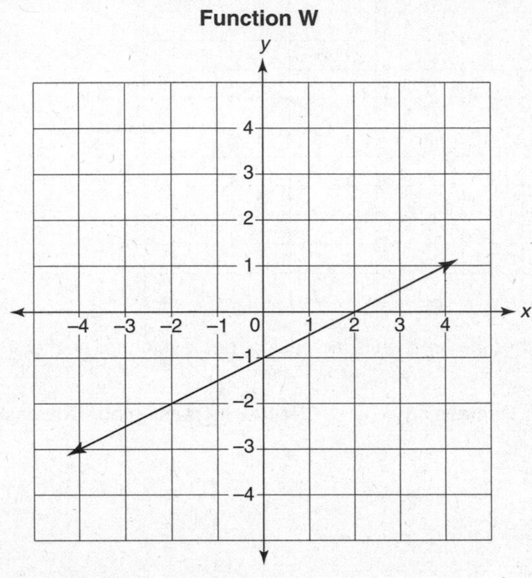


**Function Z**

| *x* | *y* |
|---|---|
| −2 | −2.5 |
| 0 | −2 |
| 2 | −1.5 |
| 4 | −1 |

Which statement comparing the functions is true?

A. The slope of Function W is less than the slope of Function Z.
B. The slope of Function W is greater than the slope of Function Z.
C. The *y*-intercept of Function W is equal to the *y*-intercept of Function Z.
D. The *y*-intercept of Function W is less than the *y*-intercept of Function Z.
E. The *y*-value when $x = -4$ for Function W is greater than the *y*-value when $x = -4$ for Function Z.
F. The *y*-value when $x = -4$ for Function W is equal to the *y*-value when $x = -4$ for Function Z.

(7) To answer this type of question you will need to ***sketch a parallelogram*** *A′B′C′D′* ***on scrap paper***. (You cannot sketch on the screen.) Then *click to select* ***all that are correct***. (There is more than one correct answer.) Your school is also permitted to supply graph paper.

Parallelogram *ABCD* is shown on the coordinate plane.

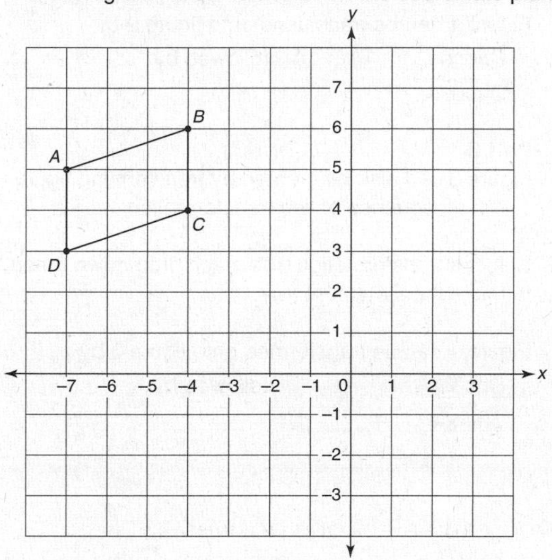


Parallelogram *A′ B′ C′ D′* (not shown) is the image of parallelogram *ABCD* after a rotation of 180° about the origin.

Which statements about parallelogram *A′ B′ C′ D′* are true? Select **each** correct statement.

A. $\overline{A'B'}$ is parallel to $\overline{B'C'}$.
B. $\overline{A'B'}$ is parallel to $\overline{A'D'}$.
C. $\overline{A'B'}$ is parallel to $\overline{C'D'}$.
D. $\overline{A'D'}$ is parallel to $\overline{B'C'}$.
E. $\overline{A'D'}$ is parallel to $\overline{D'C'}$.

(8) Some examples have **drop-down menus** to help you select the correct answer: This is a typical **Part A** and **Part B** type question. This particular example requires four separate answers.

Three congruent figures are shown on the coordinate plane.

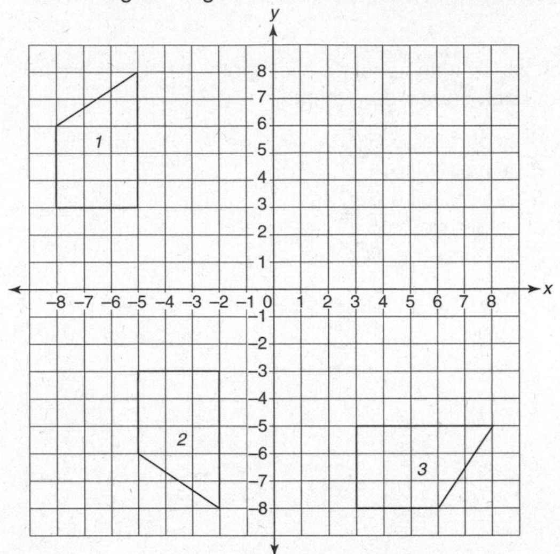


**Part A**

Select a transformation from each drop-down menu to make the statement true.

Figure 1 can be transfused onto figure 2 by

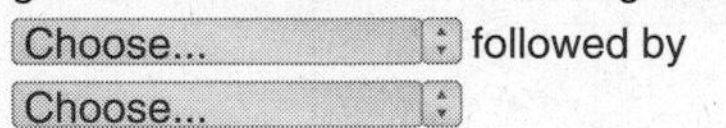


Choose...

**Part B**

Figure 3 can also be created by transforming figure 1 with a sequence of two transformations.

Select a transformation from each drop-down menu to make the statement true.

Figure 1 can be transformed onto figure 3 by

Choose... followed by

Choose...

When you click on the top **"Choose"** in **Part A** you can select one answer.

Three congruent figures are shown on the coordinate plane.

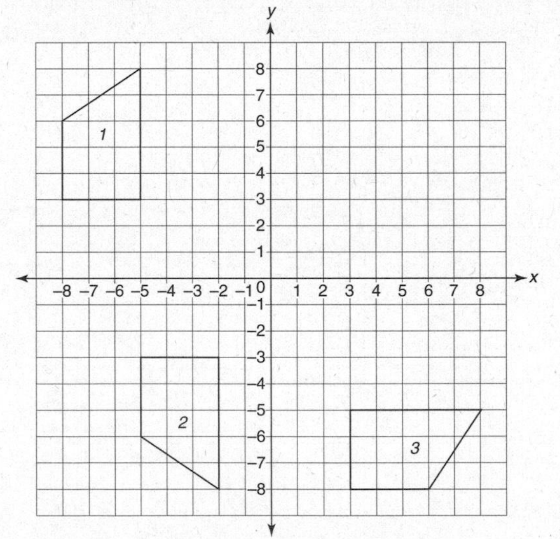


**Part A**

Select a transformation from each drop-down menu to make the statement true.

Figure 1 can be transformed onto figure 2 by

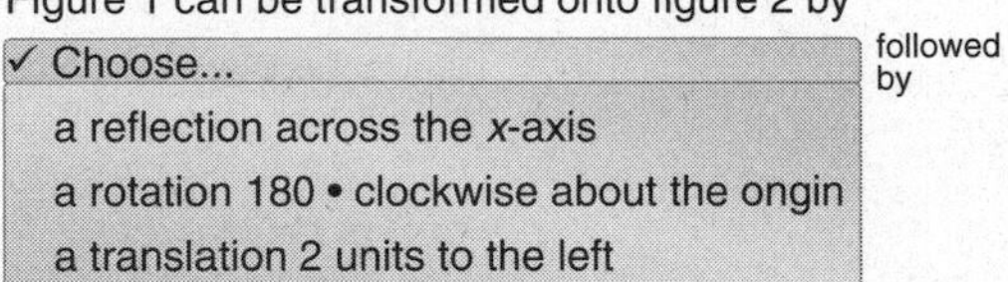


Then click on the bottom **"Choose"** in **Part A** for the second part of this answer. You would continue and do something similar for **Part B**.

(9) In this type of problem you click on the ***LINE TOOLS*** on the left to make points and lines on a graph. For example, click on the ***Line s*** tool; then click on ***the graph*** plot the point of intersection with the $y$-axis. This point (0, –3) will appear on the graph. Next, click on ***Line s*** again, and click on a point on ***the graph*** to show the slope (1 up and 3 to the right); the new point (3, –2) will appear, and the line $y = \frac{1}{3}x - 3$ will automatically be drawn for you. Each line on the graph will appear in a different color to match the tool.

The equation of line $s$ is $y = \frac{1}{3}x - 3$.
The equation of line $t$ is $y = -x + 5$.
The equations of lines $s$ and $t$ form a system of equations. The solution to the system of equations is located at point $P$.

To graph a line, select line $s$ and plot two points on the coordinate plane. A line will be drawn through the points.
In similar fashion, select line $t$ and plot two points on the coordinate plane. A line will be drawn through the points.
Select point $P$ and plot the point on the coordinate plane.

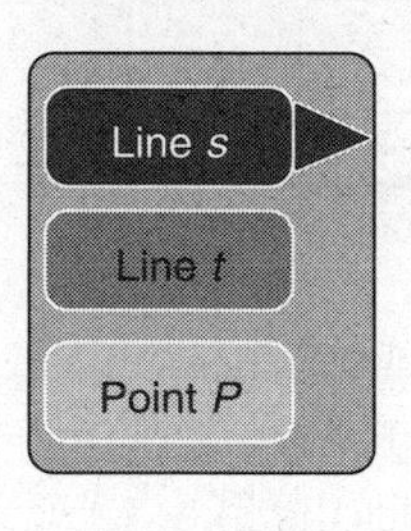


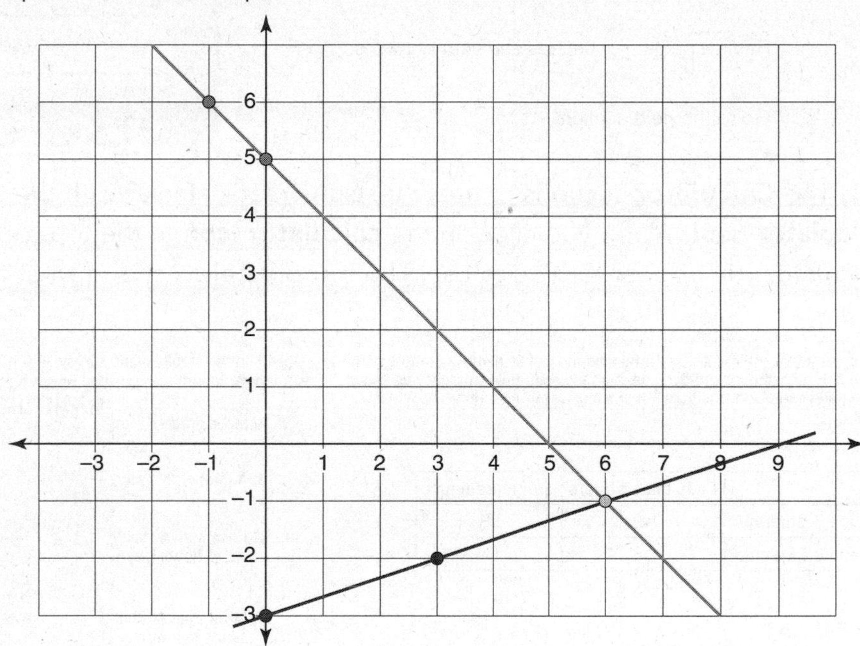

(10) In some examples you will be asked to **click on a number** on a number line to show your answer.

Select the point on the number line that **best** approximates the location of $\sqrt{14}$.

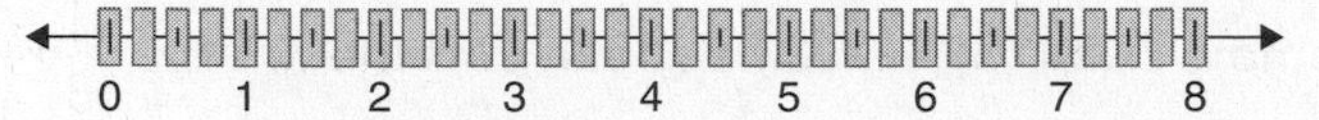


When you complete a section of a test you are able to **Review** your work, and to **make corrections**. Once you complete this and go on to a new section you may not return to an earlier section.

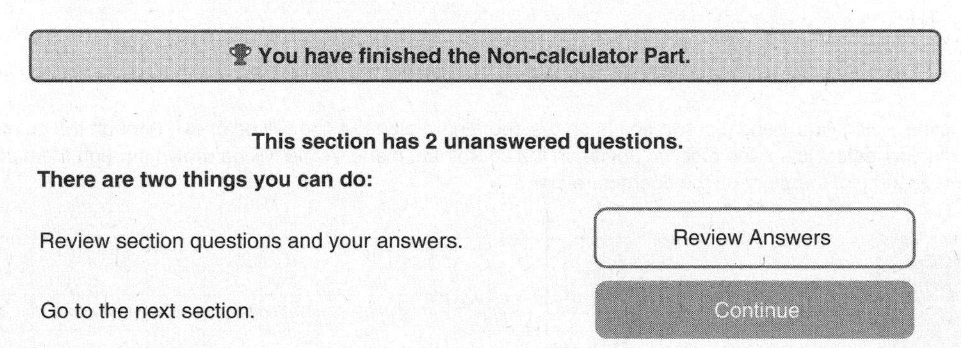


(11) The **Calculator** sections of these tests may look similar with the addition of a **calculator tool**. When you click on the **calculator icon** in the top tool bar a working calculator will appear for you to use. The sample below shows what this looks like.

The table shows the results of a random survey of students in grade 7 and grade 8. Every student surveyed gave a response. Each student was asked if he or she exercised less than 5 hours last week or 5 or more hours last week.

| | Less than 5 hours | 5 or more hours |
|---|---|---|
| Grade 7 Students | 49 | 63 |
| Grade 8 Students | 58 | 51 |

Based on the results of the survey, which statements are true? Select **each** correct statement.

A. More grade 8 stud

B. A total of 221 stud

C. Less than 50% of more hours last w

D. More than 50% of hours last week.

E. A total of 107 grac

Calculator x Close

0.

shift | 1/x | log | ln | $x^n$
hyp | sin | cos | tan
M+ | Min | MR | $x^2$ | √
( | ) | % | CE | AC
EXP | 7 | 8 | 9 | ÷
Π | 4 | 5 | 6 | x
rand | 1 | 2 | 3 | –

## Material from Grade 6 and Grade 7

Although this book has been written as a grade 8 review book focusing on the grade 8 Common Core State Standards (CCSS) and expectations of the Grade 8 PARCC exams, you will see some grades 6 and 7 material in the instruction, practice sections, and tests.

Some material has been included as pre-requisites that often need review in grade 8. Other material has been included because it will be tested on the Grade 8 Math PBA and EOY.

## Score and Test Results

School districts and teachers should receive the results of both the March/April, PBA test, and May/June, EOY test at the same time, before the end of the school year. School districts may use these scores to determine the best placement for students going into the 9th grade.

## Helpful Resources

Become familiar with the PARCC Grade 8 Mathematics Assessment Reference Sheet on page 359 so you will be able to use it as you take the tests in this book. When taking the exam online, the information will be available to you on the screen or on a separate sheet of paper.

## Calculators

### GRADES 6-8 CALCULATOR POLICY*

- PARCC math assessments for grades 6–7 will allow for an online four-function calculator with square root.
- PARCC math assessments for grade 8 will allow for an online scientific calculator.
- PARCC math assessments are to be divided into calculator and non-calculator sessions.
- No tablet laptop (or PDA) or phone-based calculators are allowed during testing.
- Calculators are not to be shared between students during testing.
- Memory on all calculators should be cleared before and after testing.
- Calculators with "QWERTY" keyboard are not permitted.

*New update may permit hand-held calculators. Check online for updates.

## Accommodations and Modifications

Refer to the online PARCC Accessibility Features and Accommodations manual made available for all students to view the additional accommodations for students with IEPs or 504 classifications. (www.parcconline.org/parcc-accessibility-features-and-accomodations-manual)

# The Number System

CHAPTER 1

## WHAT DO PARCC 8 NUMBER SYSTEM QUESTIONS LOOK LIKE?

## Multiple-Choice Question (MC)

### Example 1:

The regular price of a pair of sneakers is $79.00. They are on sale for 15% off. If Jose buys them on sale, how much change will he get from $70?

○ **A.** $9.00 ○ **C.** $3.85

○ **B.** $2.85 ○ **D.** $67.15

**Strategies and Solutions**

Read the *question part* of the example twice to be sure you understand what is being asked.

First, determine the sale price.
$79 × 0.15 (the % off) = $11.85
79.00 − 11.85 = $67.15 (sale price) or 79 × .85 (the % he would pay) = $67.15 (sale price)

Now, determine the change he would receive from $70.00.
70.00 − 67.15 = $2.85 (Watch your computation; a careless error might make you select $3.85.)

The correct answer is **B** $2.85.

## What a New Multiple-Choice Question Might Look Like

### Example 2:

(No calculator permitted.)

A team jacket costs $79.00.

This week it is on sale with an 8% discount.

**Part A** What would the approximate savings be on one jacket?

○ **A.** $6.40 ○ **B.** $6.80 ○ **C.** $7.00 ○ **D.** $7.30

**Part B** If we bought two jackets *at the sale price*, what would be the total cost?

○ **A.** >$150 ○ **B.** =$150 ○ **C.** <$150

**Part C** If we had $175 and bought *two jackets on sale*, our change would be at least $5.00. Do you agree? Check Yes ..✔... or No ..✔... .

Yes ...... No ......

Notice that with these new multiple-choice questions, no calculator is permitted.

**Solutions**

**Part A** Since it says approximate savings, you are expected to round $79 to 80 for easy estimating.

$$80 \times .08 = 6.40 \quad \mathbf{\$6.40}$$

**Part B** If you use exact numbers the sale price would be

$$\mathbf{\$79} - 6.32 = \$72.68 \quad \text{and} \quad 72.88 \times 2 = \$145.36$$

If you estimated, the sale price would be about

$$\mathbf{\$80} - 6.40 = \$73.60 \quad \text{and} \quad 73.60 \times 2 = \$147.20$$

Either way, two jackets would cost less than $150. Choice ***C*** <$150 is correct.

**Part C** No

$$150 - 147.20 = 2.80$$

$$\mathbf{2.80 < 5.00}$$

*(2.80 is less than 5.00)*

## Constructed-Response Question (CR)

### Example 3:

(No calculator permitted.)

Mr. Wieland gave his class a homework assignment that included responding on the Internet to a blog about that night's homework. Of Mr. Wieland's 120 students, 90 responded by 8:00 P.M.

What percentage of his students responded by 8:00 P.M.?

Answer: ..................

**Solution**

$$\frac{\text{Favorable outcome}}{\text{Total}} = \frac{\text{90 students responded}}{\text{120 students total}}$$

$$\frac{90}{120} \text{ reduces to } \frac{9\not{0}}{12\not{0}} = \frac{9}{12} = \frac{3}{4} = 0.75 = 75\%$$

Answer: **75%**

## Example 4:

(No calculator permitted.)

Last week, Kelsey worked 40 hours at $10.00 per hour. Then she worked 10 hours overtime on Saturday. She gets paid time-and-a-half for her overtime hours.

What was her total salary that week?

Answer: ________________

**Solution**

40 hrs. × $10 = (40)(10) = $400

10 hrs. × ($10 + 5) = (10)(15) = $150

Total salary: $550

Answer: **$550.00**

You'll notice that in these CR (constructed-response) questions you are NOT permitted to use a calculator, so the numbers used are not complicated.

# Performance-Based Assessment Questions (PBA)

## Example 5:

Nancy wants to buy a television. The regular price is $325. The store is having a special sale and is reducing all television sets by 8% each Monday for the next five weeks.

**Part A** If Nancy can spend $250, when will she be able to buy a television?
**Part B** How much will it cost her?
**Part C** How much more will she save if she waits one more week?

### Strategies and Solutions

Underline or circle important information in the question. Make a table to help you organize data. Remember, many PBA questions can be organized with a chart or table. Show your work even if you use a calculator.

If the TV is reduced by 8% each week, the cost will be 92% of the previous week's price.

(100% – 8% = 92%)

| Original Price | Show Work | $325.00 |
|---|---|---|
| 1st Monday | $325 \times 0.92$ | $299.00 |
| 2nd Monday | $299 \times 0.92$ | $275.08 |
| 3rd Monday | $275.08 \times 0.92$ | $253.07 |
| 4th Monday | $253.07 \times 0.92$ | $232.83 |
| 5th Monday | $232.83 \times 0.92$ | $214.20 |

**Part A** On the 4th Monday, Nancy will be able to buy a television.

**Part B** It will cost her $232.83.

**Part C** If she waits one week, she will save $18.63 more.

| | |
|---|---|
| Week 4 price: | $232.83 |
| Week 5 price: | –214.20 |
| More savings: | $18.63 |

## OUR NUMBER SYSTEM (NUMBER SETS)

### Integers

Before we begin, let's review some vocabulary you should remember. The first set of numbers people used to count their possessions was the *natural numbers*, or *counting numbers*.

- Counting numbers = {1, 2, 3, 4, 5,..., 100, 101, 102,...}
- Whole numbers — counting numbers and zero = {0, 1, 2, 3,..., 100, 101, 102,...}
- Positive numbers — whole numbers like 24 or 531
- Negative numbers — whole numbers like –6 or –3,000
- Integers — positive and negative whole numbers

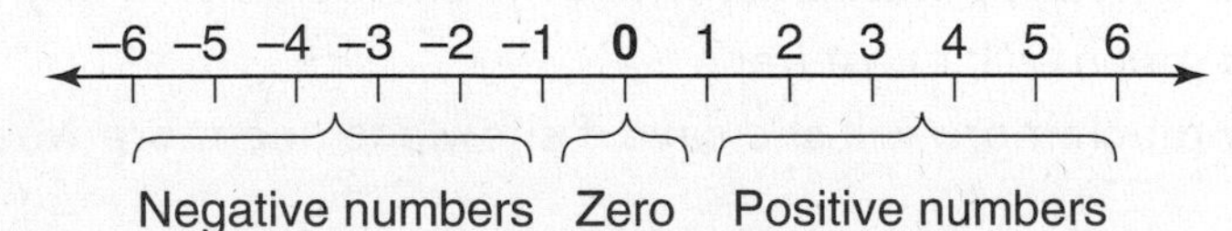


*(Note: Zero is neither negative or positive.)*

- Even numbers — –2,..., 4, 6,..., 212,..., 254,... the last digit is always divisible by 2
- Odd numbers — –5, –3,..., 7, 9,..., 273,..., 209,... the last digit is **not** divisible by 2

## PRACTICE: Compare and Order Integers

(For answers, see page 229.)

**Hint:** Draw and label a number line to help you "see" the order.

*Each question has more than one correct answer.*

**1.** Select all of the following integers that are in order from least to greatest.

- [ ] **A.** –2, 2, 6
- [ ] **B.** 5, –1, –8
- [ ] **C.** –10, –4, 0, 3
- [ ] **D.** 0, –2, –3
- [ ] **E.** 0, –1, –3

**2.** Select all of the following integers that are in order from greatest to least.

- [ ] **A.** –4, –6, –8
- [ ] **B.** 4, –3, –5
- [ ] **C.** –5, –2, 0, 1
- [ ] **D.** –3, –2, 0, 1
- [ ] **E.** 5, 1, –6, –8

**3.** Which of the following integers are in order from greatest to least?

- [ ] **A.** 6, 4, –8, –10
- [ ] **B.** –4, –6, 0, 5
- [ ] **C.** –3, 0, 2, 5
- [ ] **D.** 5, –1, –6, –8
- [ ] **E.** –10, –8, 4, 6

## Absolute Values (from Grade 6.NS.7C)

Another term you should understand is *absolute value*. The absolute value of a number is the distance of that number from 0 on the number line. The absolute value of a number is written as $|n|$. $|5|$ is read as "the absolute value of five;" $|-6|$ is read as "the absolute value of negative six." Because the absolute value of a number represents distance, it is always positive. $|-6| = 6$.

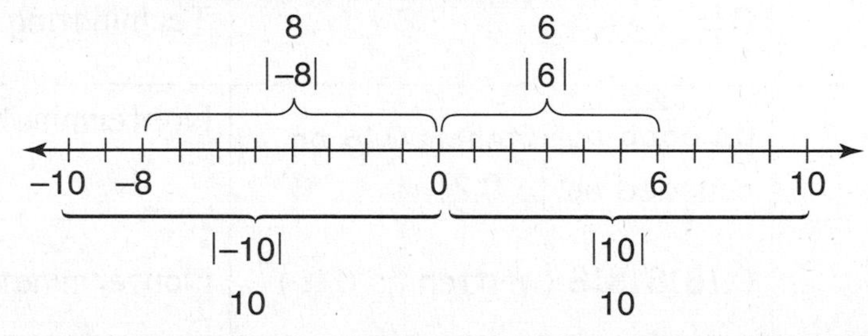


- The distance from 0 to –10 is the same length as the distance from 0 to +10.
- The distance from –8 to zero is greater than the distance from zero to +6; $|-8| > |6|$.

### Examples

**A.** $|-6|$ The absolute value of –6 could represent the number of yards the quarterback ran back before throwing a pass. He ran 6 yards.

**B.** $|-15|$ The absolute value of −15 could represent the number of feet Pete swam under water. He swam 15 feet under water.

**C.** $|-12|=|12|$ Whether I drive 12 miles south, or 12 miles north, I still drive 12 miles.

## PRACTICE: Absolute Values

(For answers, see page 229.)

**1.** Which is the longer distance, $|-16|$ or $|5|$?

○ **A.** $|-16|$ ○ **B.** $|5|$

**2.** What value is greater $|-20|$ or $|-6|$?

○ **A.** $|-6|$ ○ **B.** $|-20|$

**3.** Which absolute value expression would best describe the distance the shark dove underwater?

○ **A.** $|30|$ ○ **C.** $|20|$

○ **B.** $|-20|$ ○ **D.** $|-10|$

## RATIONAL AND IRRATIONAL NUMBERS

What are rational and irrational numbers?

A *rational number* is any number that can be expressed as a ratio of two integers.

### Examples

4 can be expressed as $\frac{4}{1}$, and 0.8 as $\frac{8}{10}$. The mixed number 3½ can be written as $\frac{7}{2}$.

| Rational Numbers | Decimal Form | Decimal Type |
|---|---|---|
| $\frac{1}{2}$ | 0.5 | Terminating |
| $\frac{1}{6}$ | $0.1\overline{6666}$ (written as $0.1\overline{6}$ or rounded up to 0.2) | Nonterminating, repeating |
| $\frac{2}{11}$ | 0.18181818 (written as $0.\overline{18}$) | Nonterminating, repeating |

Any number that *cannot* be expressed as a ratio of two integers or as a repeating or terminating decimal is called an *irrational number*.

| Irrational Numbers | Decimal Form | Decimal Type |
|---|---|---|
| $\sqrt{3}$ | 1.7320508... | Nonterminating, nonrepeating |
| $\pi$ | 3.1415926... | Nonterminating, nonrepeating |

## PRACTICE: Rational and Irrational Numbers

(For answers, see pages 229–230.)

Use a calculator to investigate the following numbers and decide if they are rational or irrational. Ask yourself, does the pattern repeat?

**1.** $\sqrt{6}$

○ **A.** rational ○ **B.** irrational

**2.** $\sqrt{8}$

○ **A.** rational ○ **B.** irrational

Use your calculator to find the square root of each given number. If the number is irrational, select the answer to the nearest hundredth.

**3.** $\sqrt{144}$

○ **A.** 71 ○ **C.** 17
○ **B.** 21 ○ **D.** 12

**4.** $\sqrt{86}$

○ **A.** 9.2736 ○ **C.** 9.3
○ **B.** 9.27 ○ **D.** 43

**5.** $\sqrt{54}$

○ **A.** 7.348 ○ **C.** 7.35
○ **B.** 7.34 ○ **D.** 7

**6.** $\sqrt{110}$

○ **A.** 10.49 ○ **C.** 10.488
○ **B.** 10.50 ○ **D.** 10.489

*Use the number line below to answer questions 7 through 14:*

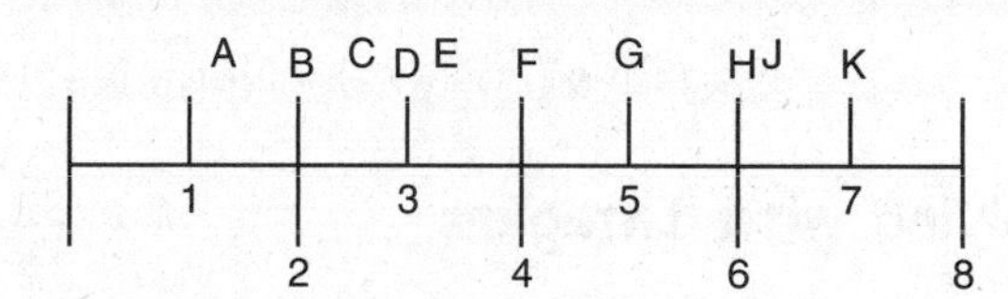


What point on the number line represents the exact or approximate value of the following? Check the correct answer.

**7.** $\sqrt{49}$ A ____ B ____ C ____ D ____ E ____ F ____ G ____ H ____ J ____ K ____

**8.** $\sqrt{4}$ A ____ B ____ C ____ D ____ E ____ F ____ G ____ H ____ J ____ K ____

**9.** $\sqrt{2}$ A ____ B ____ C ____ D ____ E ____ F ____ G ____ H ____ J ____ K ____

**10.** $\sqrt{25}$ A ____ B ____ C ____ D ____ E ____ F ____ G ____ H ____ J ____ K ____

**11.** $\sqrt{9}$ A ____ B ____ C ____ D ____ E ____ F ____ G ____ H ____ J ____ K ____

**12.** $\pi$ A ____ B ____ C ____ D ____ E ____ F ____ G ____ H ____ J ____ K ____

**13.** $\pi + 3$ A ____ B ____ C ____ D ____ E ____ F ____ G ____ H ____ J ____ K ____

**14.** $\sqrt{7}$ A ____ B ____ C ____ D ____ E ____ F ____ G ____ H ____ J ____ K ____

# NUMERICAL OPERATIONS

## Adding and Subtracting Integers

**Remember these rules!**

- Positive + Positive = Positive $6 + 3 = 9$
- Negative + Negative = Negative $-4 + -6 = -10$
- Positive + Negative or Negative + Positive $6 - 8 = -2$ or $-8 + 6 = -2$

  (Take the difference of the two numbers and use the sign of the larger number.)

# MULTIPLYING AND DIVIDING POSITIVE AND NEGATIVE NUMBERS

Positive × Positive = Positive (Positive)(Positive) = Positive; $(8)\left(\frac{1}{2}\right) = \frac{8}{2} = 4$

Negative × Negative = Positive (Negative)(Negative) = Positive; $(-8)\left(-\frac{1}{2}\right) = \frac{-8}{-2} = 4$

Positive × Negative is the same as (Positive)(Negative) = Negative; $(2)(-8.5) = -17$

Negative × Positive (Negative)(Positive) = Negative; $(-2)(8.5) = -17$

## Examples of Multiplying with Integers

**A.** $3 \times 12 = 36$ I have 3 boxes of donuts and each box has 12 donuts.

**B.** $-4 \times 2 = -8$ Mary owes $4 to two different friends. She owes a total of $8.

**C.** $(10)(-120) = -1,200$ If we rented a storage facility for 10 months at $120 per month it would cost us $1,200.

**D.** $(-6)(-3) = 18$ Six months ago you started a diet and lost 3 lbs. per month. Six months ago you were 18 lbs. heavier.

**E.** A swimmer dove 10 feet under water three times. Which expression represents this?

**A.** $(-3)(-10)$ **B.** $3|-10|$ **C.** $(3)(10)$ **D.** $-3|10|$

The answer here is **B**.

Since multiplication is the inverse operation of division, the sign rule remains the same.

- $\frac{\text{Positive}}{\text{Positive}} = \text{Positive}$ Positive ÷ Positive = Positive; $\frac{24}{2} = 12$ or $24 \div 2 = 12$
- $\frac{\text{Negative}}{\text{Negative}} = \text{Positive}$ Negative ÷ Negative = Positive; $\frac{-15}{-3} = 5$ or $-15 \div -3 = 5$
- $\frac{\text{Positive}}{\text{Negative}} = \text{Negative}$ $\frac{18}{-2} = -9$ or $\frac{\text{Negative}}{\text{Positive}} = \text{Negative}$; $\frac{-18}{9} = -2$

## Examples of Dividing with Integers

**A.** $16 \div 2 = 8$ If Ryan has 16 cans of oil and put them evenly into 2 boxes, he would have 8 cans in each box.

**B.** $10 \div -2 = -5$ If Diane had \$10 and the books at the library sale cost \$2 each, she could buy 5 books.

**C.** $-\$30 \div 5 = -\$6$ Dan is at the zoo for 5 hours. He spends \$30 all together. Dan spent an average of \$6 per hour.

**D.** $-45 \div -5 = 9$ If Kevin wants to lose 45 pounds and planned on losing 5 pounds each month, he would reach his goal in 9 months.

*Study the following chart to remind yourself how to add, multiply, or divide when you are working with positive and negative integers.*

## A Review of Operations with Integers (from Grade 6)

| Adding Integers | | Subtracting Integers | Multiplying & Dividing Integers | |
|---|---|---|---|---|
| Same sign<br>+ + − − | Integers with the different signs | Add the opposites. | Same sign<br>+ + − − | Integers with the different signs |
| Add the numbers and take the original sign. | Take the difference of the two numbers and the sign of the larger number. | Follow rules for addition. | Positive | Negative |
| 6 + 4 + 5 = 15 | −9 + 6 = −3 | (+5) − (−16) = 5 + 16 = 21 | (3)(4) = 12<br>(−6)(−2) = 12<br>−(−12) = 12 | (3)(−4) = −12<br>(−6)(2) = −12 |
| −4 − 3 − 2 = −9 | −2 + 7 = 5 | −30 − (−20) = −30 + 20 = −10 | 21 ÷ 3 = 7<br>(−21) ÷ (−3) = 7 | (21) ÷ (−3) = −7<br>(−21) ÷ (3) = −7 |

## PRACTICE: Adding, Subtracting, Multiplying, and Dividing Integers and Fractions (A Review from Grade 4)

(For answers, see pages 230–231.)

Perform the operations shown, and simplify each expression. No calculator is needed here!

**1.** $-4-(-9)=$

- ○ **A.** $-13$
- ○ **B.** 13
- ○ **C.** 5
- ○ **D.** $-5$

**2.** $6-(-3)=$

- ○ **A.** 9
- ○ **B.** $-9$
- ○ **C.** $-3$
- ○ **D.** 3

**3.** $(-3)(5.2)=$

- ○ **A.** 2.2
- ○ **B.** $-15.6$
- ○ **C.** 15.6
- ○ **D.** 4.9

**4.** $(-2)(-5)=$

- ○ **A.** $-7$
- ○ **B.** 7
- ○ **C.** $-10$
- ○ **D.** 10

**5.** $25 \div 5 =$ ------

**6.** $-30 \div -6 =$ ------

For questions 7–10, please show your answers in the boxes.

**7.** $-\frac{1}{2}+\frac{2}{3}+\frac{3}{4}=$ [ ]

**8.** $8-2\frac{2}{5}=$ [ ]

**9.** $-4\frac{3}{4}+2\frac{1}{5}=$ [ ]

**10.** $16-\frac{1}{2}(8)-\frac{3}{4}(16)=$ [ ]

## EXPONENTS, SQUARE ROOTS, AND CUBE ROOTS (NEW AND IMPORTANT!)

In mathematics, *exponents* are used as a short way of saying a number is multiplied by itself a certain number of times.

- Instead of writing $5 \times 5$, we can write $5^2$, which means 5 times itself 2 times ($5^2 = 25$).
- Instead of writing $-10 \times -10$, we can write $(-10)^2$, which means $-10$ times itself 2 times $(-10)(-10) = 100$. $(-10)^2 = 100$.

We read these numbers two different ways.

- We can say $5^2$ is *five squared,*
- or we can say $5^2$ is *five raised to the second power.*
- We can say $(-8)^2$ is *negative eight squared or negative eight raised to the second power.*

### Examples

**A.** Instead of writing $2 \times 2 \times 2$, we can write $2^3$, which means 2 times itself 3 times ($2^3 = 8$).

We can say that $2^3$ is *two cubed*, or we can say that it is *two raised to the third power.*

**B.** Instead of writing $1 \times 1 \times 1$, we can write $1^3$, which means 1 times itself 3 times ($1^3 = 1$).

We can say that $1^3$ is *one cubed*, or we can say that it is *one raised to the third power.*

A number written in *exponential form* has a base and an exponent. The following are samples of numbers written in exponential form:

$$(-5)^2 \quad 4^3 \quad 6^2 \quad 2^4$$

- In the term $(-5)^2$, the $-5$ is the base and the 2 is the exponent.
- In the term $4^3$, the 4 is the base and the 3 is the exponent.

**Note:** Any number except zero raised to the zero power equals 1.

$$59^0 = 1, \quad 2{,}980{,}000 = 1, \quad (1/5)^0 = 1$$

Note that $0^0$ is undefined.

## PRACTICE: Exponents

(For answers, see page 231.)

**1.** Write *six squared* in exponential form.

- ○ **A.** $2^6$
- ○ **B.** $6^2$
- ○ **C.** $6 \times 2$
- ○ **D.** $(6)(6)$

**2.** Which statement means the same as $5^3$?

- ○ **A.** five squared
- ○ **B.** three cubed
- ○ **C.** five cubed
- ○ **D.** five times three

**3.** Which expression = $3^4$?

- ○ **A.** $3 \times 3$
- ○ **B.** $4 \times 3$
- ○ **C.** $4 \times 4 \times 4$
- ○ **D.** $3 \times 3 \times 3 \times 3$

**4.** Which represents the largest number? (Show your work; do not guess.)

- ○ **A.** $4^2$
- ○ **B.** $(4)(2)$
- ○ **C.** $24^0$
- ○ **D.** $8^2 - 4^2$

**5.** Which expression is equal to $9^2$?

- ○ **A.** $3^2 + 3^2$
- ○ **B.** $(3^3)(6)^0$
- ○ **C.** $(9)(3)(3)$
- ○ **D.** $(2)(2)(9)$

**6.** How would you write two cubed?

\-------------------------------------------

**7.** How would you write negative eight squared?

\-------------------------------------------

**8.** Write $8 \times 8 \times 8$ in exponential form.

\-------------------------------------------

**9.** Which of the following expressions is not equivalent to 1/36?

- ○ **A.** $6^{-2} \times 6^4$
- ○ **B.** $6^{-1} \times 6^{-1}$
- ○ **C.** $6^{-3} \times 6$
- ○ **D.** $6^3 \times 6^{-5}$

**10.** Answer the following statements based on the expression $3^2 \times 3^{-4}$.
Check Yes __✔__ or No __✔__ for each.

**A.** Simplified, this is equivalent to $3^{-8}$. Yes ------ No------

**B.** This expression is equivalent to $\frac{1}{9}$. Yes ------ No------

**C.** The following is an equivalent expression: $3^{-5} \times 3^3$. Yes ------ No------

**D.** An equivalent expression is $9 \times \frac{1}{3^4}$. Yes ------ No------

**E.** Another way to write this is $(9)(-3)^4$. Yes ------ No------

**F.** Another way to write this expression is $3^{2-4}$. Yes ------ No------

## Using a Calculator

Many calculators have an $x^2$ key so you can easily *square a number*. Try these on your calculator:

- Find the value of $4^2$. Use your calculator. Even though this is an easy one (you probably know automatically $4^2 = 16$), try it on your calculator to see which buttons you need to press and in what order. Remember this so you can work with more difficult numbers easily.
- Find the value of $29^2$. Use your calculator again and remember how your calculator works.

The correct solution is 841.

- Try a few more to practice pressing the correct buttons on your calculator.

What is the value of $(-16)^2$? What is the value of $42^3$? What is the value of $(19.85)^2$?

$(-16)^2 = 256$ $42^3 = 74{,}088$ $(19.85)^2 = 394.0225$

If you ask other students in your class how their calculators work, you may find that other models work differently. This is one reason it is important for you to use the same calculator you practice with when you take the official test.

Now let's work in reverse!

The following are some *perfect square numbers:*

1, 4, 9, 16, 25, 36, 49, 64, 81, 100

Perfect square numbers can be represented as perfect square shapes. Remember that the formula for the *area of a square* is *length* × *width*. See the following examples and diagrams.

- A square with length 1 and width 1 has an area of 1 square unit; 1 is a perfect square number.
- A square with length 3 and width 3 has an area of 9 square units; 9 is a perfect square number.
- A square with length 5 and width 5 has an area of 25 square units; 25 is a perfect square number.
- A square with length 7 and width 7 has an area of 49 square units; 49 is a perfect square number.

These represent *perfect square* numbers:

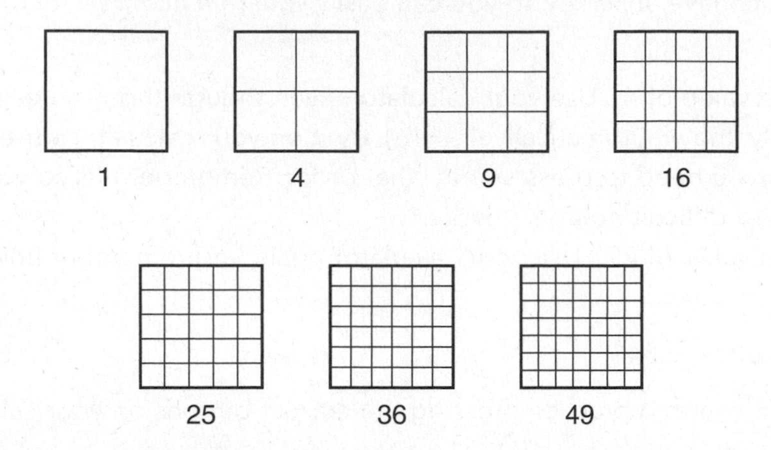


You should be familiar with these perfect square numbers and their *square roots*.

| Perfect square number | The square root is | Because | |
|---|---|---|---|
| 1 | $\sqrt{1} = 1$ | $1 \times 1 =$ | $1^2 = 1$ |
| 4 | $\sqrt{4} = 2$ | $2 \times 2 =$ | $2^2 = 4$ |
| 9 | $\sqrt{9} = 3$ | $3 \times 3 =$ | $3^2 = 9$ |
| 16 | $\sqrt{16} = 4$ | $4 \times 4 =$ | $4^2 = 16$ |
| 25 | $\sqrt{25} = 5$ | $5 \times 5 =$ | $5^2 = 25$ |
| 36 | $\sqrt{36} = 6$ | $6 \times 6 =$ | $6^2 = 36$ |
| 49 | $\sqrt{49} = 7$ | $7 \times 7 =$ | $7^2 = 49$ |
| 64 | $\sqrt{64} = 8$ | $8 \times 8 =$ | $8^2 = 64$ |
| 81 | $\sqrt{81} = 9$ | $9 \times 9 =$ | $9^2 = 81$ |
| 100 | $\sqrt{100} = 10$ | $10 \times 10 =$ | $10^2 = 100$ |
| 121 | $\sqrt{121} = 11$ | $11 \times 11 =$ | $11^2 = 121$ |
| 144 | $\sqrt{144} = 12$ | $12 \times 12 =$ | $12^2 = 144$ |
| 169 | $\sqrt{169} = 13$ | $13 \times 13 =$ | $13^2 = 169$ |
| 196 | $\sqrt{196} = 14$ | $14 \times 14 =$ | $14^2 = 196$ |
| 225 | $\sqrt{225} = 15$ | $15 \times 15 =$ | $15^2 = 225$ |

## Using a Calculator

Most calculators have a *square root key* so you easily can find the square root of any number. Try these on your calculator. Practice on your own and remember the steps that work for your calculator. Directions for some calculators follow. Yours may work differently.

- Press [4] then [second], and finally [$\sqrt{\ }$]. You'll see the number 2.
- Press [8] [1], then [second], and finally [$\sqrt{\ }$]. You'll see the number 9.
- Press the buttons [3] [6] [1], then [second], and finally [$\sqrt{\ }$]. Did you get the number 19? The square root of 361 = 19. Check for yourself. Multiply $19 \times 19$.

The following are some *perfect cube numbers:*

$2^3 = 2 \times 2 \times 2 = 8$ $\quad 3^3 = 3 \times 3 \times 3 = 27$ $\quad 4^3 = 4 \times 4 \times 4 = 64$ $\quad 5^3 = 5 \times 5 \times 5 = 125$

If you think of cubed numbers as geometric figures, you will see that a perfect cubic number represents the *volume of a cube*.

$$\text{Volume} = \text{Length} \times \text{Width} \times \text{Depth}$$

- If a cube had length = 2, width = 2, depth = 2, its volume would $= 2 \times 2 \times 2 = 8$ cubic units.

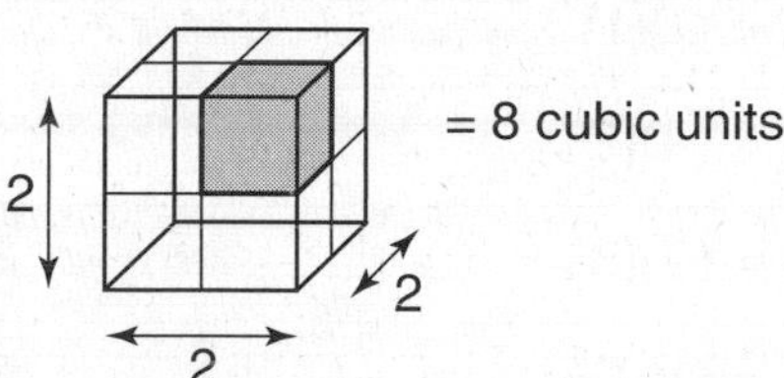


- If a cube had length = 4, width = 4, depth = 4, its volume would $= 4 \times 4 \times 4 = 64$ cubic units.

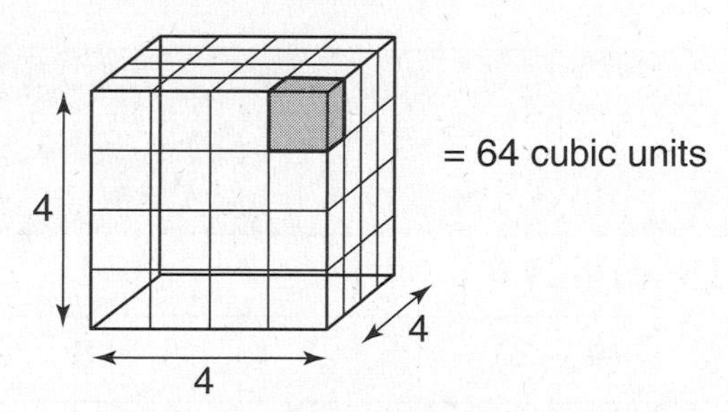


## Examples

**A.** What number times itself will equal 16? Answer: 4
Because $4 \times 4 = 16$ or $4^2 = 16$

**B.** What number times itself will equal 100? Answer: 10
Because $10 \times 10 = 100$ or $10^2 = 100$

In mathematics we say this another way. We say, "What is the *square root* of 100?"

**C.** What is the *square root* of 16? Answer: 4
Because $4 \times 4 = 16$ or $4^2 = 16$

**D.** What is the *square root* of 100? Answer: 10
Because $10 \times 10 = 100$ or $10^2 = 100$

**E.** What is the *cube root* of 8? Answer: 2
Because $2 \times 2 \times 2 = 8$ or $2^3 = 8$

Still, in mathematics we show this another way. We use a symbol to represent the words *the square root of*. The symbol is $\sqrt{\ }$. Examples: $\sqrt{25}=5$, $\sqrt{81}=9$, $\sqrt{144}=12$.

**F.** $\sqrt{16}$ is read as *the square root of* 16. Answer: $\sqrt{16}=4$
Because $4 \times 4 = 16$ or $4^2 = 16$

**G.** $\sqrt{100}$ is read as *the square root of* 100. Answer: $\sqrt{100}=10$
Because $10 \times 10 = 100$ or $10^2 = 100$

You should know these often-used *perfect square numbers* and *perfect cube numbers*.

| | Work shown | Perfect cube positive integers | |
|---|---|---|---|
| $1^3$ | $(1)(1)(1) = (1)(1) =$ | **1** | $\sqrt[3]{1}=1$ |
| $2^3$ | $(2)(2)(2) = (4)(2) =$ | **8** | $\sqrt[3]{8}=2$ |
| $3^3$ | $(3)(3)(3) = (9)(3) =$ | **27** | $\sqrt[3]{27}=3$ |
| $4^3$ | $(4)(4)(4) = (16)(4) =$ | **64** | $\sqrt[3]{64}=4$ |
| $5^3$ | $(5)(5)(5) = (25)(5) =$ | **125** | $\sqrt[3]{125}=5$ |

| | Work shown<br>Remember the rules:<br>$(+)(+) = +$ and $(+)(-) = -$ | Perfect cube negative integers | |
|---|---|---|---|
| $(-1)^3$ | $(-1)(-1)(-1) = (1)(-1) =$ | **−1** | $\sqrt[3]{-1}=-1$ |
| $(-2)^3$ | $(-2)(-2)(-2) = (4)(-2) =$ | **−8** | $\sqrt[3]{-8}=-2$ |
| $(-3)^3$ | $(-3)(-3)(-3) = (9)(-3) =$ | **−27** | $\sqrt[3]{-27}=-3$ |
| $(-4)^3$ | $(-4)(-4)(-4) = (16)(-4) =$ | **−64** | $\sqrt[3]{-64}=-4$ |
| $(-5)^3$ | $(-5)(-5)(-5) = (25)(-5) =$ | **−125** | $\sqrt[3]{-125}=-5$ |

The rules stay the same when you see these numbers in fractions, too.

Sample 1: $\frac{2}{3^3} = \frac{2}{27}$

Sample 2: $\frac{2}{(-4)^3} = \frac{2}{-64} = \frac{1}{-32}$

Sample 3: $\frac{5^2}{10^2} = \frac{25}{100} = \frac{1}{4}$

Sample 4: $\frac{2^2}{\sqrt{64}} = \frac{4}{8} = \frac{1}{2}$

*Remember that (–2)(–2) = 4 and (2)(2)= 4.

## PRACTICE: Perfect Square and Perfect Cube Numbers

(For answers, see page 231.)

Do these **multiple-choice** questions for some quick practice.

**1.** $\sqrt{100} =$

- ○ **A.** 5
- ○ **B.** 20
- ○ **C.** 10

**2.** If $x = \sqrt{121}$, then $x$ could =

- ○ **A.** 6
- ○ **B.** 11
- ○ **C.** 12

**3.** If $y = \sqrt{\frac{1}{16}}$, then $y$ could =

- ○ **A.** $\frac{1}{8}$
- ○ **B.** $\frac{1}{4}$
- ○ **C.** $\frac{1}{6}$

Now do some of these **short constructed-response** questions for more practice.

**4.** If $x^2 = 81$ then $x$ could equal ______

**5.** $(-\sqrt{81})(-\sqrt{81})(2) =$ ______

**6.** If $x^3 = 64$ then $x$ equals ______

These examples may have **more than one correct answer**. Select all that are true.

**7.** If $x = \sqrt{81}$ then $x$ could equal

- ☐ **A.** 3
- ☐ **B.** –3
- ☐ **C.** 9
- ☐ **D.** 34
- ☐ **E.** $2^5$
- ☐ **F.** –6

**8.** If $a = \sqrt{144}$ then $a$ could equal

- ☐ **A.** 72
- ☐ **B.** –72
- ☐ **C.** 12
- ☐ **D.** –21
- ☐ **E.** $4^3$
- ☐ **F.** $\sqrt{81} + 3$

**9.** 64 =

- ☐ **A.** $4^2$
- ☐ **B.** $4^3$
- ☐ **C.** $(\sqrt{16})(4^2)$
- ☐ **D.** $(-\sqrt{16})(-\sqrt{16})(4)$
- ☐ **E.** 4
- ☐ **F.** 16

## PRACTICE: Square Roots and Cube Roots

(For answers, see page 232.)

1. What number times itself will equal 25?

2. What number times itself will equal 81?

3. What is the square root of 36?

4. What is the square root of 144?

5. What is $\sqrt{49}$?

6. What is $\sqrt{100}$?

7. What do you think $\sqrt{25}$ plus $\sqrt{100}$ equals?
   - ○ **A.** $25 + 100$
   - ○ **B.** $5 + 10$
   - ○ **C.** $5^2 + 10^2$
   - ○ **D.** $50 + 10$

8. What do you think $\sqrt{36} + \sqrt{25}$ equals?
   - ○ **A.** $6 + 50$
   - ○ **B.** $36 + 25$
   - ○ **C.** $6^2 + 5^2$
   - ○ **D.** $11$

9. Now we'll combine both ideas. What is $4^2 + \sqrt{25}$?
   - ○ **A.** $16 + 5$
   - ○ **B.** $8 + 10$
   - ○ **C.** $16 + 7$
   - ○ **D.** $8 + 5$

10. Try one more combination. What is $2^3 + \sqrt{49}$?
    - ○ **A.** $6 + 7$
    - ○ **B.** $5 + 7$
    - ○ **C.** $8 + 7$
    - ○ **D.** $8 + 13$

11. Which of the following is NOT equal to 7?
    - ○ **A.** $\sqrt{1} + \sqrt{36}$
    - ○ **B.** $1^2 + 2^2 + 2$
    - ○ **C.** $\sqrt{49}$
    - ○ **D.** $\sqrt{100} - 1^2$

12. The following equal 16 except
    - ○ **A.** $2^3 + (4)(2)$
    - ○ **B.** $\sqrt{4}$
    - ○ **C.** $5^2 - 3^2$
    - ○ **D.** $6^2 - (10)(2)$

## More Practice with Radicals (Square Roots and Cube Roots)

(For answers, see page 232.)

Find each square root or cube root.

1. $\pm\sqrt{81}$ ------ 2. $\pm\sqrt{144}$ ------ 3. $-\sqrt{121}$ ------

The following are multiple-choice questions that have more than one correct choice.

Check (√) all equivalent expressions.

**4.** $\sqrt[3]{\frac{1}{8}} =$

------**A.** $\frac{1^0}{2^1}$ ------**B.** $\frac{1}{4}$

------**C.** 0.5 ------**D.** $\sqrt{\frac{25}{100}}$

**5.** $\sqrt{\frac{9}{144}} =$

------**A.** $\frac{3}{72}$ ------**B.** $\frac{3}{12}$

------**C.** $\frac{1}{4}$ ------**D.** $\frac{1}{16}$

## More Practice with Multiple-Choice Questions (Square Roots and Cube Roots)

(For answers, see page 232.)

**1.** $\sqrt[2]{49} = 7$ then $\sqrt[2]{49} + 4^2 =$

○ **A.** 16 ○ **C.** 49

○ **B.** 23 ○ **D.** 15

**2.** $\sqrt{144} =$

○ **A.** 14 ○ **C.** 12

○ **B.** 1.4 ○ **D.** 72

**3.** $5^{-2} =$

○ **A.** –25 ○ **C.** $\frac{1}{5}$

○ **B.** 25 ○ **D.** $\frac{1}{25}$

**4.** $\frac{1}{16} =$

○ **A.** $4^{-2}$ ○ **C.** $\frac{1}{8^{-2}}$

○ **B.** $4^2$ ○ **D.** $8^{-2}$

**5.** $\frac{1}{81} =$

○ **A.** $9^2$ ○ **C.** $9^{-2}$

○ **B.** $\frac{1}{9}$ ○ **D.** $-9^2$

**6.** Consider the expression $\sqrt{72}$ and answer the following statements.
Check Yes --✔-- or No --✔--.

**A.** This represents an irrational number. Yes ------ No------

**B.** This represents a perfect square number. Yes ------ No------

**C.** This has the same value as $3\sqrt{8}$. Yes ------ No------

**D.** When simplified, this is a negative integer. Yes ------ No------

**E.** The most exact way to write $\sqrt{72}$ would be in decimals. Yes ------ No------

## Mixed Practice with Square Roots and Cube Roots

(For answers, see page 233.)

**1.** Solve for $x$, if $x^2 = 25$ $x =$ ------

**2.** Solve for $y$, if $y^3 = 8$ $y =$ ------

**3.** Solve for $a$, if $a^2 = 100$ $a =$ ------

**4.** Solve for $m$, if $m^3 = 27$ $m =$ ------

**5.** If $x^3 = 64$, then $x =$

- ○ **A.** 2
- ○ **B.** 4
- ○ **C.** 8
- ○ **D.** 16

**6.** If $y^3 = 8$, then $y =$

- ○ **A.** 1
- ○ **B.** 2
- ○ **C.** 24
- ○ **D.** 64

## ESTIMATING SQUARE ROOTS

Some numbers are not perfect squares. For example the number 30 is not a perfect square, but we can still estimate its square root. Remembering these *perfect-square numbers* will help you here.

### Examples

**A.** $\sqrt{30}$ is between the perfect squares $\sqrt{25}$ and $\sqrt{36}$. Therefore, $\sqrt{30}$ is between 5 and 6.

**B.** $\sqrt{90}$ is between the perfect squares $\sqrt{81}$ and $\sqrt{100}$. Therefore, $\sqrt{90}$ is between 9 and 10.

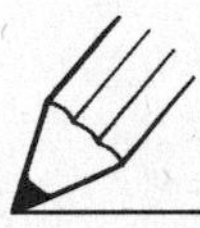

## PRACTICE: Estimating Square Roots

(For answers, see page 233.)

**1.** $\sqrt{50}$ is between

- ○ **A.** 3 and 4
- ○ **B.** 2 and 3
- ○ **C.** 6 and 7
- ○ **D.** 7 and 8

**2.** $\sqrt{15}$ is closest to the number

- ○ **A.** 3
- ○ **B.** 4
- ○ **C.** 5
- ○ **D.** 6

**3.** Which of the following square roots is between 5 and 6?

- ○ **A.** $\sqrt{10}$
- ○ **B.** $\sqrt{20}$
- ○ **C.** $\sqrt{30}$
- ○ **D.** $\sqrt{40}$

**4.** Which statement is true about $\sqrt{115}$? It is

- ○ **A.** less than 10
- ○ **B.** almost 8
- ○ **C.** between 10 and 12
- ○ **D.** greater than 20

**5.** $\sqrt{100}$ is equal to ALL of the following except

- ○ **A.** (2)(5)
- ○ **B.** 100 divided by 2
- ○ **C.** $\sqrt{4} + 2^3$
- ○ **D.** $4^2 - 6$

**6.** Consider the expression $\sqrt{49}$ and answer the following statements. Check Yes ✔ or No ✔.

**A.** This represents a rational number.
Yes ______ No______

**B.** When simplified, the answer is a perfect square number.
Yes ______ No______

**C.** This has the same value as 7.
Yes ______ No______

**D.** This has the same value as $2^2 + \sqrt{9}$.
Yes ______ No______

**E.** $\sqrt{49} \geq 7^2$ Yes ______ No______

**The following are multiple-choice questions that have more than one correct choice.**

**Select all that are true.**

**7.** Which statements are true about $\sqrt{38}$?

- [ ] **A.** It is a perfect square number.
- [ ] **B.** It is <7.0.
- [ ] **C.** It is equivalent to 6.
- [ ] **D.** It is > the value of $\pi$.
- [ ] **E.** It is a perfect number.

**8.** Check all that are true about the integer 36.

- [ ] **A.** It is equivalent to $(-6)^2$.
- [ ] **B.** It is < (6)(–6).
- [ ] **C.** It is equivalent to $(2^2)\ [(-18) \div (-2)]$.
- [ ] **D.** It is equivalent to (–72) (–1/2).
- [ ] **E.** It is (–72).

## SCIENTIFIC NOTATION

Before reviewing scientific notation, let's review the number 10 written with different exponents. The following table shows some examples of this, which we call *powers of ten*.

| Powers of 10 | Meaning | Value | Power | Number of Zeros |
|---|---|---|---|---|
| $10^1$ | 10 | 10 | Power of 1 | One zero |
| $10^2$ | $10 \times 10$ | 100 | Power of 2 | Two zeros |
| $10^3$ | $10 \times 10 \times 10$ | 1,000 | Power of 3 | Three zeros |
| $10^4$ | $10 \times 10 \times 10 \times 10$ | 10,000 | Power of 4 | Four zeros |

- Multiply by powers of 10

  $33 \times 10^5 = 33 \times 10 \times 10 \times 10 \times 10 \times 10 = 33 \times 100{,}000 = 3{,}300{,}000.$

- Multiply by powers of 10 mentally

  $25 \times 10^9$ *Think!* This is just 25 with 9 zeros added on to the end: 25,000,000,000.

3,300,000 and 25,000,000,000 are written in the *standard form* for large numbers. However, very large numbers become confusing and very difficult to write and to read. Scientific notation is a shorthand way of writing very large and very small numbers.

| *Standard Form* | *Scientific Notation Form* |
| --- | --- |
| 9,670,000,000,000 | is written as $9.67 \times 10^{12}$ |

By the definition of *scientific notation*, there can be only one digit to the left of the decimal point. In this number, the first digit must be greater than or equal to 1 but less than 10. So we move the decimal point and then count how many places the decimal point was moved (count all of the zeros and the 6 and 7) 9,670,000,000,000 and you have 12 (the number of the exponent).

## Examples of Very Large Numbers

**A.** 540,000,000 (Remember there is really a decimal point at the end.) In scientific notation form, this number is written as $5.4 \times 10^8$. We moved the decimal point 8 places to the left.

**B.** 24,000,000,000 In scientific notation form, this number is written as $2.4 \times 10^{10}$.

**C.** 2,000,000,000 In scientific notation form this number is written as $2 \times 10^9$.

## Examples of Very Small Numbers

Notice that in these situations the 10 is a negative exponent.

**A. 0.0000056** In scientific-notation form this number would be written as

**$5.6 \times 10^{-6}$** This really means $\frac{5.6}{1{,}000{,}000}$

See for yourself. Type **5.6** into your calculator, then type ÷ then type **1000000**. (There are 6 zeros here.) You should see the original 0.00000056 in the calculator window. Right?

**B.** $0.000000000194 = 1.94 \times 10^{-10}$

**C.** $0.425 = 4.25 \times 10^{-1}$

Now we'll work in reverse. We'll begin with a number in scientific notation form and change it to a number in standard form.

**D.** $6 \times 10^7$ Means 6 times $10 \times 10 \times 10 \times 10 \times 10 \times 10 \times 10$ or 60,000,000 (7 zeros).

**E.** $3.8 \times 10^5$ We must move the decimal point 5 places; but we don't have the necessary zeros. Think 3.8 is the same as 3.80000; so $3.8 \times 10^5 = 380{,}000$ so we can add the zeros we need.

## PRACTICE: Scientific Notation

(For answers, see pages 233–234.)

**1.** Fill in the missing parts in the table.

| Standard Form | How Many Places Will You Move the Decimal Point? | Scientific Notation Form |
|---|---|---|
| 650,000,000,000,000 | 14 | |
| 2,000,000,000 | | $2.0 \times 10^9$ |
| | | $4.5 \times 10^{11}$ |
| | 9 | $3.25 \times 10^9$ |
| 12,000,000,000 | | |

**2.** In space, light travels about 9,450,000,000,000 kilometers per year. This is called a light year. Write this number in scientific notation form.

○ **A.** $9.45 \times 10^{12}$ ○ **C.** $945 \times 10^{10}$

○ **B.** $94.5 \times 10^{11}$ ○ **D.** $9 \times 10^{12}$

**3.** The number 650 billion is equal to 650,000,000,000. Write this number in scientific notation form.

○ **A.** $650 \times 10^{10}$ ○ **C.** $6.5 \times 10^{11}$

○ **B.** $65 \times 10^{11}$ ○ **D.** $6.5 \times 10^{10}$

**4.** The air pressure at the bottom of the deepest ocean is said to be $1.1 \times 10^8$ pa (pascal). What number below is equivalent to this number?

○ **A.** 1.0000001

○ **B.** 110,000,000

○ **C.** 110.000,000

○ **D.** 1.10000000

**5.** In this example, Rachel says the answer is **A**; Barbara says it is **B**. Who is correct? Explain your answer.

280,000,000 = ?

○ **A.** $28 \times 10^7$ ○ **B.** $2.8 \times 10^8$

**6.** Nikoli and Su-Yong are having a discussion in their math class. Nikoli says you can use *scientific notation* to write very small numbers (like the measurement of the thickness of one strand of hair). Su-Yong says you can use scientific notation to write very large numbers (like the distance from earth to the moon).

**Part A:** Who is correct?

○ **A.** Nikoli

○ **B.** Su-Yong

○ **C.** Both

○ **D.** Neither

**Part B:** Change each number below from *scientific-notation form* to *standard form.*

$2.5 \times 10^2 =$ ------

$6.2 \times 10^{-2} =$ ------

**Part C:**

Explain how you can tell just by looking that $1.2 \times 10^2$ definitely represents a larger quantity than $3.6 \times 10^{-2}$? (Write your answer in the box below.)

| |
|---|
| |

**Part D:**

- Is $8 \times 10^{-2}$ equivalent to a *whole number* or to a *fraction less than one?* Answer: It is equivalent to ------------------------------------------------
- Is $8 \times 10^{-2}$ *more* or *less* than $\frac{1}{2}$? Answer: It is ------ than $\frac{1}{2}$.
- When the number 9,200,000 is written in *scientific-notation form* will it have a positive or negative exponent? Answer: It will have a ------ exponent.

## Scientific Notation Continued at a Grade 8 Level (8.EE.1, 3, 4)

Review: Scientific notation is a different way of writing very large and very small numbers. Numbers in scientific notation form are usually easier to compute when multiplying, dividing, adding, or subtracting.

> Here is a very large number in standard form: 342,000,000,000,000

> Here is that same very large number in scientific notation: $3.42 \times 10^{14}$

> Here is a very small number in standard form: 0.00000045

> Here is that same very small number in scientific notation: $4.5 \times 10^{-7}$

## COMPUTATING WITH SCIENTIFIC NOTATION

## Multiplying and Dividing Numbers in Scientific Notation Form

When you multiply and divide numbers in standard form you line up the digits by place value. When you multiply and divide numbers in scientific notation form you work with their *exponents* too.

Since the bases (the 10s) are the same this is quite easy.

### Multiply the Decimal Numbers, Add the Exponents

| Sample (1) | Sample (2) | |
|---|---|---|
| $(8.5 \times 10^3)(1.1 \times 10^5)$ | $(2.6 \times 10^{-2})(3.5 \times 10^8)$ | |
| $(8.5 \times 1.1)(10^{3+5})$ | $(2.6 \times 3.5)(10^{-2+8})$ | |
| $9.35 \times 10^8$ | $9.1 \times 10^6$ | scientific notation |
| 935,000,000 | 9,100,000 | standard form |

### Divide the Decimal Numbers, Subtract the Exponents

Sample: (3) $\frac{8.32\times10^7}{1.3\times10^5}=\left(\frac{8.32}{1.3}\right)(10^{7-5})=\mathbf{6.4 \bullet 10^2}$

Sample: (4) $\frac{2.5326\times10^8}{5.4\times10^3}=\frac{(2.5326)}{(5.4)}\times10^{8-3}=.469\times10^5$

Now we need to put the solution in correct scientific notation form

$$0.469 \times 10^5 = 4.69 \times 10^{5+(-1)} = \mathbf{4.69 \times 10^4}$$

Practice the multiplication and division problems below. Show all steps. You may use a calculator for multiplying or dividing the decimal numbers. Put your final answer in the correct scientific notation form.

**1.** $(6.5 \times 10^2)(2.1 \times 10^5)$ = (______)(______) × (10______ + ______) = ______

Put your answer in the correct scientific notation form: ______

**2.** $(9.2 \times 10^{-3})(4.8 \times 10^{-4})$ = (______)(______) × (10______ + ______) = ______

Put your answer in the correct scientific notation form: ______

**3.** $\frac{3.98\times10^5}{2.4\times10^3}=$ **4.** $\frac{4.5\times10^{-9}}{3.2\times10^4}=$

Study these solutions to the above practice examples and compare them to the work you showed and your final answers.

**1.** $(6.5 \times 10^2)(2.1 \times 10^5) = (6.5)(2.1) \times (10^{2+5}) = 13.65 \times 10^7 = 1.365 \times 10^8$

**2.** $(\mathbf{9.2} \times 10^{-3})(\mathbf{4.8} \times 10^{-4}) = (\mathbf{9.2})(\mathbf{4.8}) \times (10^{-3-4}) = 44.16 \times 10^{-7} = 4.416 \times 10^{-6}$

**3.** $\frac{3.98\times10^5}{2.4\times10^3}=\frac{3.98}{2.4}(10^{5-3})\approx1.6583\times10^2$

This answer is already in *scientific notation form.*

**4.** $\frac{4.5\times10^{-9}}{3.2\times10^4}=\frac{4.5}{3.2}(10^{-9-4})\approx1.406\times10^{-13}$

**ADDING and SUBTRACTING numbers in SCIENTIFIC NOTATION form can be more complex than multiplying or dividing.**
When you add and subtract numbers in standard form you line up the digits by place value. When you add and subtract numbers in scientific notation form you work with the *exponents* too.

## ADDING AND SUBTRACTING NUMBERS IN SCIENTIFIC NOTATION FORM

When you add or subtract regular fractions, you must have common denominators first.

**Step 1:** Make **common denominators**: $\frac{3}{5}+\frac{7}{10}=?\quad \frac{3}{5}+\frac{7}{10}=\frac{6}{10}+\frac{7}{10}=\frac{13}{10}$

**Step 2: Simplify** the fraction: $\frac{13}{10}=1\frac{3}{10}$

In a similar way, when you add or subtract numbers in scientific notation form, they must have the same exponent value.

**Step 1:** Rewrite with the **same exponents.**

$1.2 \times 10^5 + 3.45 \times 10^4 = ?$ $\quad$ $12.0 \times 10^4 + 3.45 \times 10^4 = ?$

**Step 2: Add** the decimal numbers: $(12.0 + 3.45) \times 10^4 = 15.45 \times 10^4$

**Step 3:** Put the answer in correct scientific notation form.

$15.45 \times 10^4$ in scientific notation form is $1.545 \times 10^{4+1} = 1.545 \times 10^5$

**Practice these addition and subtraction problems. Show all steps. You should not need a calculator. Put your final answer in the correct scientific notation form.**

**1.** $1.54 \times 10^4 + 1.22 \times 10^4 = (\_\_\_ + \_\_\_) \times 10^4 = \_\_\_ \times 10^4$

**2.** $8.6 \times 10^3 + 9.920 \times 10^2 =$

**3.** $9.3 \times 10^3 - 3.2 \times 10^2 =$

**Study these solutions to the above practice examples and compare them to the work you showed and your final answers.**

**1.** $1.54 \times 10^4 + 1.22 \times 10^4 = (1.54 + 1.22) \times 10^4 = 2.76 \times 10^4$

In this example both numbers had the same exponents so you just add the decimal numbers and keep the $10^4$. Your final answer is already in the correct scientific notation form.

**2.** $9.6 \times 10^3 + 920 \times 10^2 = 8.6 \times 10^3 + 92.0 \times 10^3 = (8.6 + 92) \times 10^3 = 100.6 \times 10^3$

Now put your answer in correct scientific notation form. $1.006 \times 10^5$

**3.** $3.2 \times 9.3 - 10^2 \times 10^3 =$

**Step 1:** Rewrite with the same exponents: $0.32 \times 9.3 - 10^3 \times 10^3 =$

**Step 2:** Subtract the decimal numbers: $(9.3 - 0.32) \times 10^3 = 8.98 \times 10^3$

## COMPARING DATA IN REAL-LIFE SITUATIONS

(For answers, see pages 234–235.)

(Yes, you may use a calculator.)

**1.** Compare the land area of two countries:

The land area of Sweden is listed as 173,732 square miles.

The land area of Saudi Arabia is listed as 864,869 square miles.

What is the ratio of the land area of Saudi Arabia to the land area of Sweden?

*(Show your work and round your final answer to the nearest whole number.)*

Answer: The land area of Saudi Arabia is about ------ times as large as the land area of Sweden.

**2.**

| Country | Land Area (2014) |
|---|---|
| Australia | 2.9 million square miles |
| Brazil | 3.3 million square miles |
| United States of America | 3.8 million square miles |
| India | 1.24 million square miles |
| Canada | 3.8 million square miles |
| Russia | 6.6 million square miles |

**Part A:** Which two countries in the table above have the greatest difference in land area?

Answer: ------------------------ and ------------------------

**Part B:** How many more times as large is the land area of the country with the largest land area than the country with the smallest land area? *(Write your answer in decimal form to the nearest tenth.)*

Answer: ------------ times greater.

**Part C:** How many times greater is the land area of the United States than the land area of India? *(Write your answer to the nearest whole number.)*

Answer: The land area of the United States is ____________ times larger than the land area of India.

**3.**

| Country | Population in 2006 (m = millions) |
|---|---|
| China | 1,304.2 m |
| South Korea | 47.7 m |
| Russia | 143.0 m |
| United States | 294.0 m |

**Part A:** Use the data from the table above for 2006.
Write the populations of China and South Korea in scientific notation.

China: ________________________ South Korea: ________________________

**Part B:** How many times as large was China's population than South Korea's?

Answer: In 2006, China's population was ________________________ times greater than South Korea's population. Write your answer in scientific notation.

**Part C:** Write your final answer in decimal form.
(Round your answer to the nearest whole number.)

Answer: In 2006 China's population was about ____________ times as large than South Korea's population. Write this answer in decimal form; round to the nearest whole number.

**4.**

| Country | Population in 2006 (millions) |
|---|---|
| China | 1,304.2 m |
| South Korea | 47.7 m |
| Russia | 143.0 m |
| United States (U.S.A.) | 294.0 m |

**Part A:** Use the data from the table above for 2006.
Write the populations of China and the U.S.A. in scientific notation.

China: ________________________ U.S.A.: ________________________

**Part B:** How many times larger was China's population than the U.S.A.'s population? (Leave this answer in scientific notation form.)

Answer: China's population was ____________ times greater than the U.S.A.'s population.

**Part C:** Write your final answer in decimal form. (Round your answer to the nearest tenth.)

Answer: In 2006 China's population was about ____________ times greater than the U.S.A.'s population.

## ALGEBRAIC ORDER OF OPERATIONS (A REVIEW)

*Please Excuse My Dear Aunt Sally*

When you get dressed you put your socks on before your shoes. In life there are some things that are best done in a certain order. In math there also is a correct order for doing some things.

We'll be working with *expressions*. Expressions are made up of numbers and operations. The numbers might be 16, –4, 3.4, $\frac{1}{2}$; operations look like ×, ÷, +, –, or ( ).

If different people were asked to *evaluate* the expression $3(4 - 6) + 12 \div 2 - 16(-1)^0 + 4^2$, they must follow certain rules of order if they are to get the same answer. Should you divide by 2 first? Maybe you should add the four-squared first. This can get confusing. Is there a rule to follow? Yes! Just remember the phrase ***P**lease **E**xcuse **M**y **D**ear **A**unt **S**ally* (***PEMDAS***) and evaluate your expressions in this order: ***P**arentheses*, ***E**xponents*, then ***M**ultiplication* and ***D**ivision* (work from left to right), and finally ***A**dd* and ***S**ubtract* (work from left to right).

### Examples

Think of the operation you would do first in each of the following expressions; then evaluate each expression. Remember PEMDAS.

**A.** $(9 - 2) \times 4 - 4$

First, work inside **P**arentheses — $7 \times 4 - 4$
Do **M**ultiplication next — $28 - 4$
Finally, **S**ubtract — $24$

**B.** $46 - 4^2 + 3$

This expression has no parentheses, so
First, work with **E**xponents — $46 - 16 + 3$
Next, **S**ubtract and **A**dd — $30 + 3$
$33$

**C.** $(28 \div 4) + 5^2(3 - 1)$

| Step | Result |
|---|---|
| First, work inside both **P**arentheses | $7 + 5^2(2)$ |
| Work with **E**xponents next | $7 + 25(2)$ |
| Then do **M**ultiplication | $7 + 50$ |
| Finally, **A**dd | $57$ |

**D.** $50 \times 2 + 12 - 2 \times 4$

| Step | Result |
|---|---|
| This expression has no parentheses or exponents, so | $100 + 12 - 8$ |
| First, do both **M**ultiplications | $112 - 8$ |
| Next, **A**dd and **S**ubtract | $104$ |

**E.** $5 + 6^2 \times 10$

| Step | Result |
|---|---|
| First, work with the **E**xponent | $5 + 36 \times 10$ |
| Then **M**ultiply | $5 + 360$ |
| Finally, **A**dd | $365$ |

**F.** $[(6 \times 5) + 3](3)$

What do you do if you have parentheses inside parentheses or parentheses inside brackets? Keep the same order of operations. Do the inside parentheses first.
$[(6 \times 5) + 3](3)$

| Step | Result |
|---|---|
| First **M**ultiply 6 and 5 | |
| Next, continue working inside | $[(30) + 3](3)$ |
| the brackets and **A**dd inside parentheses | $[33]\,(3)$ |
| Last, **M**ultiply | $99$ |

**G.** $(5 + 3 \times 20) \div 13 + 3^2$

| Step | Result |
|---|---|
| Work inside **P**arentheses first | $(5 + 3 \times 20) \div 13 + 3^2$ |
| **M**ultiply inside parentheses | $(5 + 60) \div 13 + 3^2$ |
| **A**dd inside parentheses next | $(65) \div 13 + 3^2$ |
| **E**xponents are next | $65 \div 13 + 9$ |
| **D**ivide | $5 + 9$ |
| **A**dd | $14$ |

## The New Look of Multiple-Choice Questions

(For answers, see page 235.)

Use the expression shown to answer the following statements.
Check Yes __✔__ or No __✔__.

$$3(5 - 2)^2 + 4(3^2) - 9 \div 3(2)$$

**A.** $-9 \div 3$ is an appropriate first step. Yes ______ No ______

**B.** $3^2$ is equivalent to 6. Yes ______ No ______

**C.** $(5-2)^2 = 5^2 - 2^2$. Yes ------ No ------

**D.** $4(3^2) = 4 \times 3 \times 3$. Yes ------ No ------

**E.** The expression $3(5-2)^2 + 4(3^2) - 9 \div 3(2)$ is equivalent to 57. Yes ------ No ------

## PRACTICE: Algebraic Order of Operations

(For answers, see pages 235–236.)

Do NOT use a calculator to practice these examples!

**1.** $27 - 18 \div 6$

What do you do first?

- ○ **A.** subtract
- ○ **B.** divide

Evaluate the expression.

- ○ **A.** 9
- ○ **B.** 24
- ○ **C.** 1.5
- ○ **D.** 25

**2.** $36 + 18 \div 2 \times 3 - 8$

What do you do first?

- ○ **A.** add
- ○ **B.** subtract
- ○ **C.** divide
- ○ **D.** multiply

Evaluate the expression.

- ○ **A.** 96
- ○ **B.** 73
- ○ **C.** 32
- ○ **D.** 55

**3.** Evaluate the expression.

$$(72 \div 9 - 2) + 2 \times 3$$

- ○ **A.** 12
- ○ **B.** 16
- ○ **C.** 24
- ○ **D.** 36

**4.** Evaluate the expression.

$$(20 + 8 \div 4) + (3^2 \times 4)$$

- ○ **A.** 19
- ○ **B.** 43
- ○ **C.** 58
- ○ **D.** 71

**5.** Evaluate the expression.

$$[2(5-1) + 3(2 \times 2)] - 4$$

- ○ **A.** 7
- ○ **B.** 16
- ○ **C.** 15
- ○ **D.** 20

**6.** Evaluate the expression.

$$(6-4)^3 - 4$$

- ○ **A.** –5
- ○ **B.** 2
- ○ **C.** 4
- ○ **D.** 12

**7.** To evaluate this expression, what should you do first?

$$(9 - 2 \times 4) + 2^3(6)$$

- ○ **A.** $2 \times 2 \times 2$
- ○ **B.** $9 - 2$
- ○ **C.** $2(6)$
- ○ **D.** $2 \times 4$

8. Add parentheses to this expression $2 + 3 \times 2 \times 5 - 1$ so its evaluation is 49.

A. $(2 + 3) \times 2 \times 5 - 1$

B. $2 + (3 \times 2) \times 5 - 1$

C. $2 + 3 \times (2 \times 5) - 1$

D. $2 + 3 \times 2 \times (5 - 1)$

9. Add parentheses to this expression $4 + 3 \times 2 \times 5 - 1$ so its evaluation is 28.

A. $(4 + 3) \times 2 \times 5 - 1$

B. $4 + (3 \times 2) \times 5 - 1$

C. $4 + 3 \times (2 \times 5) - 1$

D. $4 + 3 \times 2 \times (5 - 1)$

10. Jaime says the evaluation of the following expression is 28; Billy says it is 24.

$$2(2 + 5 \times 2)$$

Who is correct? Explain your answer and show all work.

## RATIO AND PROPORTION (From Grade 7.RP And 7.G1)

A fraction is a ratio. Two ratios connected with an equals sign is called a *proportion*. You can understand the concept of proportion by looking at some drawings.

### Examples

**Part A.** **Proportion:** In Figure A, the person is not in proportion to the house. When you compare the height of the person to the door, the person looks too big. In Figure B, the person seems to be in proportion to the door. The person and the house seem to be in the correct proportion.

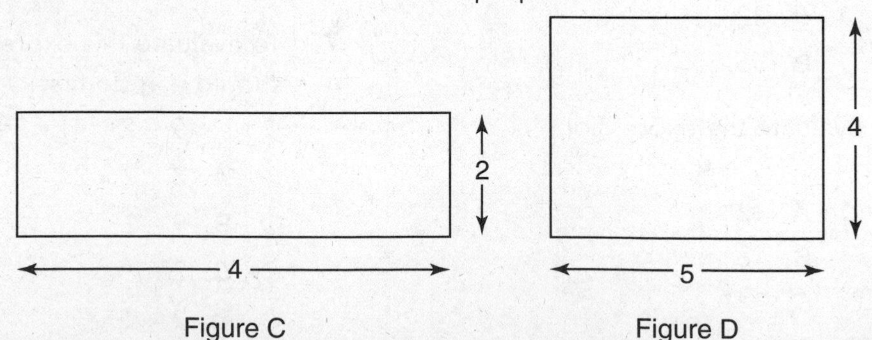

**Part B.** Refer to Figures C and D at the bottom of page 44. These shapes are *not* in proportion. In Figure C, the rectangle has length = 4 and height = 2. In Figure D, the rectangle has length = 5 and height = 4.

Figure C: $\frac{\text{Length } 4}{\text{Height } 2}$  Figure D: $\frac{\text{Length } 5}{\text{Height } 4}$  $\frac{4}{2} \neq \frac{5}{4}$

These shapes are *not* in proportion.

**Part C.** Ratio and proportion are often used in drawings of maps or architectural drawings. For example, if the architect's plans are in the scale of $\frac{1}{3}$, it means that 1 inch on the plans equals 3 feet in real life. How long is a garage in reality if the plans show it to be 8 inches long?

$\frac{1}{3} = \frac{8}{x}$ In this proportion, both numerators represent the scale from the plans (1 inch and 8 inches). The denominators give you the actual size in real life (3 feet and $x$ feet). We can solve this problem by cross multiplying

$\frac{1}{3} \times \frac{8}{x}$  $\frac{\text{inches}}{\text{feet}}$  $(1)(x) = (3)(8)$

$x = 24$ feet

So, if 1 inch represents 3 feet, then 8 inches represent 24 feet. The garage is 24 feet long.

**Part D.** The builder is showing plans of a new housing development to a future buyer. The buyer notices that the plans show that the scale used is $\frac{1}{4}$ inch = 25 feet. Look at the scale drawing below; this represents the lot for one single-family house.

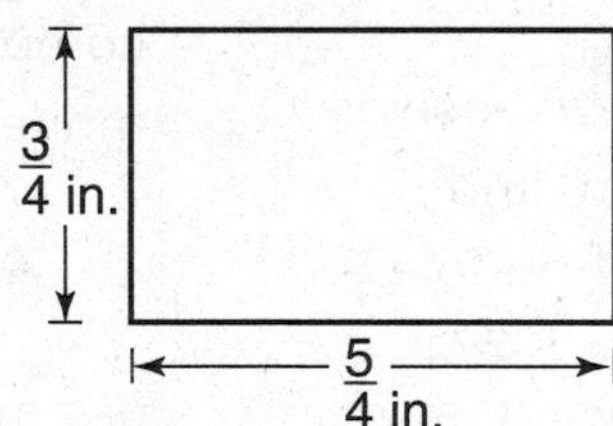


What expression would give you the correct area in square feet of this lot?

**A.** $125 \times 75$  **B.** $\frac{3}{4} \times \frac{5}{4}$  **C.** $3 \times 25$  **D.** $75 \times 5$

Since $\frac{1}{4} = 25$ ft, the width = $\frac{3}{4}$ inch on paper and 75 feet in reality and the length is $1\frac{1}{4}$ inches or $\frac{5}{4}$ inches on paper and 125 feet in reality.

Area of a rectangle = Length × Width = 125 × 75

Solution ***A*** is correct.

**E.** You need 1 cup of pancake mix to make 6 pancakes. How many cups of mix will you need to make 18 pancakes? Which is the correct proportion to solve this problem?

**A.** $\frac{1}{6} = \frac{18}{x}$  **B.** $\frac{1}{6} = \frac{x}{18}$  **C.** $\frac{6}{1} = \frac{x}{18}$  **D.** $\frac{6}{1} = \frac{12}{x}$

$$\frac{1 \text{ cup mix}}{6 \text{ pancakes}} = \frac{x \text{ cups mix}}{18 \text{ pancakes}} \qquad (1)(18) = (x)(6)$$

$$18 = 6x$$

$$3 = x$$

(3 cups of mix will give you 18 pancakes)

The correct answer is **B** $\frac{1}{6} = \frac{x}{18}$; $6x = 18$, $x = 3$

## PRACTICE: Ratio and Proportion

(For answers, see page 236.)

**1.** If there are 14 girls in the class and 11 boys, what is the ratio of girls to the total number of students in the class?

○ **A.** $\frac{25}{14}$

○ **B.** $\frac{14}{25}$

○ **C.** $\frac{14}{11}$

○ **D.** $\frac{11}{14}$

**2.** If 4 out of 7 people in a New Jersey town use Brand *X* detergent, find the approximate number of people that use Brand *X* if there are 5,271 people in the town.

○ **A.** 1,757

○ **B.** 3,012

○ **C.** 5,250

○ **D.** 9,975

**3.** At college, Will gets discounted meals; he spends \$2.00 for breakfast, \$3.00 for lunch, and \$7.00 for dinner each weekday. If he makes \$100 each week at his library job, write a ratio that shows the amount of money spent on food Monday through Friday as compared to the total amount of money he makes each week.

○ **A.** $\frac{12}{100}$

○ **B.** $\frac{60}{100}$

○ **C.** $\frac{72}{100}$

○ **D.** $\frac{84}{100}$

**4.** Charleen is enlarging the figure shown so that the base of the new figure changes from 9 to 18 centimeters.

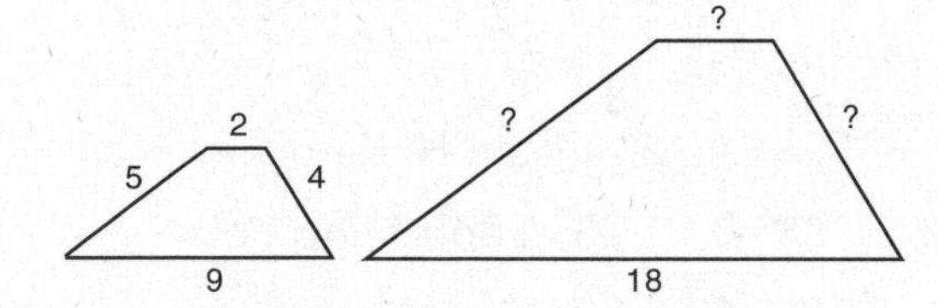


Note: Figures are not drawn to scale.

What is the perimeter of the enlarged figure?

- ○ **A.** 29 cm
- ○ **B.** 38 cm
- ○ **C.** 40 cm
- ○ **D.** 42 cm

**5.** If triangles are similar, then their sides are in proportion. Look at these two *similar* triangles. The small triangle has a height of 3 feet. How tall is the larger triangle?

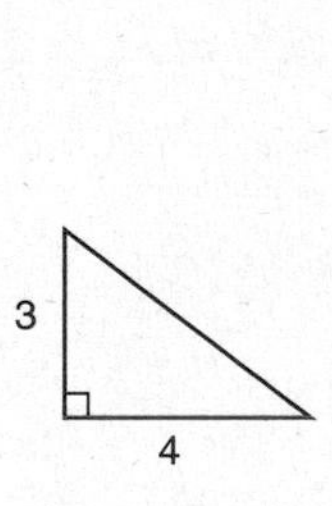


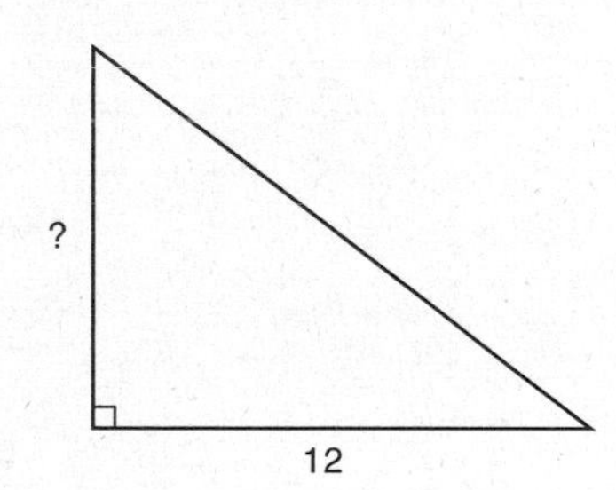


- ○ **A.** 4 feet
- ○ **B.** 6 feet
- ○ **C.** 8 feet
- ○ **D.** 9 feet

**6.** Trey wants to design a model car. The diameter of the real tire is 28", and the diameter of the hubcap is 15". If he makes a drawing where the tire is 7" in diameter, which proportion would help him find the diameter of the hubcap in his drawing?

- ○ **A.** $\frac{28}{15} = \frac{x}{7}$
- ○ **B.** $\frac{28}{15} = \frac{7}{x}$
- ○ **C.** $\frac{15 + x}{28 + 7}$
- ○ **D.** $\frac{\text{Real tire}}{\text{Real hubcap}} = \frac{\text{Drawn hubcap}}{\text{Drawn tire}}$

**7.** Jaime Escalante became a well-known math teacher from California because of his ability to inspire his students to succeed even though they thought they could not. Solve the following proportion example to find out the name of the movie made about this group.

$$\frac{8}{2} = \frac{m}{2.5}$$

- ○ **A.** $m = 10$ *Stand and Deliver*
- ○ **B.** $m = 6.4$ *Blackboard Jungle*
- ○ **C.** $m = 20$ *Beverly Hills High*
- ○ **D.** $m = 8.25$ *California Black Hawks*

**8.** Solve the following proportion example for $c$.

$$\frac{6}{c} = \frac{0.3}{9}$$

- ○ **A.** $c = 0.2$
- ○ **B.** $c = 2.0$
- ○ **C.** $c = 180$
- ○ **D.** $c = 18.0$

**9.** In real life, Mike's ladder is 24 feet tall. At the base, it is leaning 5.4 feet away from the wall of his house. In the smaller drawing, the ladder is only 6 inches tall. How far away from the house should it be drawn to keep the picture in proportion to the real thing?

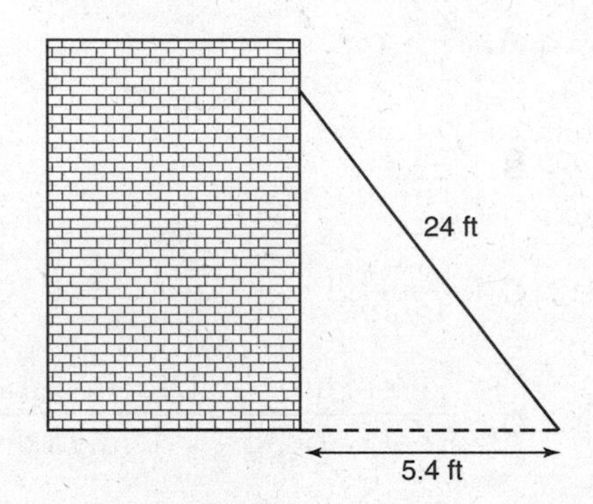


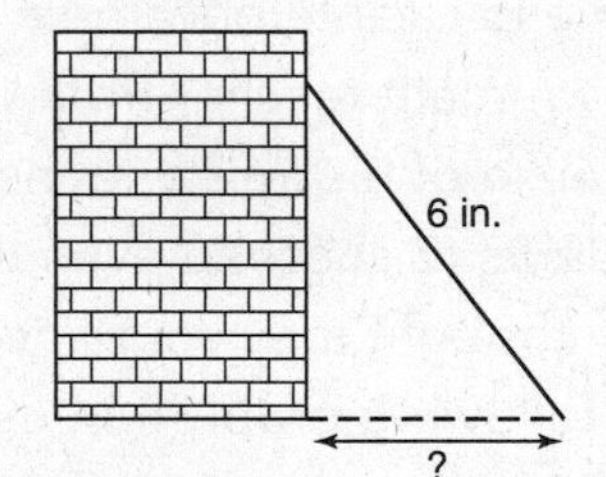


Note: Figures are not drawn to scale.

- ○ **A.** 0.135"
- ○ **B.** 1.35"
- ○ **C.** 0.266"
- ○ **D.** 2.66"

**10.** Solve this proportion problem to find out the name of the architect who designed the John F. Kennedy Library in Boston, Massachusetts.

$$\frac{5}{j} = \frac{12}{3} \quad j = ?$$

- ○ **A.** 20 Jin H. Kinoshita
- ○ **B.** 1.25 Leoh Ming Pei
- ○ **C.** 0.80 An Wang
- ○ **D.** 8.0 Maya Ying Lin

## PRACTICE: SCR Non-calculator Questions

Each question is worth 1 point. No partial credit is given.

(For answers, see pages 236–237.)

**1.** Using the correct algebraic order of operations, what would be the first step you should do to simplify this expression?

$$10 + 8 \div 2 + 3(5 - 2) - 4^2$$

Answer: ______________________

**2.** To add the fractions below, Jessica says you should use 12 as the lowest common denominator, but Sarah says that 18 is the lowest common denominator. Who is correct?

$$\frac{1}{6}+\frac{2}{9}$$

Answer: -----------

**3.** The center of the moon is 238,900 miles away from the center of Earth. Write these miles in scientific notation.

Answer: -----------

**4.** Put these numbers in numerical order starting with the smallest number.

$3^2$ $\pi$ $\sqrt{9}$ $2^3$ $\sqrt{3}$

Answer: -----------

**5.** What is 0.85 written as a fraction?

Answer: -----------

**6.** On an official New Jersey map, the scale shows 0.5 inches for every 10 miles. How far apart are two towns if the map shows them as 2.5 inches apart?

Answer: -----------

**7.** If your dad receives an average of 40 e-mails each day at work, how many e-mails can he expect to receive during a month? (For this example, we'll let 4 weeks = 1 month and each week = 5 workdays.)

Answer: -----------

Use the menu below to answer questions 8 and 9.

**LUNCHEON MENU**

(Special Today: NO SALES TAX)

| Item | Price |
|---|---|
| Hamburger plain | $2.50 |
| Cheeseburger | $3.00 |
| Chicken nuggets | $3.25 |
| Hot dog plain | $2.00 |
| Hot dog with chili | $2.50 |
| French fried potatoes | $1.50 |
| Soda | $1.00, $1.50, $2.00 (sm., med., large) |
| Juice | $1.50, $2.00 (sm., med.) |

**8.** If Max had $5.00, would he have enough money to buy a hamburger, fries, and a small soda?

Answer: -----------

**9.** Jose has a $10 bill. How much change would he receive if he bought chicken nuggets, fries, and a medium juice?

Answer: -----------

Original
drawing

**10.** Abigail was selected to draw a picture for the school's Art and Literary magazine. The magazine picture must be no wider than 8 inches. If Abigail uses a scale factor of 3, how wide should she make her original drawing?

**A.** 24 inches wide

**B.** 27 inches wide

**C.** 32 inches wide

**D.** 36 inches wide

Answer: ______________________

## PRACTICE: Performance-Based Assessment Questions

(For answers, see pages 237–239.)

**1.** A New York City parking garage charges $40.00 to park for the whole day, or $6.00 for the first hour and $4.00 for each additional half-hour.

**Part A:** If Bernice and her mom were taking a trip to the New York City Planetarium, they would need to park for 3 hours. Which plan should they use? Show your work and explain your answer.

**Part B:** Next month they plan to go to the theater and a dinner in New York City and expect to be there for 7 hours. Show how the $40.00 per-day rate would be the best choice.

**2.** Sam's brother, Daryl, works at an automobile factory and makes $12.00 per hour. Since the company was not selling as many cars lately, they decreased each person's wages by 8%.

**Part A:** What will Daryl's new hourly wage be?

**Part B:** If Daryl worked a typical 40-hour week, how much less will his weekly wages be with this 8% reduction?

**3.** Brett decided to have a big breakfast at the Ridge Diner. Below is a copy of the diner's breakfast menu.

**THE RIDGE DINER MENU**

Juice
(orange, grapefruit, tomato, or cranberry)
$1.20

Eggs any style
One $1.25 Two $2.25 Three $3.25

Toast or Plain Muffin
$1.20

Coffee or Tea with refill
$1.25

Milk
Small $.95 Large $1.25

Side Dishes
Sausage or Bacon $1.50
Homefries $1.00

Morning Special
(5:00 A.M.–11:00 A.M., MONDAY-FRIDAY)
Juice, toast or muffin, coffee or tea, 2 eggs any style, sausage or bacon, and homefries
$6.95

**Part A:** How much money would Brett save if he ordered the Morning Special instead of buying each item separately?

**Part B:** How much would his total bill be with a 6% sales tax and a tip of $1.50? (Round your answer up to the nearest penny.)

**Part C:** How much change would he get from $10.00?

**4.** In a recent survey about how many people attended a local amusement park, it was determined that 3 out of 8 customers had also attended the park the week before.

**Part A:** What percentage of customers attended the park the week before?

**Part B:** If 4,000 people attended the park this week, how many of these people can we estimate also attended the week before?

**5.** Mr. B. takes excellent care of his car. He gets the oil changed every 5,000 miles, and has the tires rotated every 10,000 miles.

**Part A:** If he drove his car for 50,000 miles, how many trips would he make to the service station for the above two services?

**Part B:** If each of these trips to the service station is 8 miles, each round-trip would then be 16 miles. If Mr. B's car gets 20 miles per gallon and gasoline costs $2.45 per gallon, how much would he spend in gas for the above trips to the service station?

6. The Sweater Barn in an upstate New York mall and the Teen Outlet in New Jersey are both having big sweater sales.

| Store | Sweater and Cardigan Set | |
|---|---|---|
| Sweater Barn N.Y. | Sweater $24 less 15% | Matching cardigan is 50% off of the sweater sale price |
| Teen Outlet N.J. | Sweater $18 less 10% | Matching cardigan is the same price as the sweater |

Mary wants to purchase three sets as birthday gifts for her granddaughters. She notices that the New York State store charges 8% sales tax on clothing and New Jersey has no sales tax on clothing.

**Part A:** What will the three sets cost at the Sweater Barn in New York?

**Part B:** What will the three sets cost at the Teen Outlet in New Jersey?

**Part C:** Which is the better buy? Explain.

Name: ________________________ Date: ________________________

# Chapter 1 Test: The Number System

35 minutes

(Use the *PARCC Grade 8 Mathematics Assessment Reference Sheet* on page 359.)

(For answers, see pages 239-241.)

**1.** Evaluate the following expression.

$$-3\left(\frac{1}{2}\right)^{-2} =$$

Answer: ________________________

**2.** Evaluate the following expression.

$$\sqrt[3]{27}+\sqrt{\frac{36}{144}}+|-20| =$$

Answer: ________________________

**3.** Which rational number could be the value of $x$ on this number line?

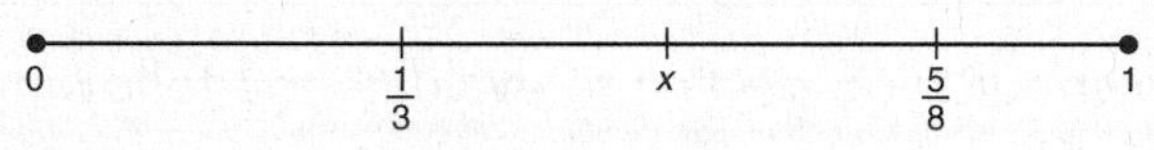


○ **A.** $\frac{1}{4}$

○ **B.** $\frac{2}{3}$

○ **C.** 0.7

○ **D.** 0.45

**4.** The number $\sqrt{115}$ is between

○ **A.** 8 and 9

○ **B.** 9 and 10

○ **C.** 10 and 11

○ **D.** 11 and 12

**5.** The following are the calorie counts of some breakfast food served at Burger Barn.

| | |
|---|---|
| BB Ultimate Breakfast platter | 1450 calories |
| Cinnamon minis (4) | 1200 calories |
| Hash browns large | 670 calories |
| Sausage and Cheese Muffin Sandwich | 300 calories |

**Part A:**

How many times more calories does The BB Ultimate Breakfast have than the Sausage and Cheese Muffin Sandwich? (Round your answer to the nearest whole number.)

Answer: It has about ______________________ times more calories.

**Part B:**

If Coach Jones wanted to order fifteen breakfasts for his team, how many times more calories would 15 BB Ultimate Breakfasts have than 15 Sausage and Cheese Muffin Sandwiches? (Round your answer to the nearest tenth.)

Answer: They would have about ______ times more calories.

**6.** Which operation should you do first to evaluate the following expression?

$$16 \times 4 \div 2 + (8 - 2)^2 + 10^2$$

○ **A.** square the 10

○ **B.** divide 4 by 2

○ **C.** multiply 16 by 4

○ **D.** subtract 2 from 8

**7.** Which of the following expressions are equivalent to $\left(\frac{-2}{5}\right)^2$?
Check Yes ✔ or No ✔.

**A.** $\left(\frac{2}{-5}\right)^2$ Yes ______ No ______

**B.** $\frac{4}{10}$ Yes ______ No ______

**C.** $-\frac{4}{10}$ Yes ______ No ______

**D.** $\frac{-4}{25}$ Yes ______ No ______

**E.** $\frac{16}{100}$ Yes ______ No ______

**8.** Which of the following statements are true about $\sqrt{144}$?

☐ **A.** $\sqrt{144}$ represents only the positive square root of 144

☐ **B.** $\sqrt{144}$ is equivalent to $4^2 - 2^2$

☐ **C.** $\sqrt{144} = \left(\frac{1}{12}\right)^2$

☐ **D.** $\sqrt{144} = 12 \text{ or } -12$

☐ **E.** $\sqrt{144}$ will always be an even number

**9.** The smaller the U-value, the better the material is for insulation. Use the chart below and select the material that is the best insulator.

| Material | U-value |
|---|---|
| Roof with no insulation | $2.2 \times 10^0$ |
| Insulated roof | $3.0 \times 10^{-1}$ |
| Single brick wall | $3.6 \times 10^0$ |
| Double brick wall, air cavity between | $5.0 \times 10^{-1}$ |
| Double brick wall filled with foam | $5.0 \times 10^{-1}$ |
| Single-glazed window | $5.6 \times 10^0$ |
| Double-glazed window with air gap | $2.7 \times 10^0$ |
| Floor without carpets | $1.0 \times 10^0$ |
| Floor with carpets | $3.0 \times 10^{-1}$ |

○ **A.** double brick wall filled with foam

○ **B.** roof with no insulation

○ **C.** double-glazed window with air gap

○ **D.** floor with carpets

**10.** A used car was priced at $7,000. The salesperson then offered a discount of $350. This means the salesperson offered a discount of ______ %.

○ **A.** 5

○ **B.** 20

○ **C.** 80

○ **D.** 95

**11.** Roni is making a large poster for the hall bulletin board. She is enlarging the figure drawn below so the side corresponding to the right side will be 9 inches long instead of 3 inches. What will be the distance around this enlarged shape (the perimeter)?

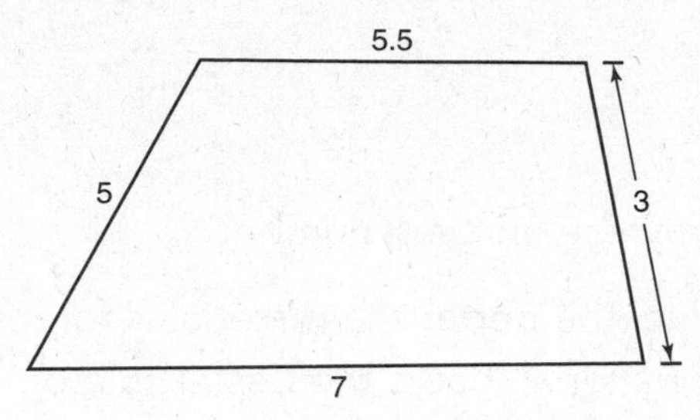


- ○ **A.** 61.5 inches
- ○ **B.** 41 inches
- ○ **C.** 20.5 inches
- ○ **D.** 189 inches

**12.** There are 650 students who were surveyed. Use the circle graph below to determine how many students selected pizza or chili as their favorite. Select all choices that are correct.

- ☐ **A.** 429 students
- ☐ **B.** 416 students
- ☐ **C.** 64% of the students
- ☐ **D.** 36% of the students
- ☐ **E.** $>(20)^2$ students

26% Burgers
56% Pizza
10% Soup
8% Chili

**13.** Which of the following expressions are equivalent to 64?

- ☐ **A.** $8^{-1} \times 8^{-1}$
- ☐ **B.** $8^{-2} \times 8^{4}$
- ☐ **C.** $8^{-3} \times 8$
- ☐ **D.** $8^{6} \times 8^{-4}$
- ☐ **E.** $8^{2} \times 8^{4}$

**14.** Based on the following expression, $4^{-3} \cdot 4^{2}$, select all that apply.

- ☐ **A.** Simplified, this is equivalent to $4^{-6}$.
- ☐ **B.** This expression is equivalent to $\frac{1}{4}$.
- ☐ **C.** The following is an equivalent expression: $4^{5} \cdot 4^{-6}$.
- ☐ **D.** An equivalent expression is $16 \div 4^{3}$.
- ☐ **E.** Another way to write this is –4.

## Performance-Based Assessment Questions

**Directions for Questions 15 and 16:** Respond fully to the PBA questions that follow. Show your work and clearly explain your answer. You will be graded on the correctness of your method as well as the accuracy of your answer.

**15.** Four local gas stations each advertise that they have the lowest prices in the area. Marco and his dad drove to each station and filled up the tanks of five different cars.

Use the chart below as a guide:

| Gas Station | Total Cost | Number of Gallons |
|---|---|---|
| Alco | $18.24 | 9.5 |
| Bright | $18.31 | 9.2 |
| Custom | $17.02 | 8.3 |
| Dixon | $16.80 | 8 |
| Extra | $18.45 | 9 |

**Part A:** Which gas station really has the lowest price? Explain.

**Part B:** If Marco purchases 20 gallons at the lowest price, how much would he save instead of going to the most expensive gas station?

**16.** There is a big sale at the local department store. Every hour their winter jackets get reduced by 5%. Susan wants to buy the jacket that is originally priced at $89.50. The sale begins at 1:00 P.M. and continues until midnight.

**Part A:** If Susan has only $70.00, will she be able to buy the jacket today?

**Part B:** If so, what will it cost?

**Part C:** What time will she be able to buy it?

Explain or show your work to describe how you arrived at your answer.

# Geometry

## WHAT DO PARCC 8 GEOMETRY QUESTIONS LOOK LIKE?

### Multiple-Choice Question (MC)

**Example 1:**

You are given a square with a side measuring 8 inches and a triangle with a base of 8 inches and a height of 16 inches. Which of the following is true?

○ **A.** Area of the square is > area of the triangle.

○ **B.** Area of the square is < area of the triangle.

○ **C.** Area of the square = area of the triangle.

○ **D.** There is not enough information to answer the question.

**Strategies and Solutions**

Draw and label diagrams.

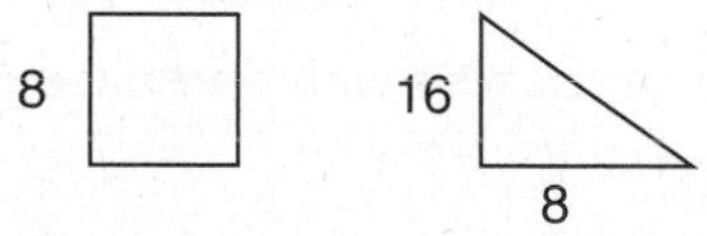

Show all work, even work done with a calculator.

Remember symbols (< means less than; > means greater than).

Area square = $s^2 = 8^2 = (8)(8) = 64$ sq. in.

Area triangle = $\frac{bh}{2} = \frac{(8)(16)}{2} = 64$ sq. in.

The correct answer is ***C***.

## Short Constructed-Response Question (SCR)

### Example 2:

(No calculator permitted.)

What is the area of the *smallest* polygon described below?

- A rectangle 18 ft long by 2 ft wide.
- A square that is 6 ft long on each side.
- A triangle that has a base of 9 ft and is 4 ft tall.

Answer: ..................

**Solution**

- Area rectangle = (length)(width) = (18)(2) = 36
- Area square = (length)(width) = (6)(6) = 36
- Area triangle $= \frac{(\text{base})(\text{height})}{2} = \frac{(9)(4)}{2} = \frac{36}{2} = 18$

The triangle has the smallest area, 18 sq. ft.

Answer: 18 sq. ft.

### Example 3:

(No calculator permitted.)

What is the area of an *isosceles trapezoid* with one base 6 cm long, the other base 10 cm long, and a height of 4 cm?

Answer: ..................

**Solution**

- Area any trapezoid = $\frac{(\text{base}_1 + \text{base}_2)(\text{height})}{2}$
- Area of this trapezoid = $\frac{(6+10)(4)}{2} = \frac{(16)(4)}{2} = (8)(4) = 32$

Answer: 32 $cm^2$

You will notice that in the SCR (short constructed-response) questions you usually are NOT permitted to use a calculator, so the numbers used are not complicated.

## Performance-Based Assessment Question

### Example 4:

John has an odd-shaped backyard. He wants to plant grass in the entire area, and he also wants to put fencing around all sides.

**Part A:** What is the area he will cover with grass seed?

**Part B:** If grass seed costs $0.05 per square foot, what would it cost him to buy enough seed for the backyard?

**Part C:** How much fencing will he need to purchase?

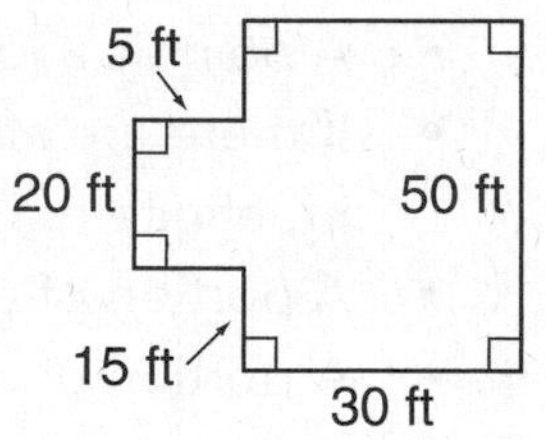


**Strategies and Solutions**

Underline or circle important information in the question.

Write formulas.

Use the *PARCC Grade 8 Mathematics Assessment Reference Sheet* on page 359.

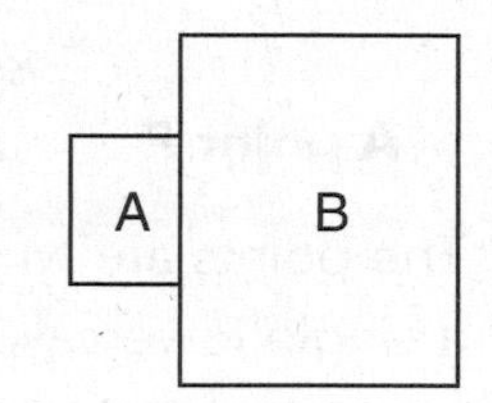


Make a chart or table to help you organize data. Remember, almost every open-ended PARCC 8 question can be organized with a chart or table.

Check computation. Does it make sense?

| Think | Plan | Do |
|---|---|---|
| Draw and divide shape | Rectangles labeled | See diagram |
| Area of A and B rectangles | Area = $l \times w$<br>$A = 20 \times 5 = 100$<br>$B = 30 \times 50 = 1{,}500$ | Total area $A + B =$ 1,600 sq. ft |
| Cost of grass seed | Total Area × Cost per square foot $1{,}600 \times .05 =$ | $80.00 |
| Perimeter for fence | Add all sides $50 + 50 + 35 + 35$ | 170 feet |

*Final Answers*:

**Part A:** He will need to cover an area of 1,600 square feet with grass seed.

**Part B:** It will cost $80 to by the grass seed.

**Part C:** He will need 170 feet of fencing.

*Geometry* really means measuring the earth. First, we'll review some basic terms that will be used throughout this section.

# POINTS, LINES, ANGLES, AND PLANES

## Points and Lines

Definitions you should know:

- A *point* is a location in space. It doesn't have any height or width.
- If there are at least two points on a plane, there is a *line*. A line contains an indefinite number of points and continues indefinitely in both directions.
- A portion of a line that has two endpoints is a *line segment*.
- A portion of a line that has only one endpoint is a *ray*.

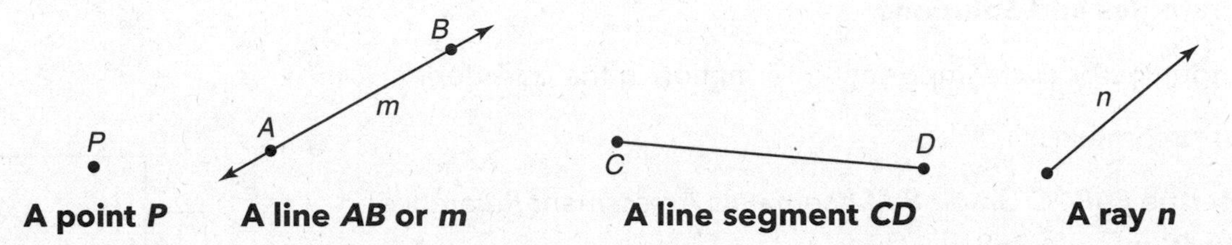


The points are written as uppercase letters, but a line can also be given a name using a single lowercase letter (line *AB* can also be called line *m*).

## Angles

When lines, segments, or rays meet or when they intersect they form angles.

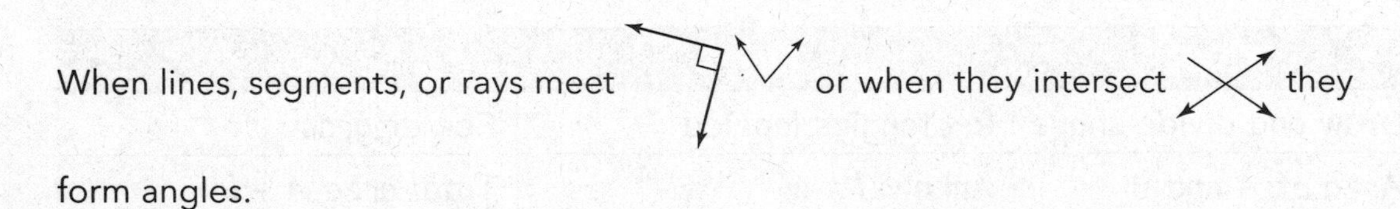

Some different types of angles you should recognize and remember:

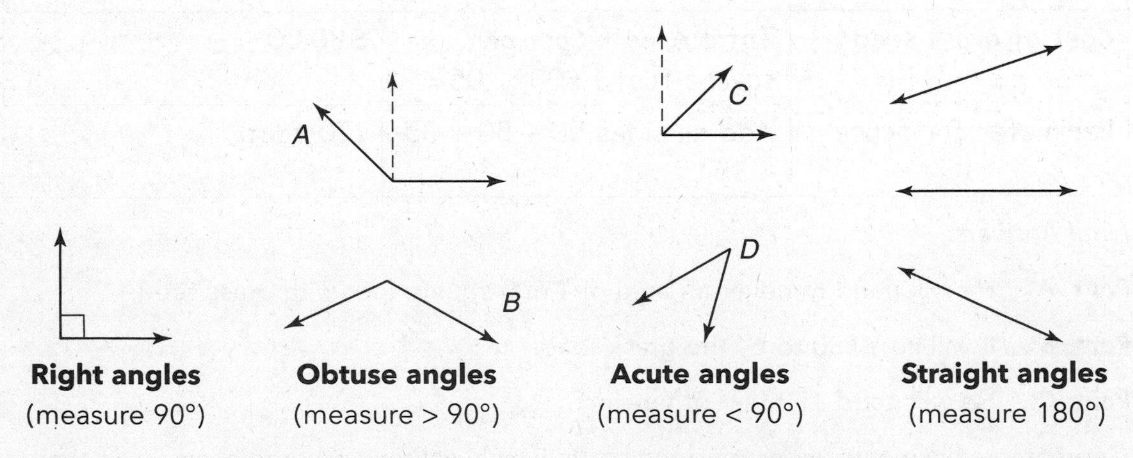


The term *vertex* is often used in geometry. A *vertex* is the endpoint that is shared by two segments that form an angle. The little symbol for a right angle is the little square near the vertex of the angle.

## Angles and Pairs of Angles with Special Names (from Grade 7.G.5)

**Complementary Angles** When the sum of two angles is 90°, the angles are called *complementary angles.*

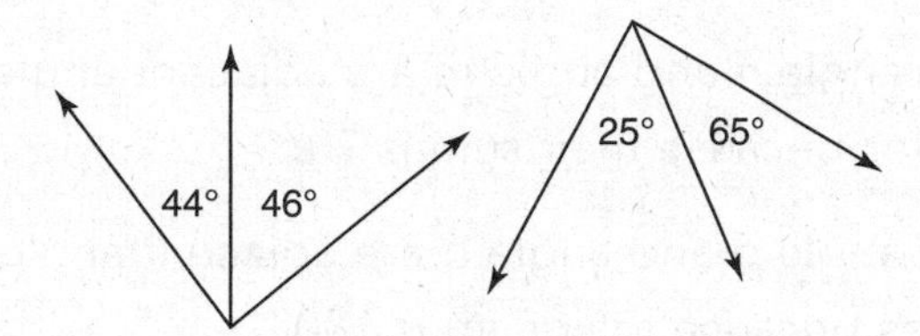

### HINT

You can remember the name "complementary" by thinking of "corner," (complementary angles look like the corner of a room). You can also write the word complementary and draw a line next to the C. You will see the number 90: Complementary.

---

**Vertical Angles** When lines, rays, or segments intersect they form four angles; their *vertical angles* are equal, ($m\angle c = m\angle e$) and ($m\angle d = m\angle f$).

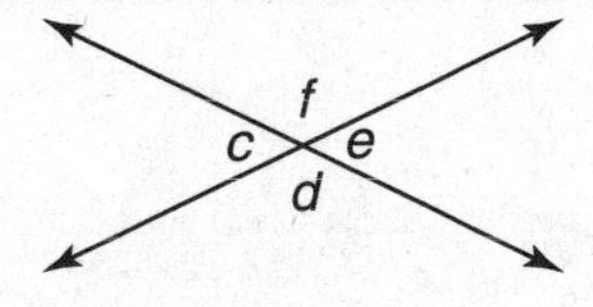

### HINT

Vertical angles create a letter *X*.

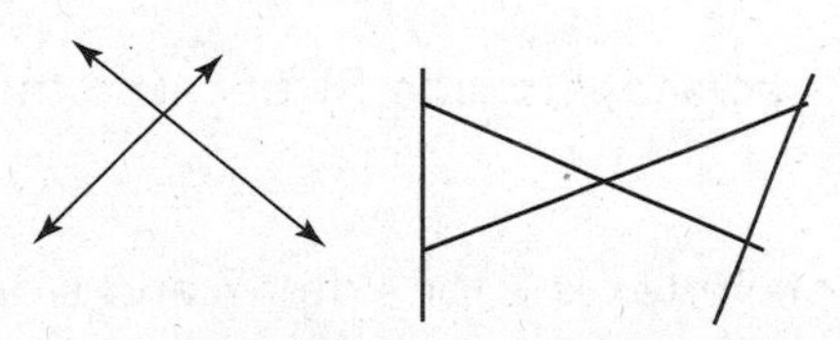

---

When we write "the measure of an angle is," we use symbols instead of words. The example $m\angle 5 = 210°$ is read as "the measure of angle five is two hundred ten degrees." A 50° angle and a 40° angle are complementary angles ($50° + 40° = 90°$). A 33° and a 57° angle are also complementary angles ($33° + 57° = 90°$). If you are told that two angles are complementary and you know the measure of one angle, you can find the measure of the other. Here are two examples.

## Adjacent Angles

Two angles are called *adjacent angles* if they share a common vertex and a common side.

In the diagram **A** below, angle *a* and angle *b* are adjacent angles.

In diagram **B** below, angle *n* and angle *m* are adjacent angles. They are also *complementary* angles because their sum is 90°

In diagram **C** below, angle *p* and angle *q* are adjacent angles. They are also *supplementary* angles because their sum is 180°.

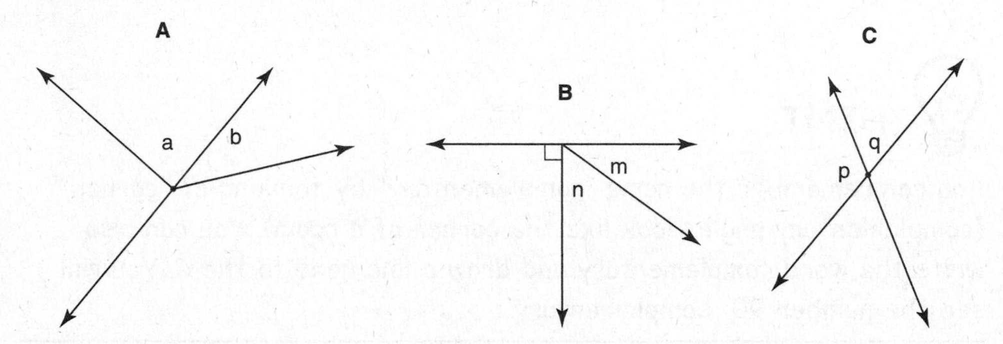


## Adjacent Sides

In a triangle, two sides sharing a common vertex are called *adjacent sides.*

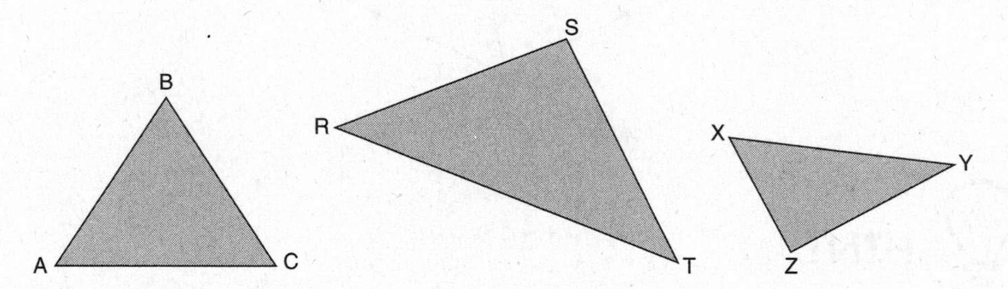


In triangle ABC, side AB is *adjacent* to side BC. The common vertex they share is **B.**

In triangle RST, side RT is *adjacent to* side ST because they share a common vertex ______?

In triangle XYZ a common vertex is Z for side XZ and side ______? Therefore sides XZ and side ______ are *adjacent* sides.

(*Answers:* vertex T, side ZY or YZ, ZY)

## Examples

**A.** Given: $\angle 5$ *and* $\angle 6$ are complementary angles, and $m\angle 5 = 10°$. What is $m\angle 6$?

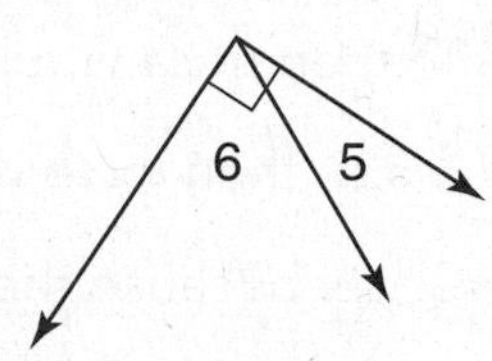


Since they should total 90° and $90° - 10° = 80°$, we know that $m\angle 6 = 80°$.

**B.** Given: $\angle 3$ *and* $\angle 4$ are complementary angles, and $m\angle 4 = 78°$. What is $m\angle 3$?

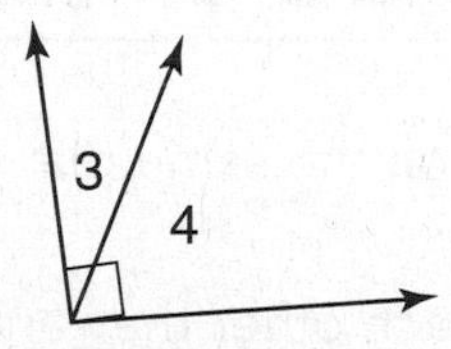


Since they should total 90° and $90° - 78° = 12°$, we know that $m\angle 3 = 12°$.

**C.** If $\angle a$ and $\angle b$ are complementary, and $m\angle a = 35°$, then $m\angle b$ must = ______° because their sum is 90°. ($m\angle b = 55°$)

**D.** If $m\angle d = 55°$ and $m\angle e = 35°$, they are complementary angles because their sum is ______. ($55° + 35° = 90°$)

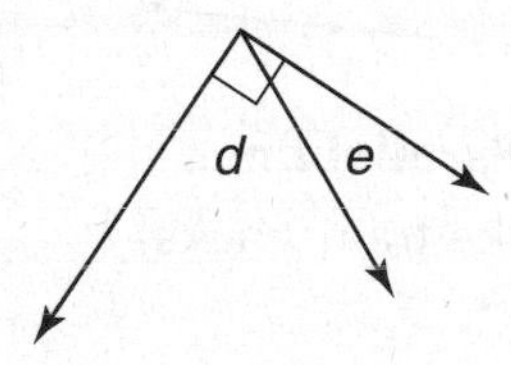


### Supplementary Angles

When the sum of two angles combine to make a straight line, they are called *supplementary* angles (their sum = 180°). In the following diagrams, angles *a* and *b* are supplementary. Angles *d* and *e* are also supplementary.

**HINT**

You can remember the word supplementary by remembering supplementary angles make a straight line.

### Examples

(Refer to the diagrams at the bottom of page 65.)

**A.** In the first diagram, if $m\angle b = 40°$, then $m\angle a$ would $= 180° - 40°$ or $140°$.

**B.** In the second diagram, if $m\angle d = 52°$, then $m\angle e$ would $= 180° - 52°$ or $128°$.

**C.** If $m\angle a = 35°$, then $m\angle b$ must = ______ because their sum is 180°. ($m\angle b = 145°$)

**D.** If $m\angle d + m\angle e = 180°$, then they are called ______ angles. (supplementary)

---

## RELATIONSHIPS OF LINES (8.G.1C)

---

- Lines are *parallel* if they are on the same plane and are the same distance apart (equidistant) from each other.
- When two lines cross over each other they are *intersecting lines*.
- When two lines cross over each other at right angles they are *perpendicular* to each other.

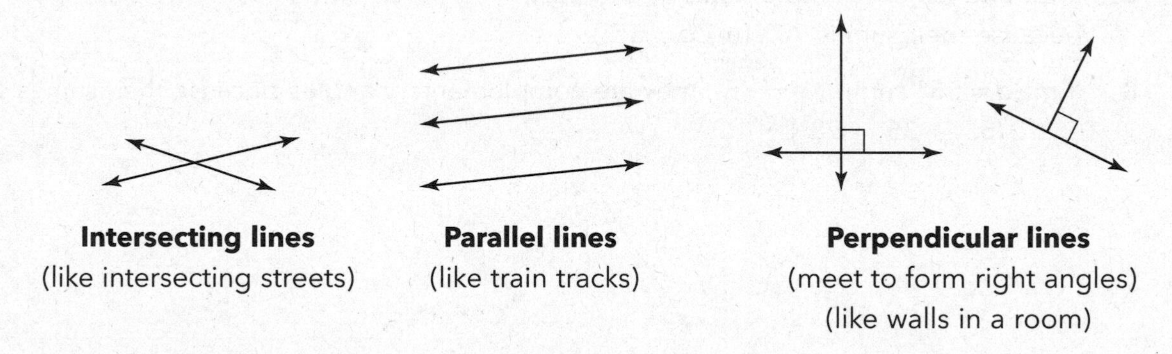

**Intersecting lines**
(like intersecting streets)

**Parallel lines**
(like train tracks)

**Perpendicular lines**
(meet to form right angles)
(like walls in a room)

### Rays with Special Names and Characteristics

An *angle bisector* divides an angle into two equal angles. Ray *AB* is an angle bisector. Ray *EF* is also an angle bisector.

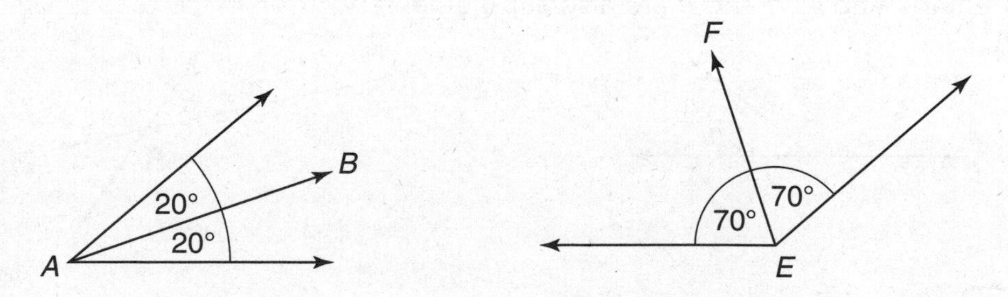

# PLANES

Besides points, lines, and angles, you should be familiar with *planes*. When we write on a sheet of paper, we are writing on one plane, but when we fold that paper, or make a three-dimensional form, then we have more planes. Below you see two different *planes*. Notice that when two planes intersect they intersect *at a line*. (See dotted line.)

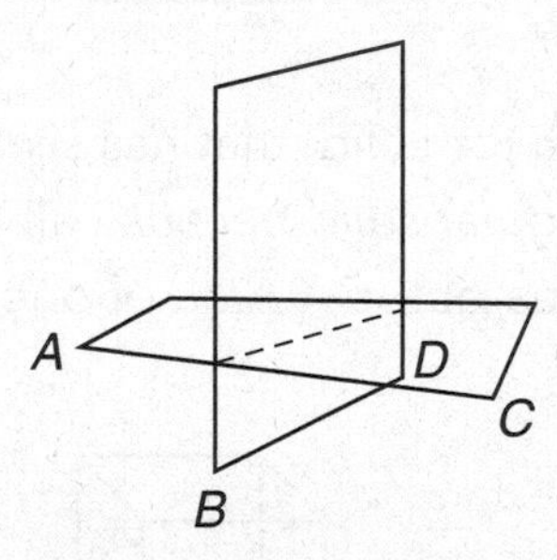


## Parallel, Perpendicular, and Intersecting Planes

- *Parallel lines* are lines on the same plane that never meet; they never intersect.
- *Parallel planes* also never meet; they never intersect.

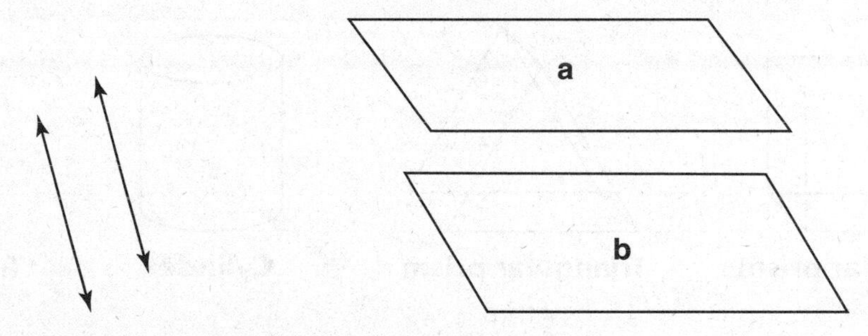


- Pages in a book look like *intersecting planes*.
- Walls in a room look like intersecting planes; since most walls meet at right angles (90°), we call these *perpendicular planes*.

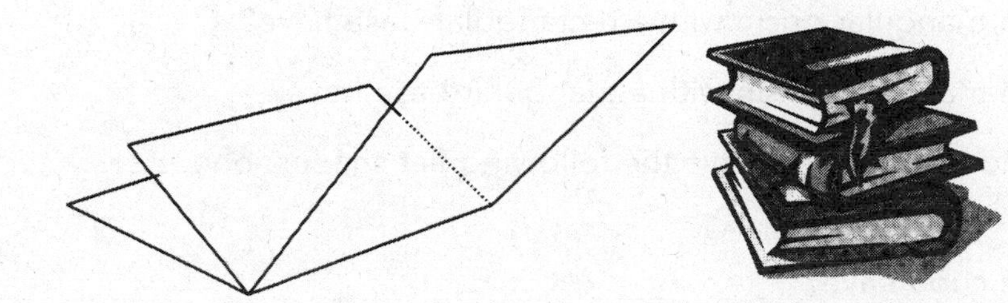

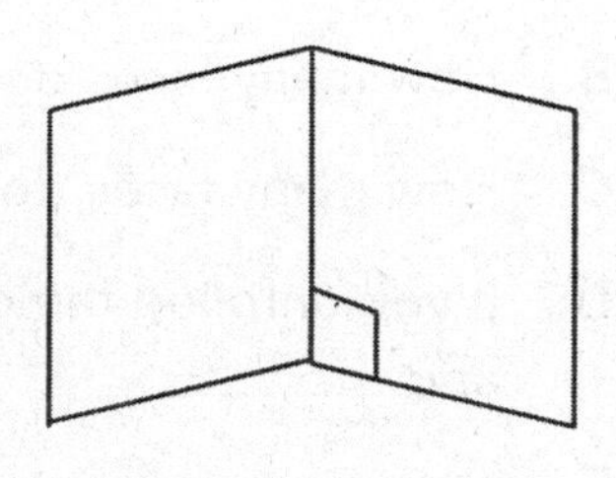

## Three-Dimensional Forms

Planes can combine to make three-dimensional forms. Three-dimensional forms have length, width, and depth.

- A tissue box is a three-dimensional form.

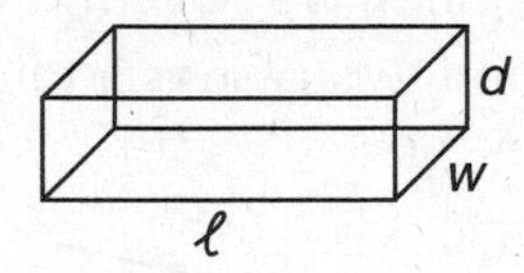


- A special rectangular form (or solid) that has six identical surfaces is called a *cube*. It is also called a *regular solid* because all of its faces have the same size and shape. (In a cube: Area of top = Area of one side = Area of front = Area of back = Area of bottom)

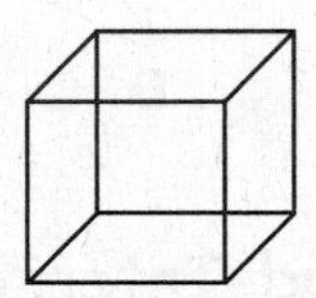

There are other three-dimensional solids. Some are drawn below. Look at these figures and answer the questions below.

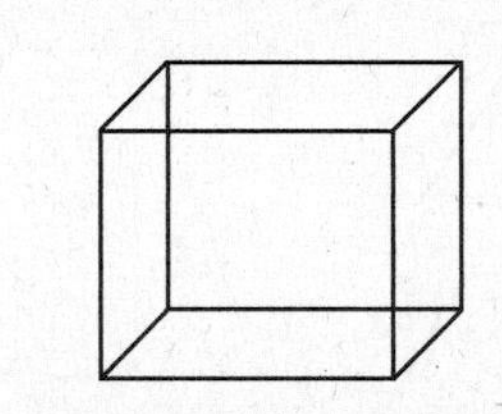

**Rectangular prism**

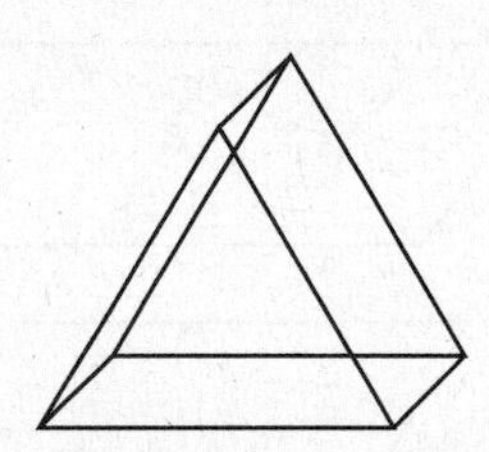

**Triangular prism** with a rectangular base

**Cylinder**

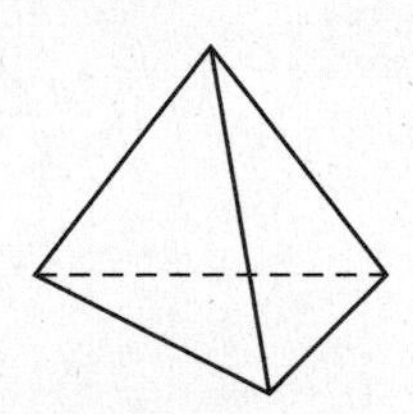

**Triangular prism** with a triangular base

**A.** How many faces does a rectangular prism have? It has a top, bottom, front, back, and two sides. Therefore it has ______ faces.

**B.** How many faces does a triangular prism with a rectangular base have?

**C.** How many faces does a triangular prism with a triangular base have?

**D.** If you unrolled the cylinder you would have the following flat shapes: one ______ and two ______.

**E.** How many faces does a cube have?

*(Answers: **A.** six **B.** five **C.** four **D.** rectangle; circles **E.** six)*

Now let's work in reverse. Look at the flat shapes below and see if you can tell what three-dimensional forms they would make when folded. These flat shapes are called *nets* (from grade 6.G.4).

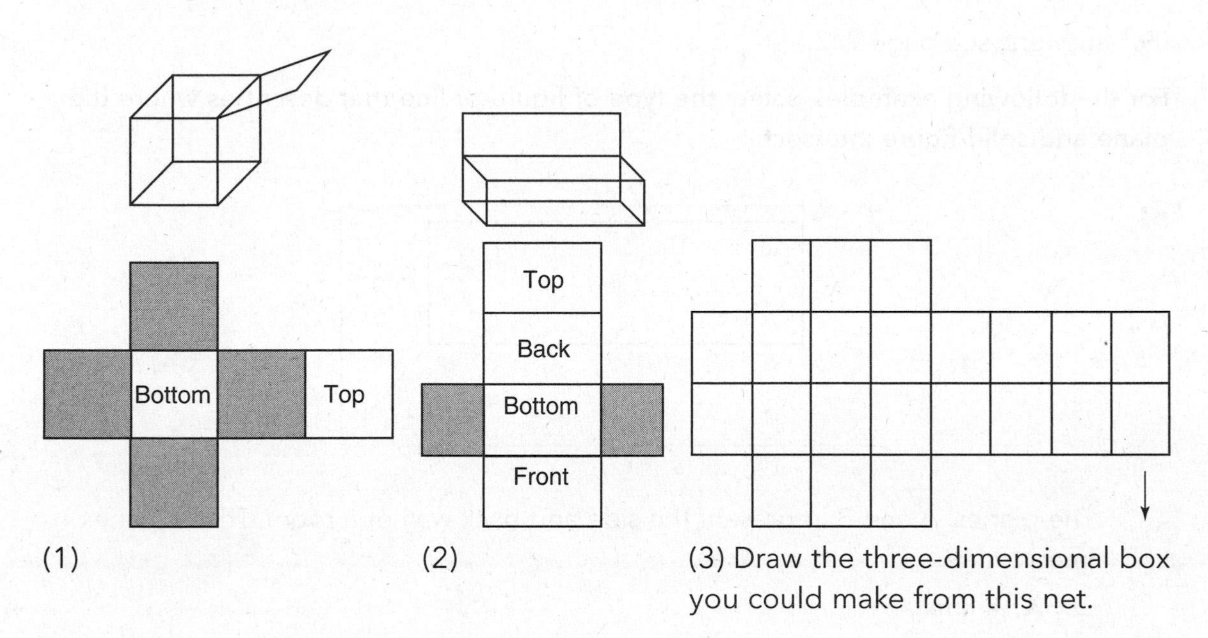


(1)

(2)

(3) Draw the three-dimensional box you could make from this net.

**Number (1)** is a cube with a top cover that flaps over from the right side.
Given: Each face is an 8 ft by 8 ft square. What is the total surface area of this cube?

Think:

a. How many faces? ------
b. Are all faces the same size? ------
c. What is the area of one face? ------
d. What is the sum of the area of all faces? (Total surface area) = ------sq. ft

**Answers (1)**

a. 6
b. yes
c. $8 \times 8 = 64$ sq. ft
d. $6 \times 64 = 384$ sq. ft

**Number (2)** is a rectangular box with a top cover that flaps over from the back.
Given: The two sides measure 5.2 in. by 5.2 in. All other faces measure 5.2 in. by 10.4 in.

What is the total surface area of this box?

------sq. in.

**Answers (2)**

$A_{sides} = 2(5.2)(5.2) = 54.08$
$A_{all\ other\ faces} = 4(5.2)(10.4) = 216.32$
Total surface area = 270.4 sq. in.

**Number (3):** Describe the box made from this net.
What are the dimensions of box number (3)?

What is the total surface area of box number (3)?

------

**Answers (3)**

Answers will vary.

## PRACTICE: Points, Lines, and Planes

(For answers, see page 242.)

**For the following examples, select the type of figure or line that describes where the plane and solid figure intersect.**

**1.**

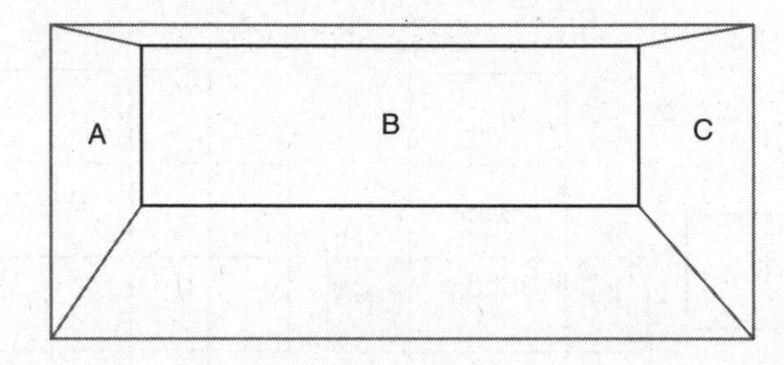


The planes, A and B, represent the side and back wall of a room. These planes intersect at:

- ○ **A.** a line
- ○ **B.** a rectangle
- ○ **C.** a point
- ○ **D.** two lines

**2.**

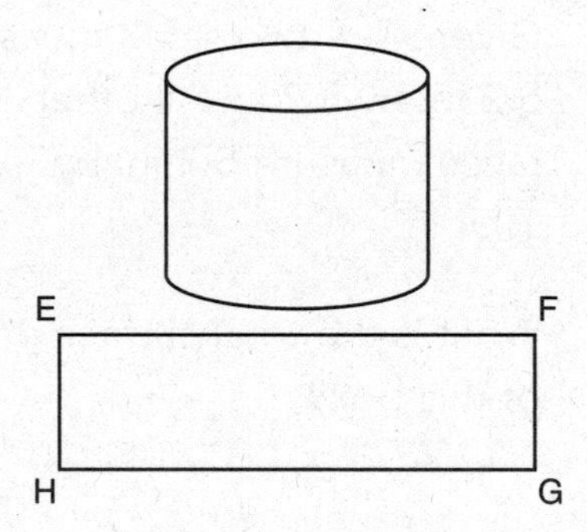


If the bottom horizontal rectangular plane ABCD is translated up to intersect with the right-cylinder shown, what shape would be formed where they intersect?

- ○ **A.** a line
- ○ **B.** an oval
- ○ **C.** a circle
- ○ **D.** a rectangle

**3.**

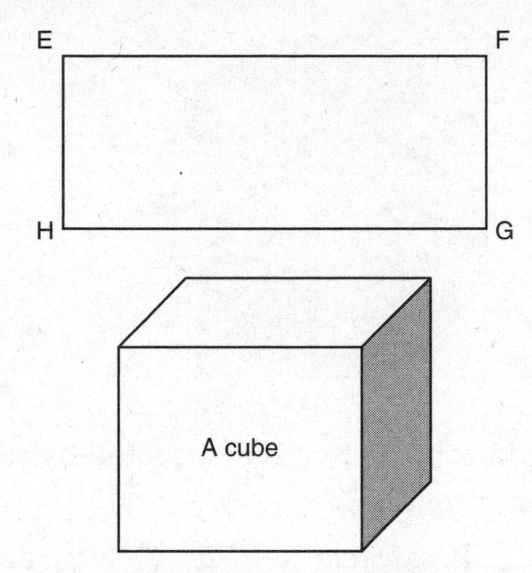


Rectangle EFGH is translated down to intersect with the middle of this three-dimensional regular right-prism. Which of the following could be used to describe their intersection? Select all that are true.

- [ ] **A.** a straight line
- [ ] **B.** a curved line
- [ ] **C.** a triangle
- [ ] **D.** a rectangle
- [ ] **E.** an oval
- [ ] **F.** a square
- [ ] **G.** a parallelogram
- [ ] **H.** a pentagon

**4.**

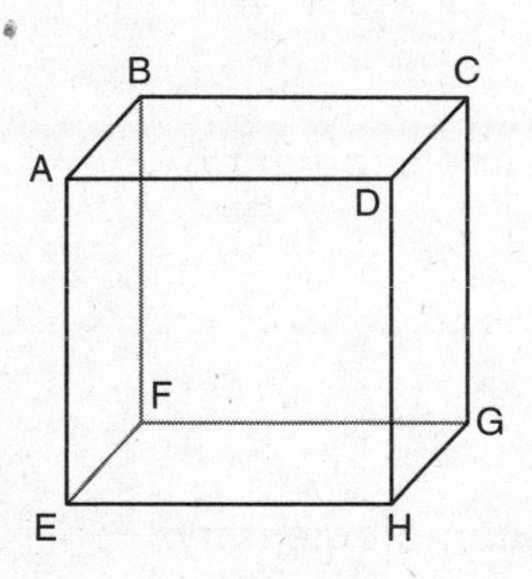


The diagram shows a cube ABCDEFGH.

- Draw the lines EC and EG.
- What is the name of the new figure ECG? Select all correct choices.

- [ ] **A.** a rectangle
- [ ] **B.** a square
- [ ] **C.** an equilateral triangle
- [ ] **D.** a right triangle
- [ ] **E.** an isosceles triangle
- [ ] **F.** a polygon

**5.**

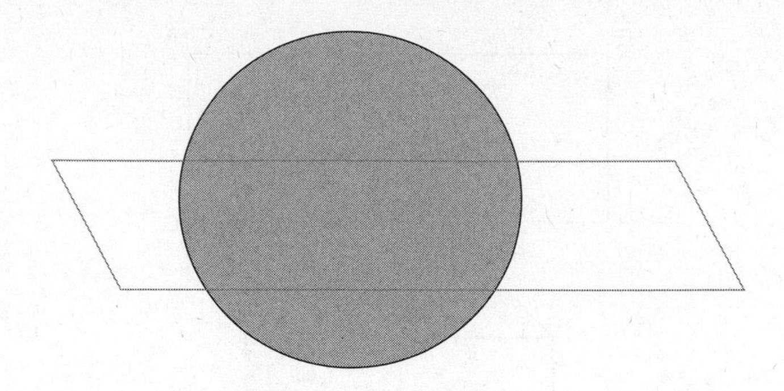

The picture is a large beach ball that is intersected by a slanted rectangle. What is the shape of their intersection?

○ **A.** an oval ○ **B.** a circle ○ **C.** a square ○ **D.** a rectangle

**6.** In $\triangle ABC$, ray $BD$ is an angle bisector of $\angle B$; therefore, we know that

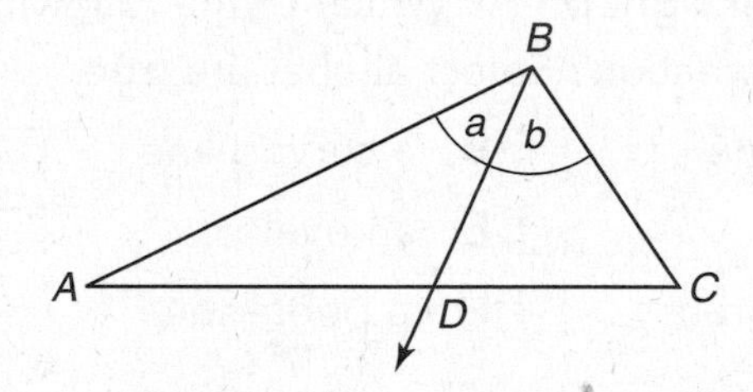


○ **A.** $m\angle A = m\angle C$

○ **B.** line segment $AD \cong$ line segment $DC$

○ **C.** $m\angle A = 90°$

○ **D.** $m\angle a = m\angle b$

**7.** Which of the following is true?

○ **A.** Vertical angles are complementary.

○ **B.** Supplementary angles total 90°.

○ **C.** Complementary angles are equal.

○ **D.** Vertical angles are equal.

**8.** In the diagram shown, if ray *NM* is an angle bisector, and $m\angle 2 = 46°$, then

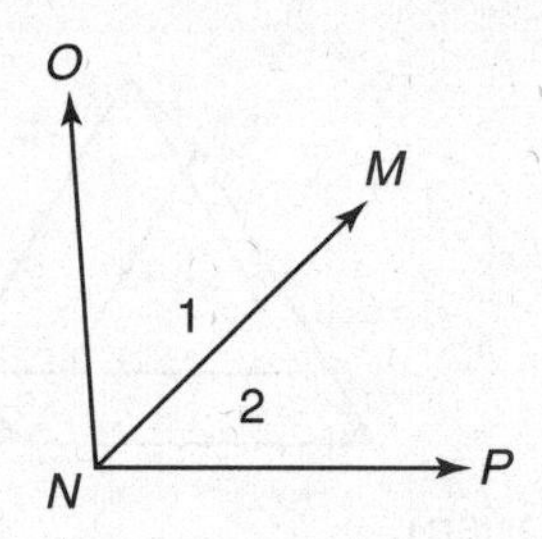


- ○ **A.** $m\angle 1 = 34°$
- ○ **B.** $m\angle 1 = 46°$
- ○ **C.** $m\angle 1 + m\angle 2 = 90°$
- ○ **D.** $m\angle 1 + m\angle 2 = 180°$

**9.** In the diagrams below, you see two rectangular prisms. Each three-dimensional form has six planes. Each plane intersects another plane at a

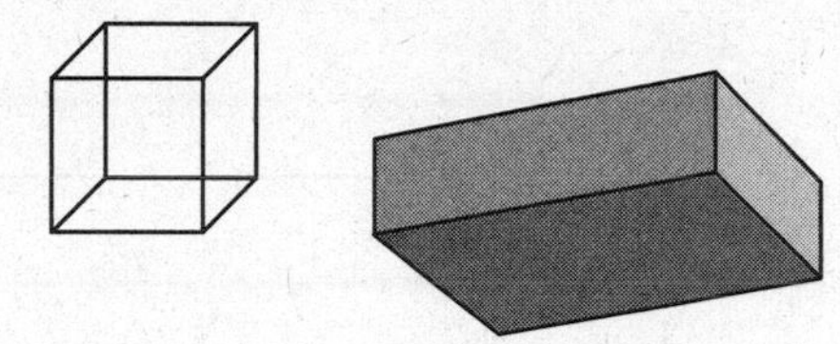

- ○ **A.** line
- ○ **B.** point
- ○ **C.** ray
- ○ **D.** bisector

**10.** If this is a rectangular prism, select all the statements that are true.

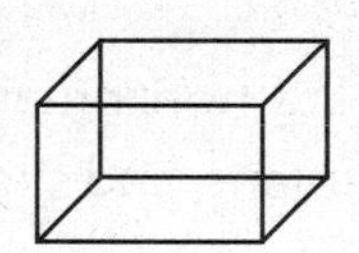

- ☐ **A.** The sides are perpendicular to the bottom plane.
- ☐ **B.** It has six faces.
- ☐ **C.** It has all right angles.
- ☐ **D.** All sides are congruent.
- ☐ **E.** The top and bottom planes are parallel.
- ☐ **F.** The front plane does not intersect the back plane.

**11.** Check all statements that correctly complete this sentence. Looking at the figure below you know that

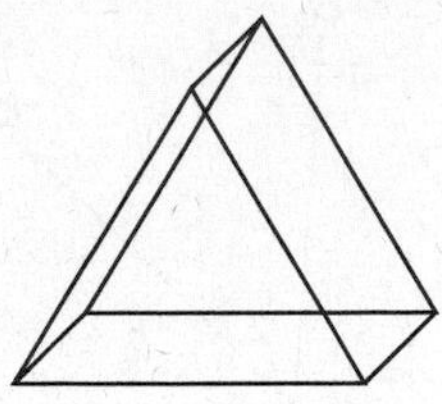

- [ ] **A.** it is a rectangular prism.
- [ ] **B.** it is a triangular prism with a rectangular base.
- [ ] **C.** the base has four right angles.
- [ ] **D.** it is a six-sided figure.
- [ ] **E.** it has four triangular faces.

**12.** Use the diagram below. Look at the line going through this two-dimensional shape. This line intersects this plane at

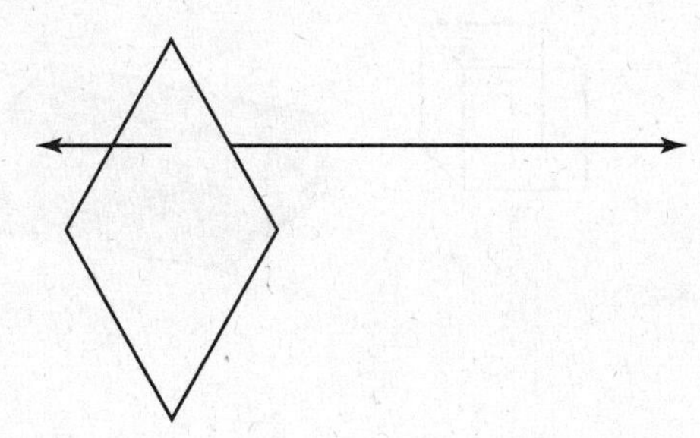

○ **A.** a point

○ **B.** a line

○ **C.** a plane

○ **D.** its vertex

**13.** Here is a cylinder on a table. If a horizontal plane cut through this cylinder, what would the intersection of the plane and the cylinder look like?

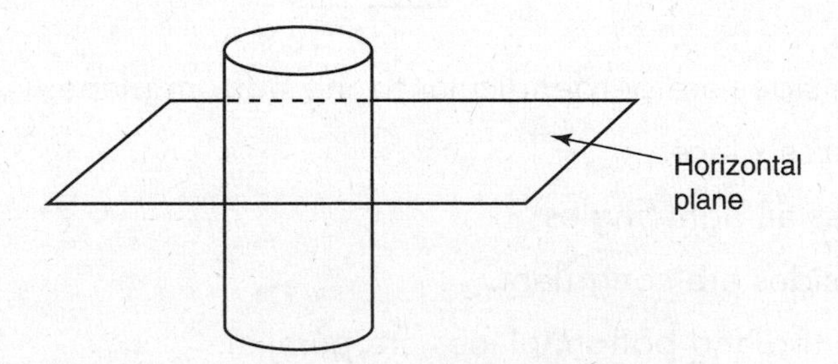


○ **A.** a square

○ **B.** an oval

○ **C.** a rectangle

○ **D.** a circle

**14.** You learned some new terms in this section. Match each term on the left with its description on the right to be sure you understand what each one means.

| | |
|---|---|
| 1. A triangle | A. All sides are the same size and all angles measure the same, like a square. |
| 2. A cube | B. Angle A and B in rectangle ABCD. |
| 3. A vertex | C. A two-dimensional flat shape. |
| 4. A regular shape | D. It has length, width, and depth. |
| 5. A three-dimensional form | E. A point where two lines meet and form an angle. |
| 6. A plane | F. They never meet; like a ceiling and a floor in a room. |
| 7. Parallel planes | G. Angle ABC and angle CBD (They share side BC). |
| 8. Complementary angles | H. Two angles that add up to 90°. |
| 9. Adjacent | I. Two angles that add up to 180°. |
| 10. Consecutive | J. A flat surface like a table top. |
| | K. A regular three-dimensional solid. |

## Parallel Lines

Two lines are *parallel* if they are on the same plane and do not intersect.

Below you can see three sets of parallel lines.

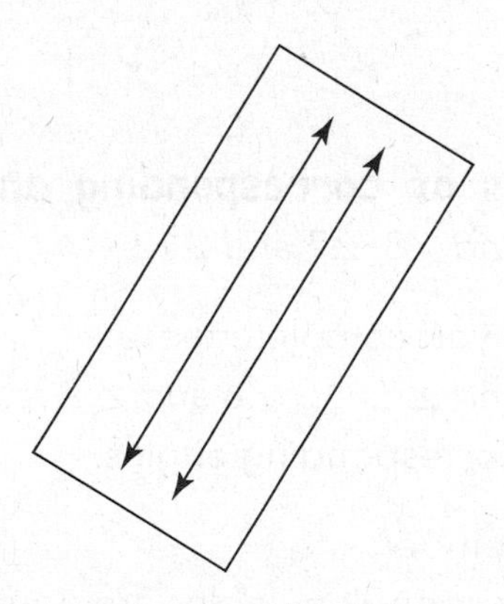

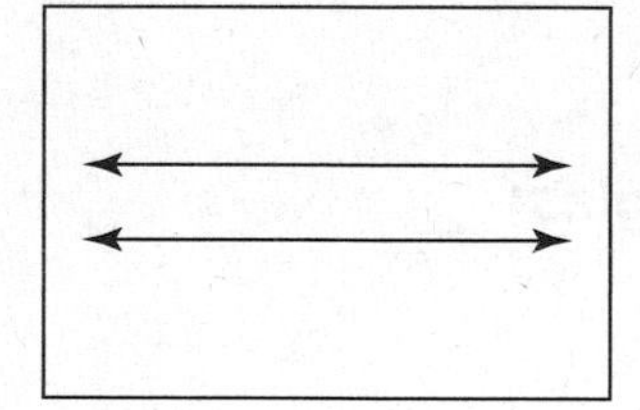

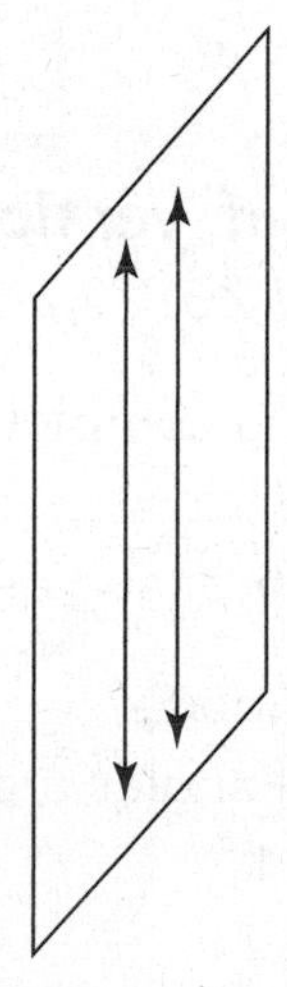

## A Transversal

For now, we'll define a *transversal* as a line that intersects two or more parallel lines.

The angles formed by these two lines have special names and special properties.

Here, we will learn about two special angles.

**Vertical Angles**

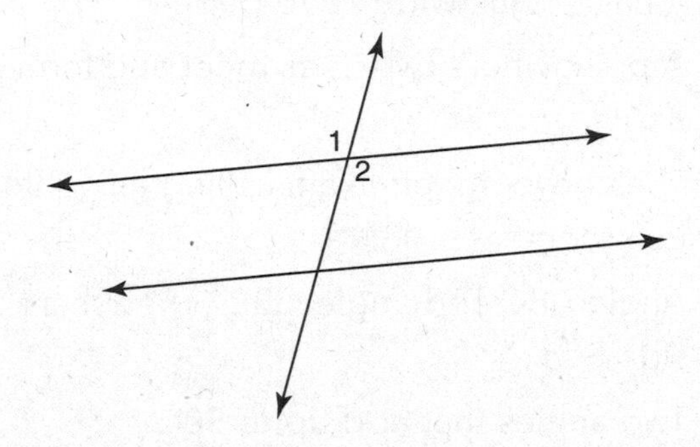


**Corresponding Angles**

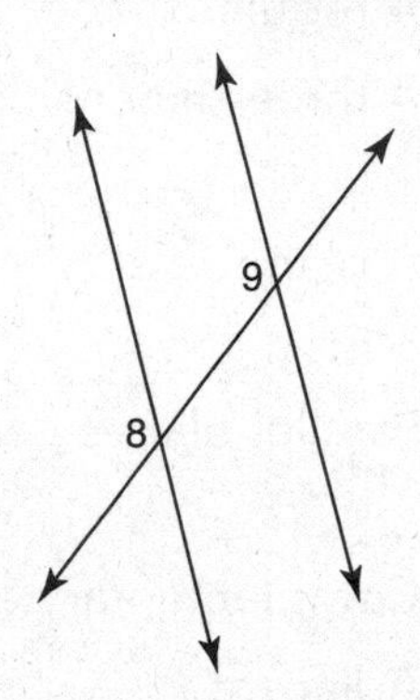


For the diagram below we will list pairs of vertical angles and pairs of corresponding angles.

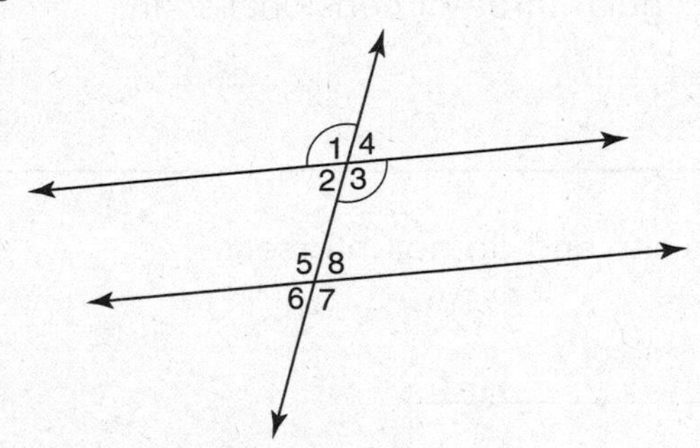


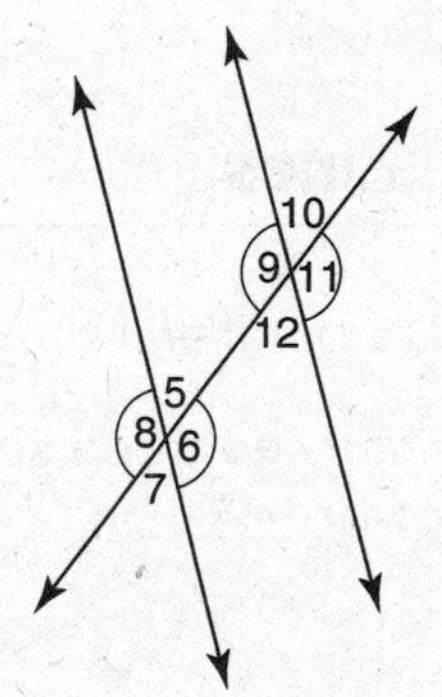


**Pairs of vertical angles:**
*∠1 and ∠3, ∠2 and ∠4*

Now you complete these:
*∠5 and ∠ ------, ∠6 and ∠ ------*
are vertical angles.

Remember:
Vertical angles are equal
∠3 = ∠1
∠7 = ∠ ?

**Pairs of corresponding angles:**
*∠9 and ∠8, ∠7 and ∠12*

Now you complete these:
*∠5 and ∠ ------, ∠6 and ∠ ------*
are corresponding angles.

Remember:
Corresponding angles are equal
∠7 = ∠12    ∠11 = ∠6
∠10 = ∠ ?

## PRACTICE: Parallel Lines Cut by a Transversal Line

(For answers, see pages 242–243.)

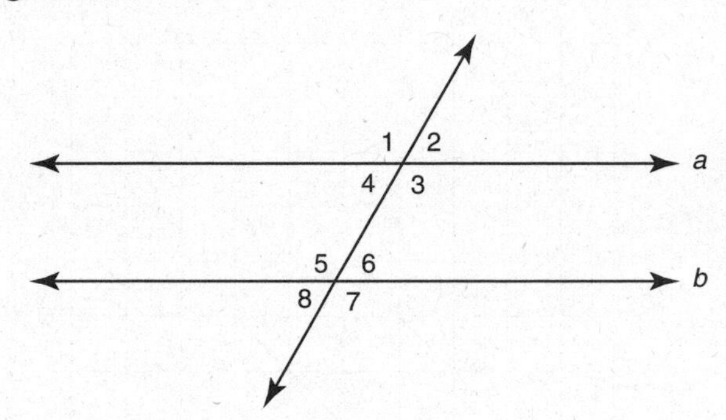


**1.** Answer the following statements based on the diagram above. Check Yes ..✔... or No ..✔.... Lines ***a*** and ***b*** are parallel lines.

**A.** ∠2 and ∠3 are vertical angles. Yes ...... No ......

**B.** ∠3 and ∠4 are supplementary angles. Yes ...... No ......

**C.** The measure of ∠2 = the measure of ∠4. Yes ...... No ......

**D.** If ∠6 measures 80° then ∠7 measures 100°. Yes ...... No ......

**E.** If ∠8 measures 80° then ∠4 measures 80°. Yes ...... No ......

**F.** ∠6 = ∠8. Yes ...... No ......

**2.** In the diagram below you are told that the measure of angle 2 is 80° and the measure of angle 8 is 85°.

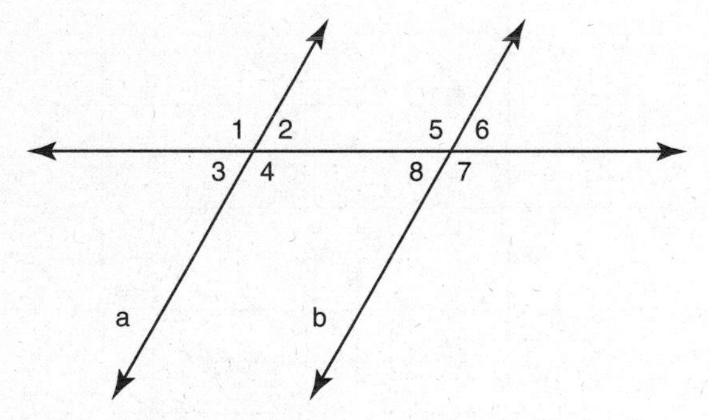


**Part A:** Is line **a** parallel to line **b**? Yes ...... No ......

**Part B:** Use the space below to explain how you know.

**3.** Use a scale factor of 2 to dilate the parallelogram ABCD. The new coordinates will be:

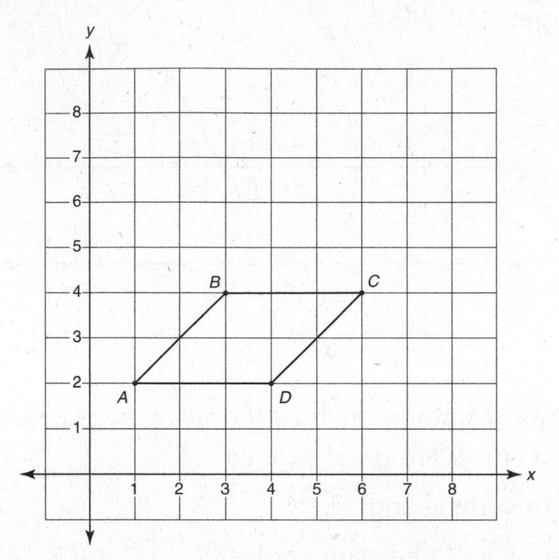


| | | | |
|---|---|---|---|
| ○ **A.** $A'(1, 4)$ | $B'(3, 8)$ | $C'(1, 8)$ | $D'(4, 4)$ |
| ○ **B.** $A'(2, 4)$ | $B'(6, 8)$ | $C'(12, 8)$ | $D'(8, 4)$ |
| ○ **C.** $A'(2, 2)$ | $B'(6, 4)$ | $C'(2, 4)$ | $D'(4, 4)$ |
| ○ **D.** $A'(2, 8)$ | $B'(6, 12)$ | $C'(2, 12)$ | $D'(8, 6)$ |

**4.** If you use a scale factor of 2 to dilate the triangle PQ the new coordinates will be:

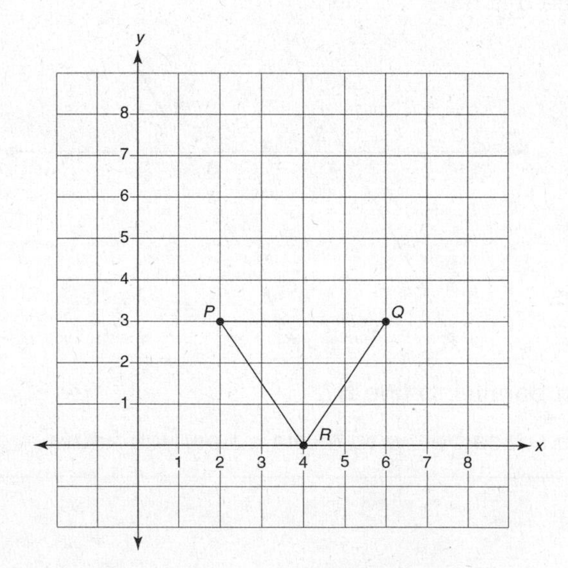


| | | |
|---|---|---|
| ○ **A.** $P'(1, 1.5)$ | $Q'(3, 1.5)$ | $R'(2, 0)$ |
| ○ **B.** $P'(1, 6)$ | $Q'(3, 3)$ | $R'(2, 0)$ |
| ○ **C.** $P'(4, 6)$ | $Q'(12, 6)$ | $R'(8, 0)$ |
| ○ **D.** $P'(-2, 3)$ | $Q'(-6, 3)$ | $R'(4, 0)$ |

**5.** On Janet's test the teacher asked this question.

How can you tell that lines ***a*** and ***b*** in the diagram below are **not** parallel?

Explain to Janet how you would answer this question.

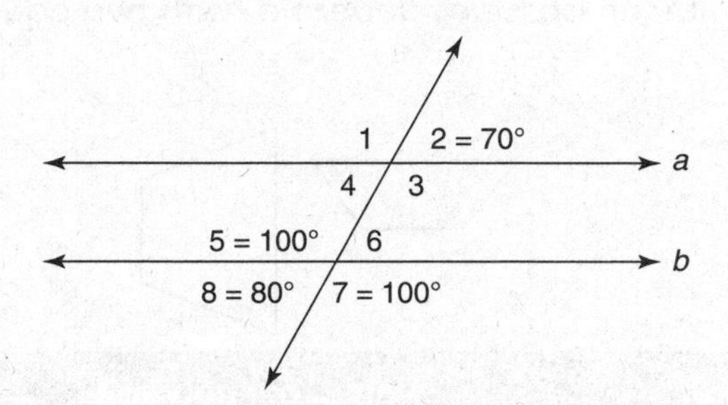


## POLYGONS: AREA OF TWO-DIMENSIONAL SHAPES

***Polygons*** *Polygon* really means many angles (*poly-* is Greek for many and *-gon* is Greek for knee or bend or angle). *Polygons* are figures with many angles and many sides. We usually identify them by the number of their sides. Some polygons are illustrated on the following pages.

***Regular Polygons*** If the lengths of the sides of a particular polygon are the same and the angles measure the same, that figure is called a *regular polygon*. Some regular polygons are shown below with their special names.

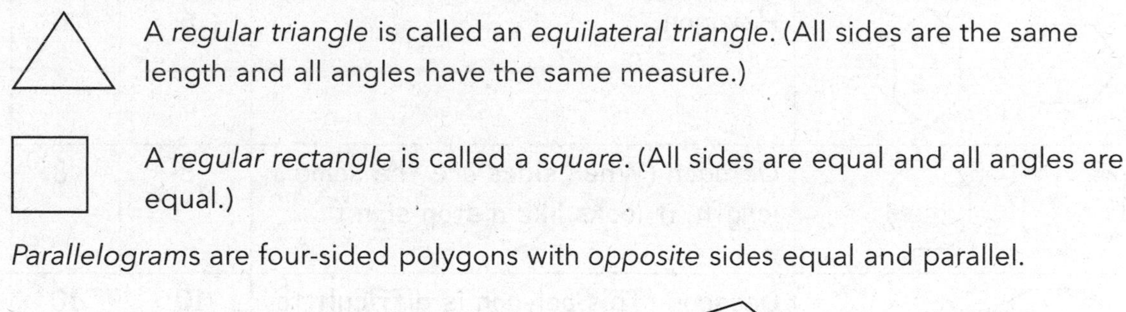

A *regular triangle* is called an *equilateral triangle*. (All sides are the same length and all angles have the same measure.)

A *regular rectangle* is called a *square*. (All sides are equal and all angles are equal.)

*Parallelograms* are four-sided polygons with *opposite* sides equal and parallel.

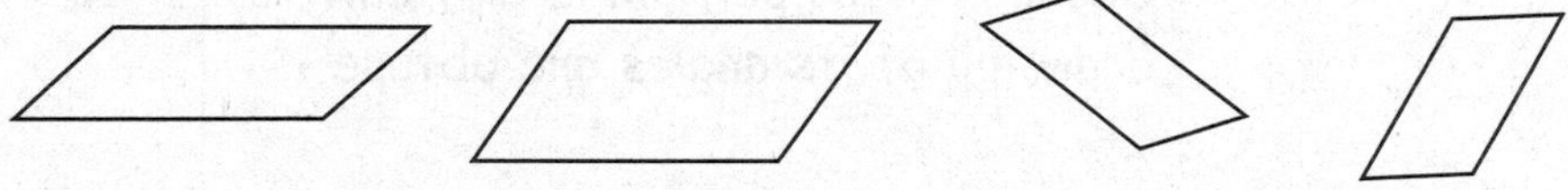

***Trapezoid*** Another polygon you should know about is a trapezoid. A *trapezoid* has four sides, but it is different from a rectangle, square, or parallelogram. A *trapezoid* has only one pair of parallel sides. A trapezoid can be a *right trapezoid* (with one or two right angles), or it can be an *isosceles trapezoid* (with two equal nonparallel sides and equal base angles).

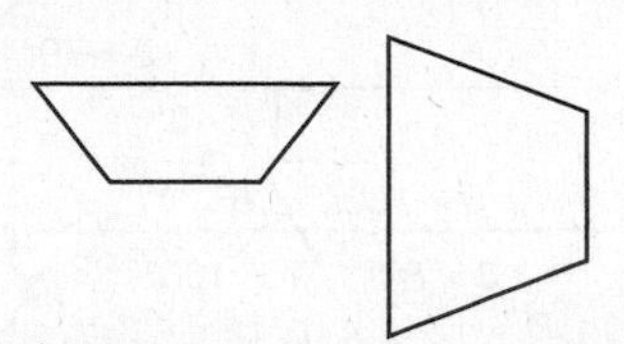

| Drawings of polygons | Name of polygon | Angles | Sides |
|---|---|---|---|
| 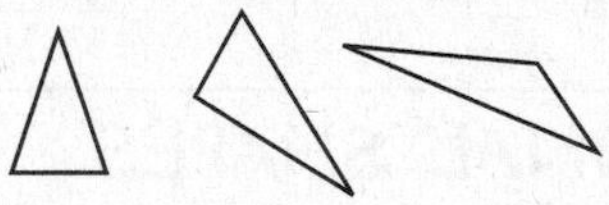 | Triangle | 3 | 3 |
| 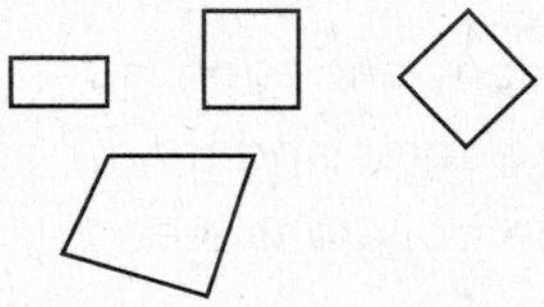 | Quadrilateral (When sides are even, it could look like a square or like a diamond.) | 4 | 4 |
| 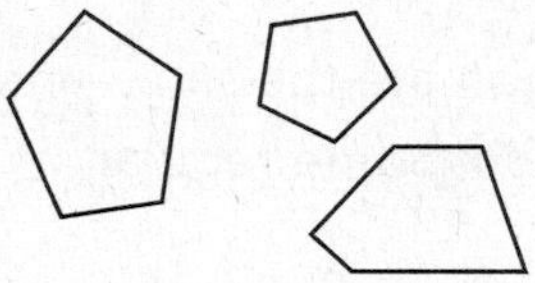 | Pentagon (When sides are the same length, it sometimes looks like a house.) | 5 | 5 |
| 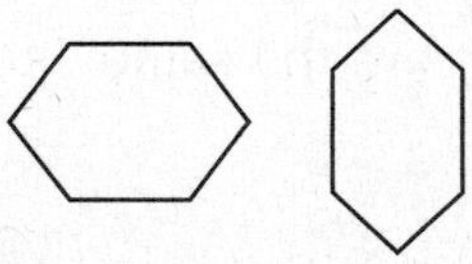 | Hexagon | 6 | 6 |
| 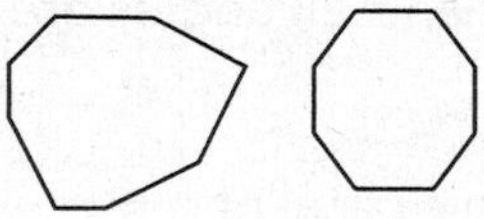 | Octagon (When sides are the same length, it looks like a stop sign.) | 8 | 8 |
| 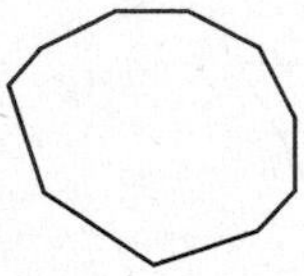 | Decagon (This polygon is difficult to draw; all of its angles are obtuse.) | 10 | 10 |

## Area of Flat Shapes (Two-Dimensional Shapes) (Some from Grade 7.G.)

When you are finding area, think of *covering* a plane figure in small squares. There are formulas to help you find the area of different shapes easily. Remember to always label your answers in square units such as 14 in.$^2$ or 14 square inches. You are working with two dimensions (length and width) so you need to use square units.

### Examples

**A.** You can count the boxes in the rectangle below and see that the area of this rectangle is 35 square units. A faster way would be to multiply the base (7 units) by the height (5 units).

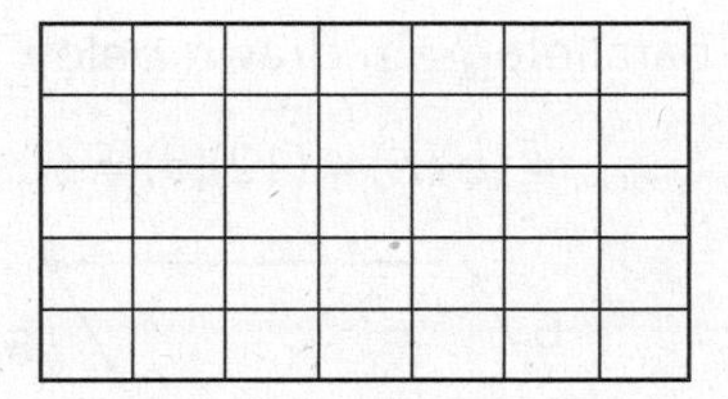

Base × Height = Area of rectangle
$7 \times 5 = 35$ square units

**B1.** The square to the left is 6 units long and 6 units high. The area of this square is (Base)(Height) or $(b)(h) = (6)(6)$ or $6^2$, which is 36 square units.

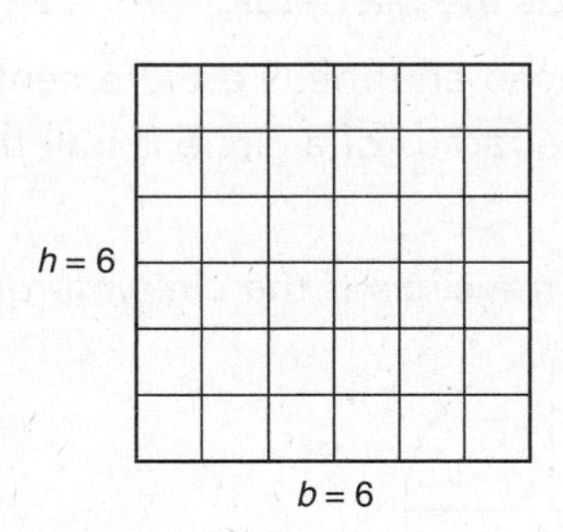


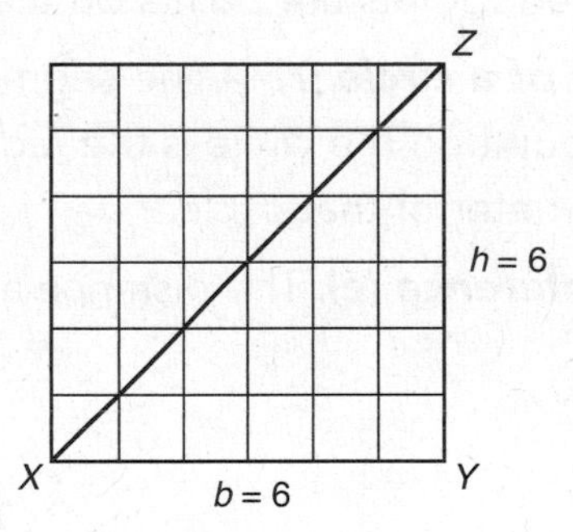


**B2.** The triangle *XYZ*, in the figure above, is half the size of the square.

$$A_{\text{triangle}} = \frac{bh}{2} = \frac{(6)(6)}{2} = \frac{36}{2} = 18 \text{ square units}$$

**C.** What is the area of a rectangle with one side 12 inches and one side 3 inches?

$$A_{\text{rectangle}} = \text{Base} \times \text{Height} = (b)(h) = 12 \times 3 = 36 \text{ square inches}$$

**D.** What is the area of a triangle with a base of 5 feet and a height (altitude) of 12 feet?

$$A_{\text{triangle}} = \frac{\text{Base} \times \text{Height}}{2} = \frac{(b)(h)}{2} = \frac{(5)(12)}{2} = \frac{(60)}{2} = 30 \text{ square feet}$$

**E.** What is the area of the obtuse triangle drawn below? (*Hint*: 4 is the height.)

$$A_{\text{triangle}} = \frac{(b)(h)}{2} = \frac{(5)(4)}{2} = \frac{(20)}{2} = 10 \text{ square units}$$

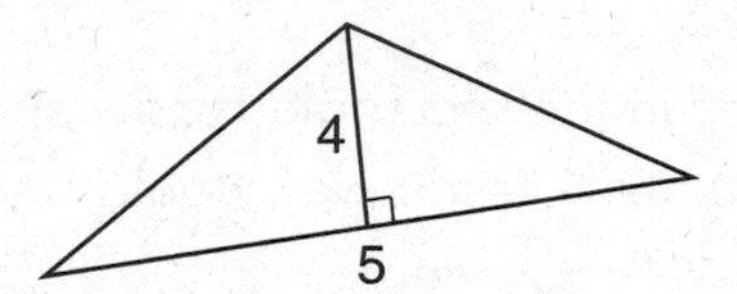


**F.** What is the area of the parallelogram drawn below? (*Notice:* 5 is the height.)

$$A_{\text{parallelogram}} = (b)(h) = (12)(5) = 60 \text{ square units}$$

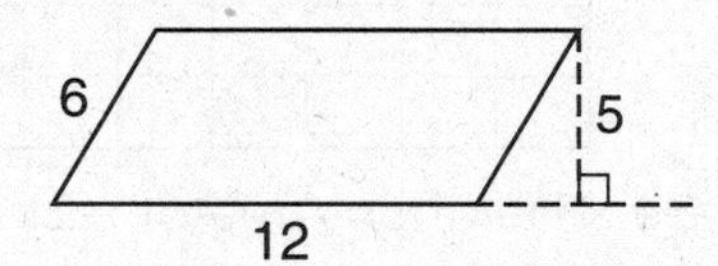


**G.** What is the approximate area of a circle with radius ($r$) = 5 cm?

***Diameter of a circle (d)***: A line segment that contains the center of the circle and whose endpoints are points on the circle is the *diameter*.

***Radius of a circle (r)***: A line segment whose endpoints are the center of the circle and a point on the circle is the *radius*. The radius of a circle is half the length of the diameter of that circle.

***Circumference (c)***: The distance around the circle is the *circumference*.

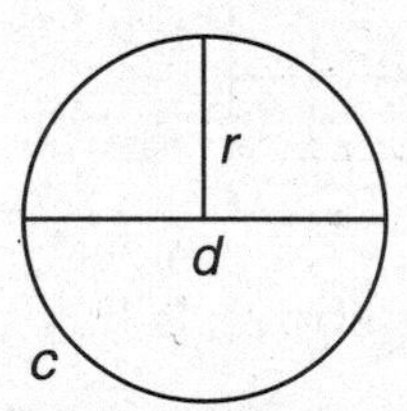


$$A_{\text{circle}} = \pi r^2 = (3.14)(5^2) = (3.14)(25) = 78.5 \text{ or } 79 \text{ sq. cm}$$

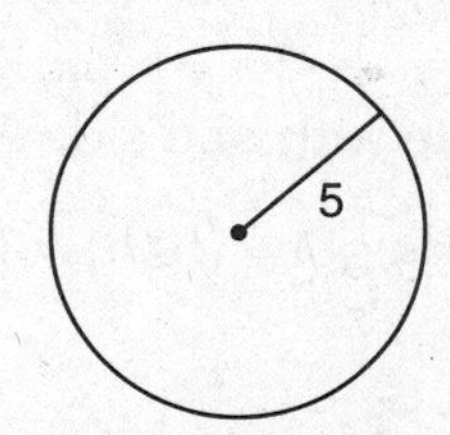


If you use a calculator, press [3.14] [×] [$5^2$], and you will see [78.5]. Or press [second] [$\pi$] [×] [5] [$x^2$]. You will see 3.141592654 × 25 = [78.53981634].

**H.** What is the approximate area of a circle with diameter ($d$) = 20 feet?

$$A_{circle} = \pi r^2 = (3.14)(10^2) = (3.14)(100) = 314 \text{ square feet}$$

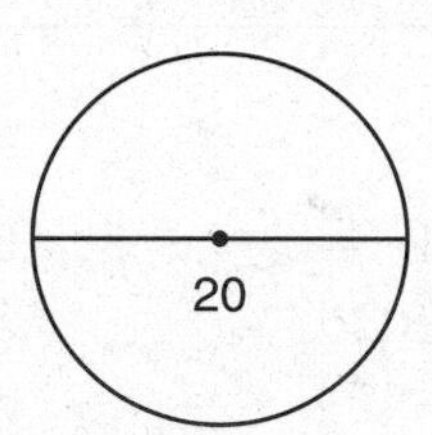


If you use a calculator, press the following keys: [3] [.] [1] [4] [×] [1] [0] [$x^2$]. You'll see 314. However, if you press π instead of 3.14 you will get a more accurate answer: [second] [π] [×] [1] [0] [$x^2$] [=]. You will see 3.141592654 × 100 = [314.1592654].

**I.** What is the area of this trapezoid?

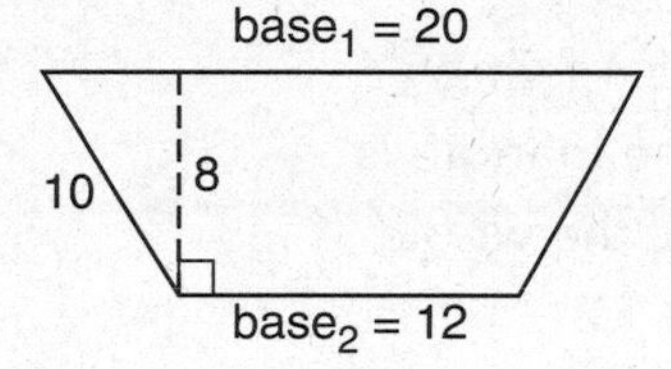


The area of a trapezoid is the average of the two bases times the height

$$\text{Area of a trapezoid} = \frac{\text{Base}_1 + \text{Base}_2}{2}(\text{Height})$$

Remember to use the altitude for the height. In this figure the height is 8.

$$\text{Area of this trapezoid} = \left(\frac{12+20}{2}\right)(8) = \left(\frac{32}{2}\right)(8) = (16)(8) = 128 \text{ square units}$$

If you can't remember the formula, (and don't have a reference sheet), use your problem-solving techniques and divide a big problem into smaller ones. Trapezoids can be divided into rectangles and triangles.

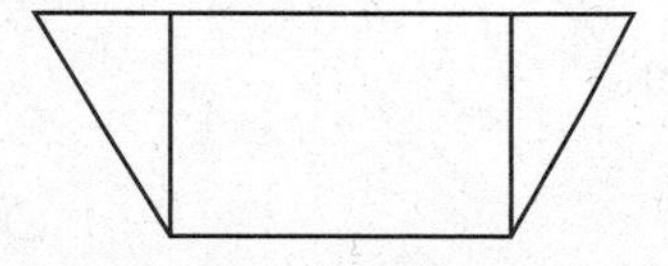

Find the area of each smaller figure and then add up the numbers to get the area of the entire trapezoid.

## PRACTICE: Area of Flat Shapes

(For answers, see page 243.)

**1.** If $AC = 14.5$ feet and $BD = 8.2$ feet, what is the area of triangle $ABC$?

------ square feet

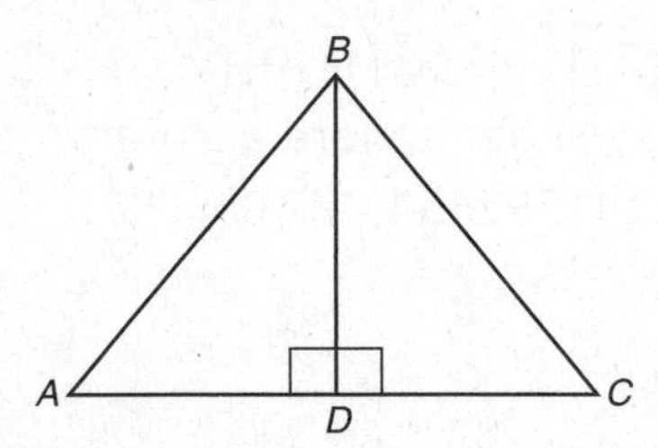


**2.** If the rectangle measures 4 feet by 6 feet, and the height of the triangle is 3 feet, what is the area of the whole shape?

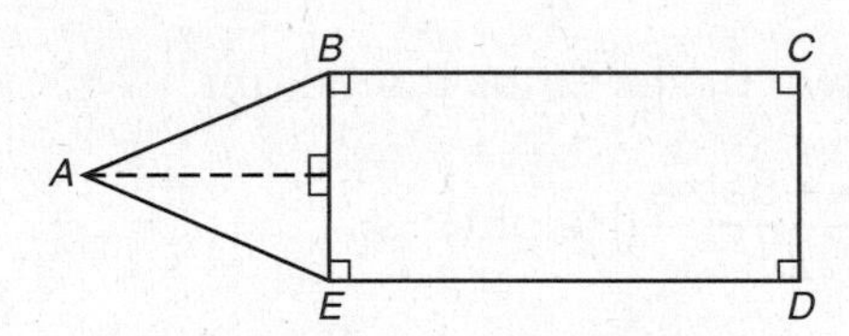


Area rectangle + Area triangle = Area whole shape

$$4 \times 6 + \frac{(3)(4)}{2} =$$

$$24 + \frac{12}{2} = 24 + 6 =$$

------ square feet

**3.** Look at the circle below to help you answer the following questions.

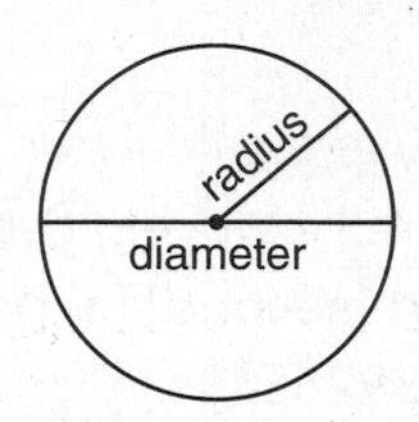


**a.** If the diameter of a circle is 15 inches, what is the *radius* of that circle?

- ○ **A.** 15 in.
- ○ **B.** 7.5 in.
- ○ **C.** 8 in.
- ○ **D.** 30 in.

**b.** If the radius of a circle is 4.3 yards, what is the *diameter* of that circle?

- ○ **A.** 4.3 yd.
- ○ **B.** 2.15 yd.
- ○ **C.** 2.6 yd.
- ○ **D.** 8.6 yd.

**c.** If the radius of a round garden is 6 feet, what is the *approximate area* of that garden?

- ○ **A.** 36 sq. ft.
- ○ **B.** 19 sq. ft.
- ○ **C.** 38 sq. ft.
- ○ **D.** 113 sq. ft.

**d.** If the diameter of a circular cover for a pool is 18 feet, what is the *approximate area* of this cover?

- ○ **A.** 19 sq. ft.
- ○ **B.** 9 sq. ft.
- ○ **C.** 113 sq. ft.
- ○ **D.** 255 sq. ft.

**e.** If I had a rectangular room that measured 12 feet wide and 18 feet long, what is the *approximate area* of the largest round rug I could fit in this room? (*Hint*: Draw and label a diagram first.)

- ○ **A.** 38 sq. ft.
- ○ **B.** 113 sq. ft.
- ○ **C.** 254 sq. ft.
- ○ **D.** 452 sq. ft.

**4.** Find the approximate area of a circle whose diameter the same as the height of the rectangle shown below. Use 3.14 for $\pi$ and round to the nearest hundredth. (No calculator needed here.)

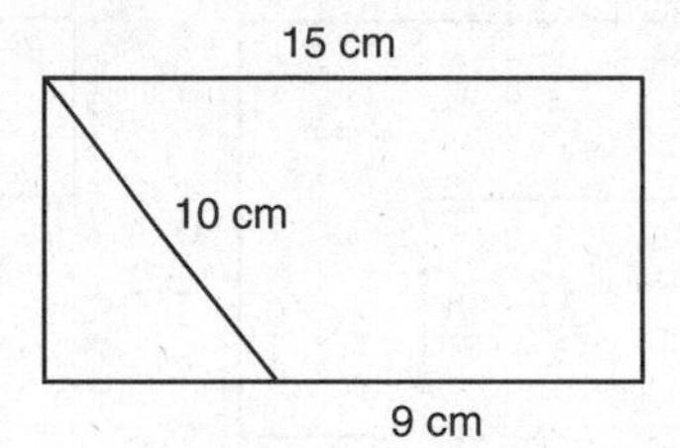


Answer: ------------------

**5.** A square has an area of 256 square meters. To the nearest whole number, what would the radius of a circle be so that the circle would have approximately the same area?

- ○ **A.** ~ 6 m
- ○ **B.** ~ 7 m
- ○ **C.** ~ 8 m
- ○ **D.** ~ 9 m

**6.** A triangle has an area of 36 square inches. What could the height and length be in this triangle? List as many integer values as you can.

## AREA OF IRREGULAR SHAPES (Some From Grade 6.G.1)

Sometimes you need to find the area of shapes that are really combinations of other geometric shapes. In Figure A, you see a shape that seems to be made up of different rectangles. Since you know how to find the area of one rectangle, you can easily find the area of the entire shape. Sometimes these are called *rectilinear shapes*.

## Examples

**A.** Figure A is the figure given. Figures B and C show different ways you could divide this irregular shape into rectangles. Now, just find the area of each rectangle and add them to find the total area of the irregular shape.

**Figure A**

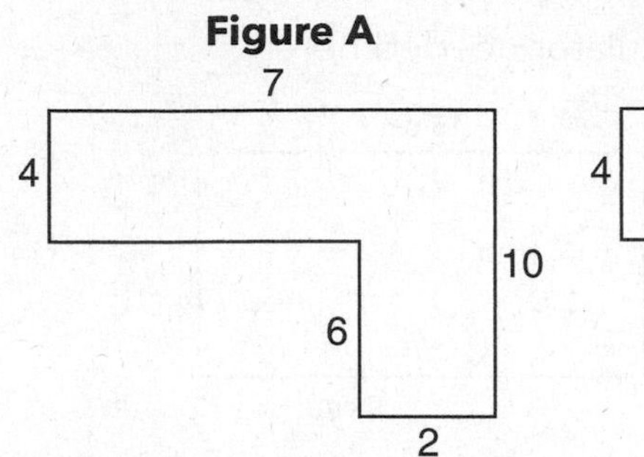


**Figure B**

7 4 10 6 2

**Figure C**

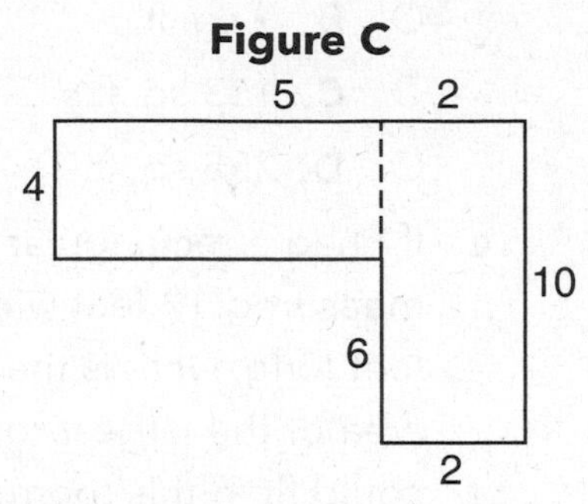


*Hint*: It is important to redraw the diagram and divide the shape into simple shapes. Then, carefully determine the length of each side you need to use.

Top rectangle
Area = 4 × 7 = 28

Smaller rectangle
Area = 6 × 2 = 12

Total area irregular shape
28 + 12 = 40 sq. units

Left rectangle
Area = 4 × 5 = 20

Right rectangle
Area = 10 × 2 = 20

Total area irregular shape
20 + 20 = 40 sq. units

**B.** Find the area of this irregular shape (composed of a rectangle and a triangle). Given: The height of the triangle is 4 feet, the base of the rectangle is 6 feet, and the height of the rectangle is 5 feet.

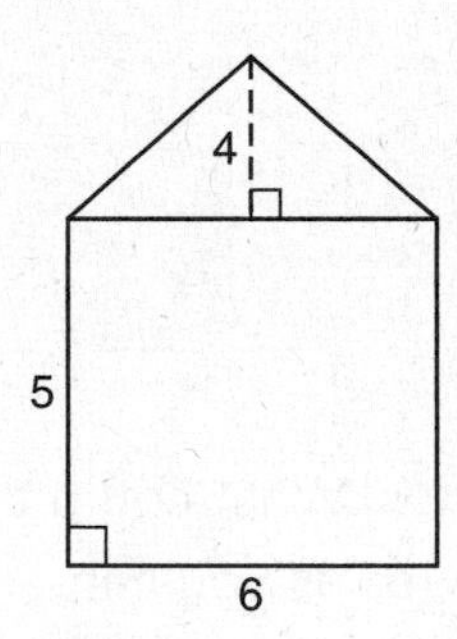


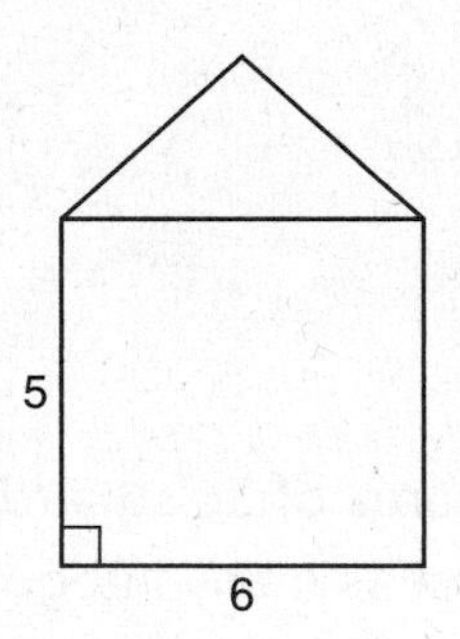


$$A_{\text{triangle}} = \frac{(b)(h)}{2} = \frac{(6)(4)}{2} = \frac{(24)}{2} \quad = 12 \text{ square units}$$

$$A_{\text{rectangle}} = (b)(h) = (6)(5) \quad = 30 \text{ square units}$$

$$A_{\text{total figure}} = A_{\text{triangle}} + A_{\text{rectangle}} \quad = 42 \text{ square units}$$

**C.** Find the approximate area of this irregular shape (composed of a rectangle and half a circle). Given: The width of the rectangle is 6 inches, and the height of the rectangle is 4 inches. From this information you can see that the *diameter* of the circle is also 4 inches (so the radius is 2 inches). Now, we'll find the area of the entire shape.

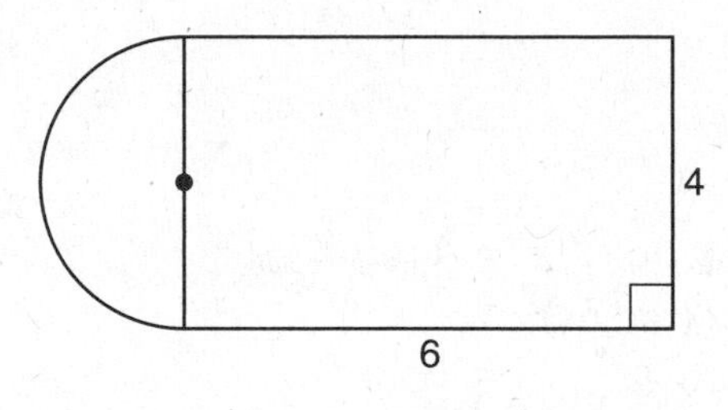


$A_{\text{rectangle}} = (b)(h) \quad = 6 \times 4 \quad = 24.00$ square inches

$A_{\text{half circle}} = (\pi r^2) \div (2) \quad = \dfrac{(3.14)(4)}{2} \quad = \underline{6.28 \text{ square inches}}$

$A_{\text{entire shape}} = A_{\text{rectangle}} + A_{\text{half circle}} \quad = 30.28$ square inches

## Area of the Shaded Region

Other times you may need to find the area of a shape within a shape. You often are asked to find the area of part of the shape. Below are four diagrams with overlapping shapes of circles, rectangles, and triangles. Here you are asked to find the area of the shaded region. Think of the white shape inside the larger shape as an empty hole (like a donut hole). Just subtract the hole from the large shape and you have the area of the shaded region.

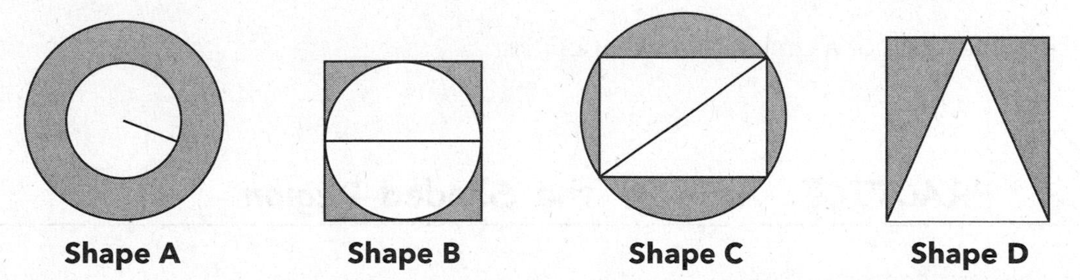


For Shape A, use the area of the larger circle minus the area of the smaller circle. For Shape B, use the area of the square minus the area of the circle. For Shape C, use the area of the circle minus the area of the rectangle. (The rectangle might also be a square.) For Shape D, use the area of the rectangle minus the area of the triangle.

### Example

Use Shape A. Diameter of large circle = 14, diameter of small circle = 10. What is the *approximate area* of the shaded region? (Use $\pi = 3.14$.)

- O **A.** 232
- O **B.** 154
- O **C.** 75
- O **D.** 6

| $A_{\text{large circle}}$ | $-\ A_{\text{small circle}}$ | $= A_{\text{shaded region}}$ |
|---|---|---|
| $(\pi) \times (\text{radius})^2$ | $(\pi) \times (\text{radius})^2$ | $=$ |
| $3.14 \times 7^2$ | $3.14 \times 5^2$ | $=$ |
| $3.14 \times 49$ | $3.14 \times 25$ | $=$ |
| $153.86$ | $-\ 78.50$ | $= 75.36$ square units |

If you use a calculator, press [π] [×] [7] [$x^2$] [−] [π] [×] [5] [$x^2$] [=]. You will see [75.39822369].

Notice that when we used 3.14 as π, our answer was only 75.36. But the calculator used a more accurate estimate for π; it used 3.141592654; therefore, the final answer with a calculator is a little higher and more accurate. However, we were able to select the closest multiple-choice answer, ***C***.

**Note**: Formulas for areas of two-dimensional shapes are on the official *PARCC Grade 8 Mathematics Assessment Reference Sheet* (on page 359).

The formulas are on the PARCC Grade 8 Mathematics Reference sheet, but you should know these formulas for area and circumference of circles from Grade 7 (7.G.4):

Area of a circle = $\pi r^2$

Circumference of a circle = $c = \pi d$ or $c = 2\pi r$

## PRACTICE: Area of the Shaded Region

(For answers, see page 244.)

**1.** Use Shape B from page 87. One side of the square = 12 centimeters. What is the area of the shaded region?

---------------------- − ---------------------- = ------------------------------

(Area of square) (Area of circle) (Area of shaded region)

**2.** Use Shape C from page 87. The square measures 6 feet on each side, and its diameter measures approximately 8.5 feet. What is the approximate area of the shaded region?

-------------------- – -------------------- = ----------------------------

(Area of circle) (Area of square) (Area of shaded region)

○ **A.** 4-5 square feet ○ **B.** 6-8 square feet

○ **C.** 15-16 square feet ○ **D.** 20-21 square feet

**3.** Use Shape D from page 87. The rectangle has a base of 5 inches and is 8 inches tall. What is the area of the shaded region?

-------------------- – -------------------- = ----------------------------

(Area of rectangle) (Area of triangle) (Area of shaded region)

**4.** Use Shape A from page 87. If the area of the large circle = 36 square units, and the area of the shaded region is 14 square units, what is the area of the small circle?

-------------------- – -------------------- = ----------------------------

(Area of large circle) (Area of small circle) (Area of shaded region)

**5.** Which figure below does NOT have an area of 100?

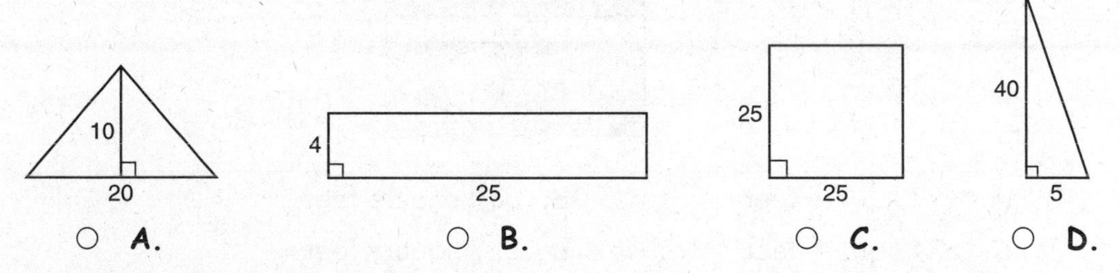


○ **A.** ○ **B.** ○ **C.** ○ **D.**

**6.** Find the area of the irregular shape below. (These are sometimes called *composite* shapes because they are *composed* of two or more geometric shapes.)

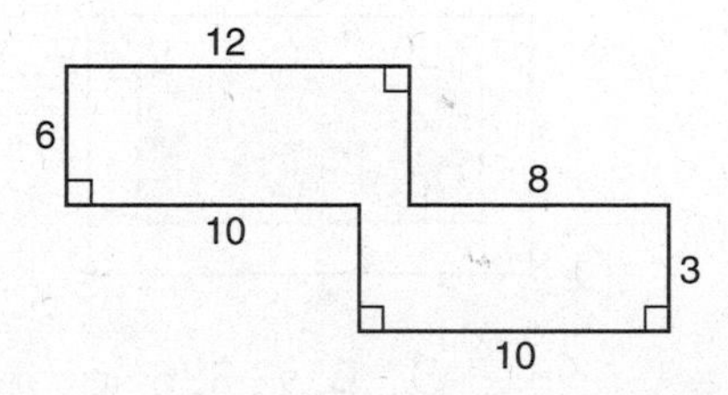


○ **A.** 96 square units ○ **B.** 90 square units

○ **C.** 102 square units ○ **D.** 55 square units

7. You are given a rectangle connected to a semicircle. If the height of the rectangle is 6 cm and the width is 4 cm, what is the approximate area of the whole shape?

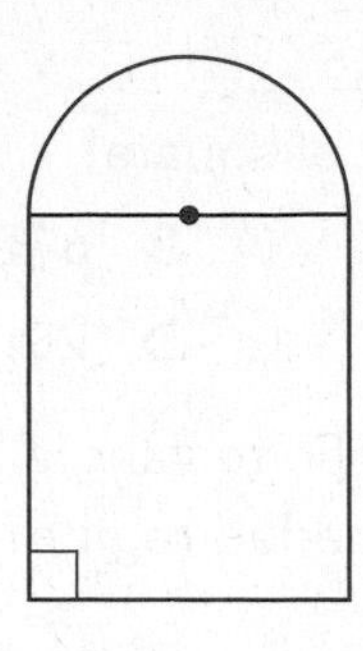

○ **A.** 24 square cm ○ **B.** 30 square cm
○ **C.** 18 square cm ○ **D.** 50 square cm

8. There is a rectangular yard that is 30 feet long and 20 feet wide. In the yard there is a circular fountain with a radius of 2 feet, and a path (as shown) that is 3 feet wide. The area left over will be covered with grass seed.

What is the area of the grass seed surface? Round your answer to the nearest whole number. (*Hint:* Label all dimensions on the diagram first.)

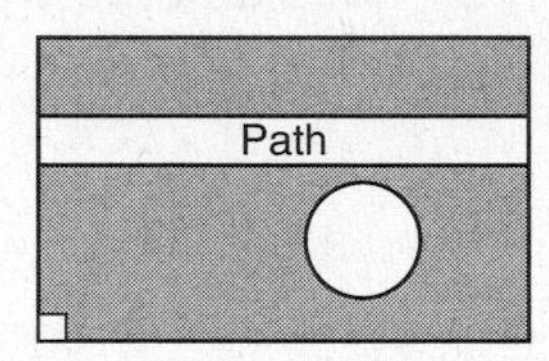


○ **A.** 497 square feet ○ **B.** 510 square feet
○ **C.** 534 square feet ○ **D.** 527 square feet

9. There is a 3-foot-wide walkway around a small park. What is the area of the park? See diagram.

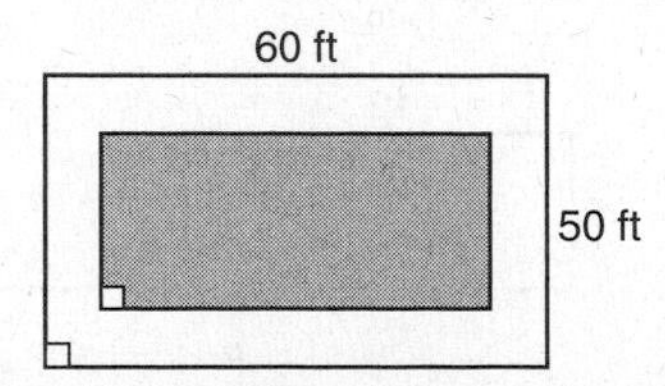


○ **A.** 220 square feet ○ **B.** 2,376 square feet
○ **C.** 2,679 square feet ○ **D.** 3,000 square feet

**10.** Which one does not have an area of 36 sq. units?

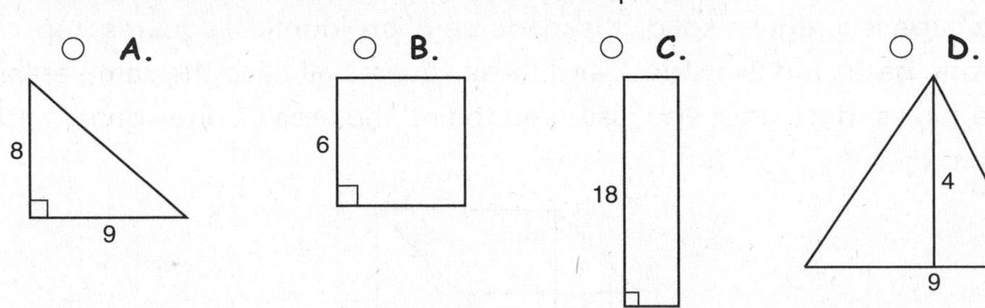


**A right triangle** **A square** **A rectangle** **An acute triangle**

**11.** What is the area of the shaded portion inside the rectangle?

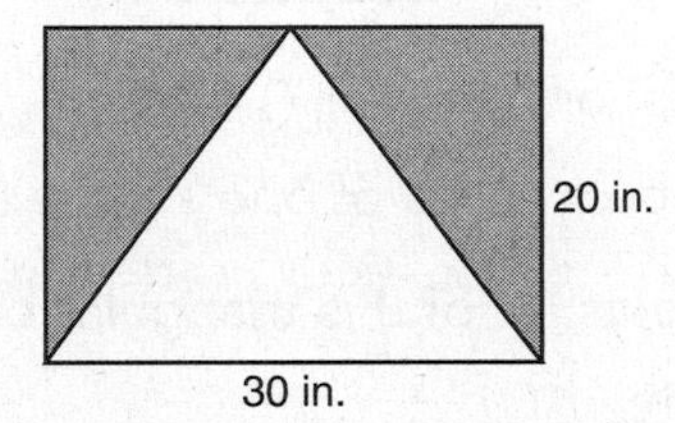


## Surface Area of Three-Dimensional Forms (Some from Grade 7.G.6)

Sometimes you are asked to think about *flat surfaces* even when you have a three-dimensional form. Don't get this confused with volume. You are still looking for the amount of *surface area*. When you find the surface area of a three-dimensional form, you just find the area of each face (top, bottom, and sides) and then add them together.

### Examples

**A.** If you were to paint the outside of this rectangular box, what is the total *surface area* you would be painting?

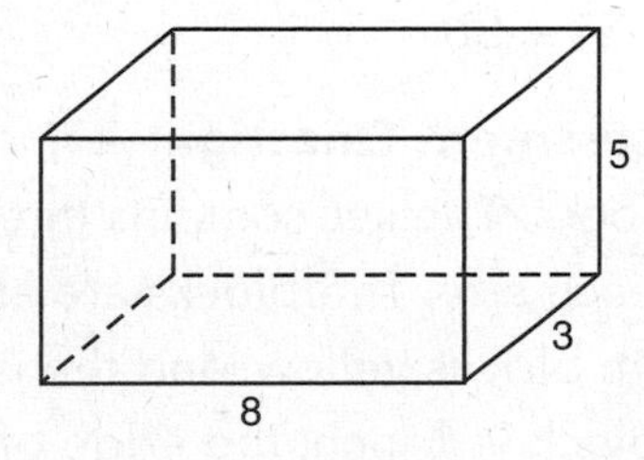


| | | |
|---|---|---|
| Area of top and bottom | $(b)(h) = (3 \times 8)(2)$ | $= 48$ |
| Area of two sides | $(b)(h) = (3 \times 5)(2)$ | $= 30$ |
| Area of front and back | $(b)(h) = (8 \times 5)(2)$ | $= \underline{80}$ |
| Total surface area | | 158 sq. units |

**B.** If you were to find the total surface area of a cube, it could be much less work. Because a cube is a regular solid, it is made up of six identical squares (top, bottom, front, back, and two sides), and these squares all have the same area. So, to find the total surface area, you just need to find the area of one square and multiply that by 6.

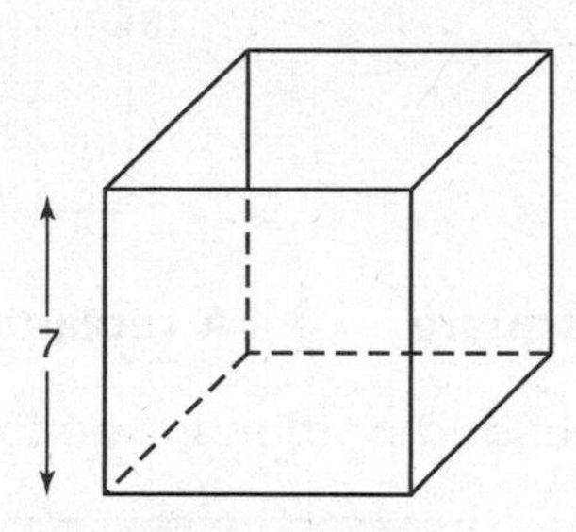


Area of one square = $(s)(s)$ or $(s^2) = (7)(7) = 49$

Surface area of entire cube = (Area of one square)(6) = (49)(6) = 294 square units

**C.** If you were to paint the outside of this triangular box, what is the total surface area you would be painting?

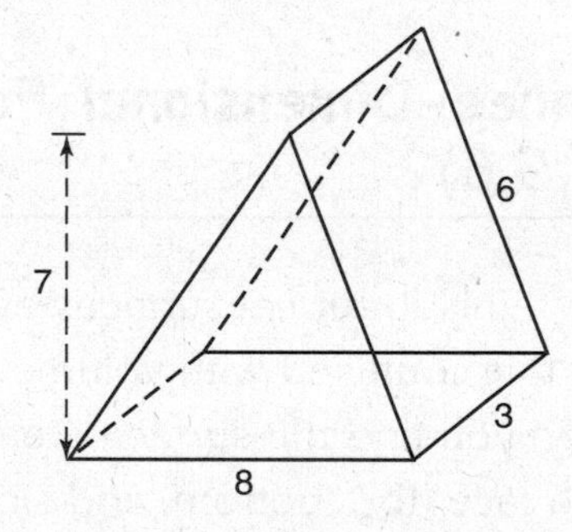


| | | |
|---|---|---|
| Area of bottom (1 rectangle) | $(b)(h) = (3 \times 8)$ | = 24 |
| Area of sides (2 rectangles) | $(2)(b)(h) = (2)(6)(3)$ | = 36 |
| Area of front and back (2 triangles) | $(2)\frac{(b)(h)}{2} = (8)(7)$ | = 56 |
| Total surface area = 24 + 36 + 56 | | = 116 sq. units |

**D. Performance-Based Assessment Question:** A nursery school teacher asks you to paint a set of wooden blocks. The set contains three different sizes of blocks. There are ten blocks of each size. The blocks are all cubes. She asks you to paint the small and the medium blocks yellow, and the large blocks red. The length of the edge of each small block is 1 inch, the edge of each medium block is 3 inches, and the edge of each large block is 4 inches.

The teacher gives you one can of yellow paint and one can of red paint. Each can will cover 1,000 square inches. Will you have enough paint for all the blocks as she asked? Explain!

This sounds complicated. But if you organize your information carefully, and label all the information, it can be easy. It is best to use a table, label each section, and show all your work.

| Blocks (Cubes) Used | Area of 1 Face | Area of 1 Cube (add, top, bottom, front, back, 2 sides = 6 faces) | Area of 10 Cubes (with work shown) | Area of 10 Cubes |
|---|---|---|---|---|
| | | | | |
| Small blocks (1 in. each edge) | (1)(1) = 1 sq. in. | (6)(1) = 6 sq. in. | (10)(6) = 60 sq. in. | 60 sq. in. |
| Medium blocks (3 in. each edge) | (3)(3) = 9 sq. in. | (6)(9) = 54 sq. in. | (10)(54) = 540 sq. in. | 540 sq. in. |
| **Total surface area of all yellow cubes** | | | | **600 sq. in.** |
| Large blocks (4 in. each edge) | (4)(4) = 16 sq. in. | (6)(16) = 96 sq. in. | (10)(96) = 960 sq. in. | 960 sq. in. |
| **Total surface area of all red cubes** | | | | **960 sq. in.** |

*Since each can of paint can cover 1,000 square inches, you would have enough paint. You need 600 square inches of yellow paint and 960 square inches of red paint.*

**E.** Find the approximate surface area of the can below to the nearest square inch. The circumference is about 6.25 inches, and the height is 5.5 inches.

First you should sketch the parts of a can the way they would look if you could lay them flat. A can (really a cylinder) is made of two circles and a rectangle.

$$A_{rectangle} = l \times w = 6.25 \times 5.5 = 34.375 \text{ square inches}$$

> **(Note: The circumference of the circle is really the same as the length of the rectangle.)**

$$A_{circle} = \pi r^2 = (3.14)\ (?)$$

Even though you don't know the radius, you do know the circumference and the value of $\pi$. That can help you find the radius.

| | | |
|---|---|---|
| $C$ | $= \pi d$ | Formula for circumference of a circle. |
| $6.25$ | $= (3.14)(d)$ | Use the values that you know. |
| $\frac{6.25}{3.14}$ | $= \frac{(3.14)(d)}{3.14}$ | Divide both sides by 3.14 to solve for $d$. |
| $1.99$ | $= d$ (diameter) | Since the radius is half the diameter, the radius = about 1. |

$$A_{circle} = \pi r^2 = (3.14)(1)(1) = 3.14 \text{ square inches}$$

The total area of the can is the area of the rectangle + the area of two circles.

$$34.375 + 2(3.14) = 40.65 \text{ or approximately } 41 \text{ in.}^2$$

## VOLUME OF THREE-DIMENSIONAL FORMS

Other times you need to think about the total space *inside* a three-dimensional form. You then will be thinking about its *volume*. Since you are working with three dimensions (width, length, and depth), you need to use cubic *units*.

## Examples

**A.** If you were to fill this rectangular solid with water, how many cubic units of water would it hold?

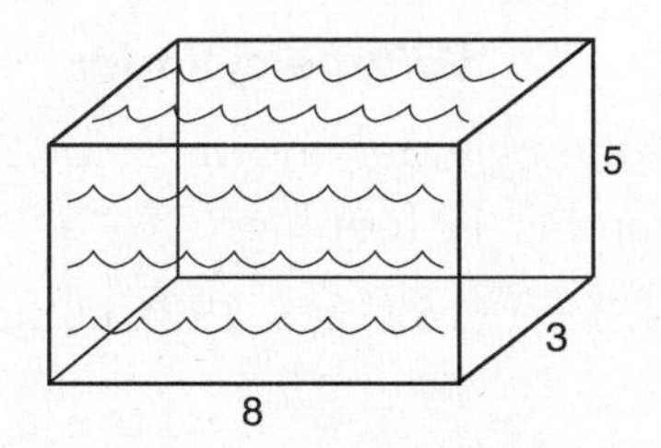


Base × Height × Depth = Volume

$b \times h \times d = V$

$V = 8 \times 5 \times 3 = 120$ cubic units

**B.** If you were to fill this rectangular cube with water, how many cubic feet of water would it hold? (*Remember*: A cube is a regular solid; all of its faces are identical.)

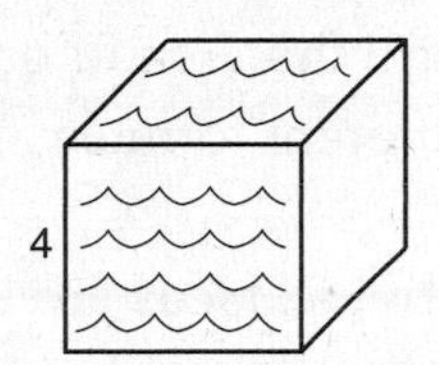


Base × Height × Depth = Volume

$b \times h \times d = V$

$V = (b)(h)(d) = (4)(4)(4) = 64$ cubic feet

**C.** The diagram below is a sphere that has a radius of 6 inches. If you were to fill this sphere with air, how many cubic inches of air would it hold? To do this we find the *volume* of the sphere.

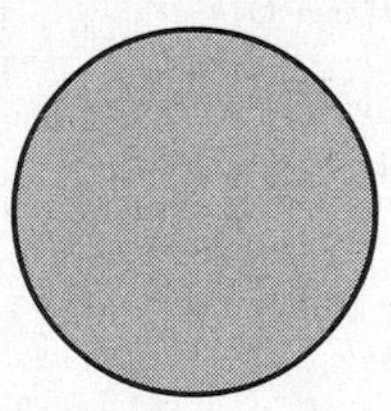

**The formula for finding the *volume of a sphere* is $\frac{4}{3}\pi r^3$.**

This formula is also on the PARCC Grade 8 Mathematics Assessment Reference Sheet.

$$\text{Volume} = \frac{4}{3}\pi r^3 = (\textit{Area of the Base})(\textit{height})$$

$$= \frac{4}{3}\pi(6)(6)(6) = 4\pi(2)(6)(6) = \pi 288 \text{ cubic inches} \sim 904 \text{ cubic inches}$$

**D.** Another solid to consider is a cylinder.

The formula for finding the *volume of a cylinder* is (*Area of the Base*) (*height*)

Since the base of a cylinder is a circle you use ($\pi r^2$)(*height*)

$$\textbf{Volume cylinder} = \pi r^2 h$$

How much liquid will fill the cylinder shown below?
It has a radius of 4 feet and is 12 feet high.

Volume $= \pi r^2 h = \pi (4^2)(12) = \pi (16)(12) = 192\pi$ cubic feet

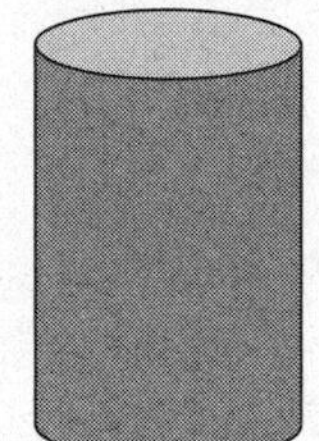


## Important Reminders

There are many other solid forms, such as triangular prisms, however, for the PARCC 8 test, we will concentrate only on the ones we have shown you here.

Remember the difference between *surface area* and *volume*!

- To find *surface area*, you find the area of each face and add them together.
- To find *volume*, you use different formulas to see how much will fit inside the solid.
- In a *regular solid*, like a cube, all faces are identical (Length = Width = Depth).

## PRACTICE: Surface Area and Volume

(For answers, see page 244.)

**1.** Given: A cube with each length equal to 3 inches. Find the *surface area* of this form (the area you would paint if you painted the exterior of the cube).

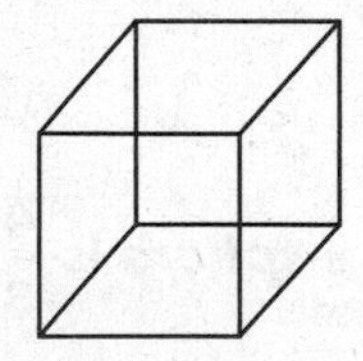

- ○ **A.** 27 sq. in.
- ○ **B.** 36 sq. in.
- ○ **C.** 54 sq. in.
- ○ **D.** 63 sq. in.

**2.** Two boxes have the same volume.

Box A: Height = 3 in.
Width = 2 in.
Length = 6 in.

Box B: Height = 4 in.
Width = 3 in.
Length = ______ in.

What is the length of Box B?

- ○ **A.** 2 in.
- ○ **B.** 3 in.
- ○ **C.** 4 in.
- ○ **D.** 9 in.

**3.** If a cube has a volume of 64 cubic centimeters, the length of one edge would =

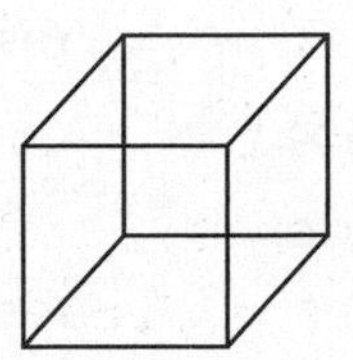

- ○ **A.** 6 cm
- ○ **B.** 4 cm
- ○ **C.** 8 cm
- ○ **D.** 16 cm

**4.** If a cube has a volume of 27 cubic feet, then the area of one face would =

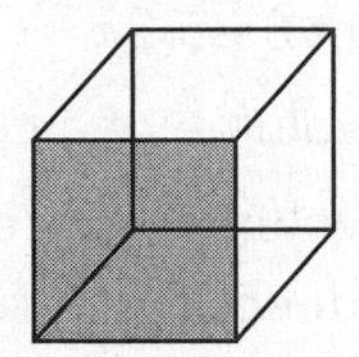

- ○ **A.** 3 sq. ft.
- ○ **B.** 6 sq. ft.
- ○ **C.** 9 sq. ft.
- ○ **D.** 12 sq. ft.

**5.** If a cube has a volume of 8 cubic feet, then what is the perimeter of one of its faces?

- ○ **A.** 4 feet
- ○ **B.** 8 feet
- ○ **C.** 12 feet
- ○ **D.** 16 feet

**6.** Find the *volume of a sphere* that has a *radius* the same length as the diagonal **"X"** of the rectangle shown below. (Use 3.14 as π and round your answer to the nearest whole number.)

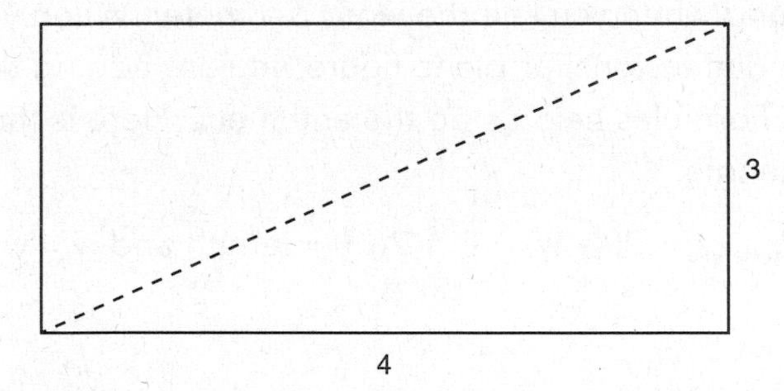


Which of the following are correct expressions for the volume of that sphere?

- ☐ **A.** $\frac{4}{3}(5^3)\pi$
- ☐ **B.** $\frac{500}{3}\pi$
- ☐ **C.** $\frac{4}{3}(125)\pi$
- ☐ **D.** $\frac{1}{2}\pi(5^3)$
- ☐ **E.** $\frac{1}{2}$

7. Which of the following is true about the volume of the listed solids? Check Yes --✔--- or No --✔---.

   A. The volume of a cylinder with radius 2 and height 8 is $32\pi$. Yes ------ No ------

   B. The volume of a cube that is 5 inches high is (25)(5). Yes ------ No ------

   C. The volume of a rectangular solid that is 2.5 ft. wide, 5 ft. deep, and 8 ft. high is (2.5) (13). Yes ------ No ------

   D. The volume of a cube that is 2 cm high is larger than the volume of a cylinder that has a radius of 2 cm and is 4 cm high. Yes ------ No ------

   E. The volume of the shorter cylinder shown below is larger than the volume of the taller cylinder. Yes ------ No ------

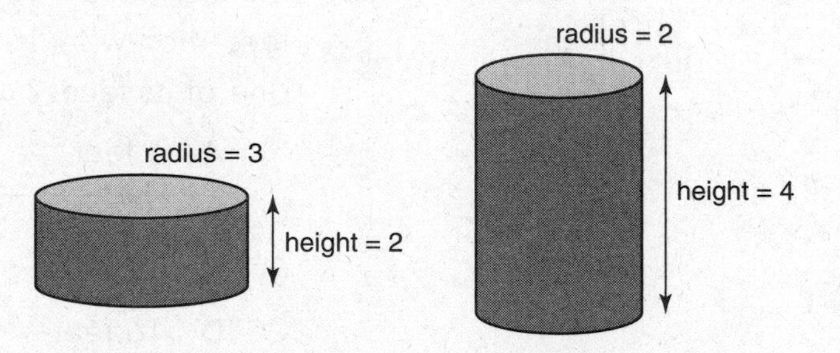


## PERIMETER

Think *distance around* when you see the word *perimeter*. When you are looking for the perimeter of a two-dimensional or plane figure, you are adding the lengths of the sides of the figure. Formulas help us do the arithmetic. Here is the formula for the perimeter of a rectangle.

$$P_{rectangle} = 2(l + w) = 2l + 2w \ (l = \text{length and } w = \text{width})$$

## Examples

**A.** Judy and Dee take a walk around their neighborhood a few times a week. They walk east 900 feet, south 1,200 feet, and then continue in a rectangular shape until they return home again. How far do they walk?

Add the four sides.

$$900 + 1{,}200 + 900 + 1{,}200 = 4{,}200 \text{ feet}$$
$$\text{Length} + \text{Width} + \text{Length} + \text{Width} = \text{Perimeter}$$
$$2l + 2w = P$$

or

$$2(900 + 1{,}200) = 4{,}200 \text{ feet}$$
$$2(\text{Length} + \text{Width}) = \text{Perimeter}$$
$$2(l + w) = P$$

This is a typical perimeter problem where we measure the distance *around* a shape.

**Hint:** It is always helpful to draw and label a diagram before solving perimeter problems.

**B.** Find the perimeter of a rectangle whose width is 3 cm and whose length is twice the width.

Draw a rectangle. Label the lengths: Width ($w$) = 3 and Length ($l$) = 6

$$2(l) + 2(w) = P_{rectangle}$$
$$2(6) + 2(3) = P_{rectangle}$$
$$12 + 6 = P_{rectangle}$$
$$18 = P_{rectangle}$$

or

$$2\,(l + w) = P_{rectangle}$$
$$2(6 + 3) = P_{rectangle}$$
$$2(9) = P_{rectangle}$$
$$18 = P_{rectangle}$$

**C.** Find the length of a side of a square whose perimeter is 64 inches. Remember, in a square, all sides are the same length. Instead of saying Perimeter = Side + Side + Side + Side, we can just say 4 times the length of one side.

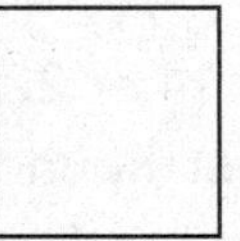

Using the formula, substitute 64 for $P$.

*Perimeter of a square*

$P_{square} = 4s$

$64 = 4s$

$16 = s$

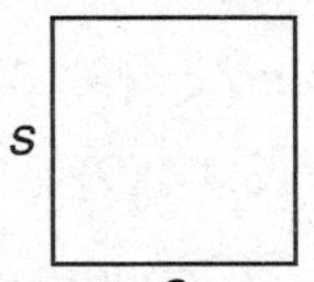


**D.** To find the perimeter of a triangle, add the lengths of the three sides together.

$$P_{triangle} = s_1 + s_2 + s_3$$

If $s_1 = 4$ and $s_2 = 5$ and $s_3 = 8$, then $P_{triangle} = 4 + 5 + 8 = 17$

**E.** Sometimes you are asked to find the perimeter of *special* triangles. The following is information you should remember about two special triangles. We use tick marks to show which sides are the same length (which sides are congruent).

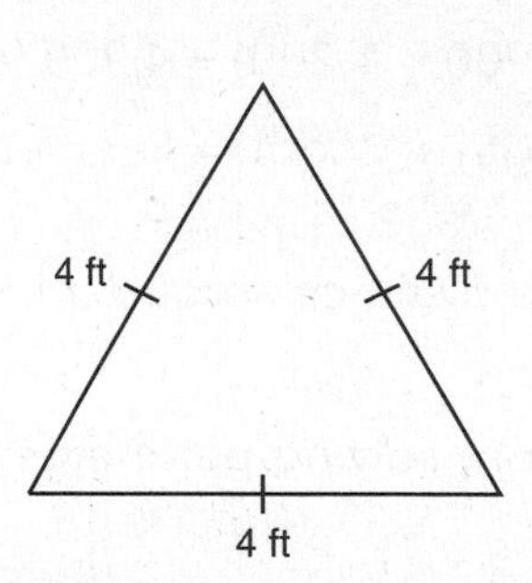


**Equilateral triangle**

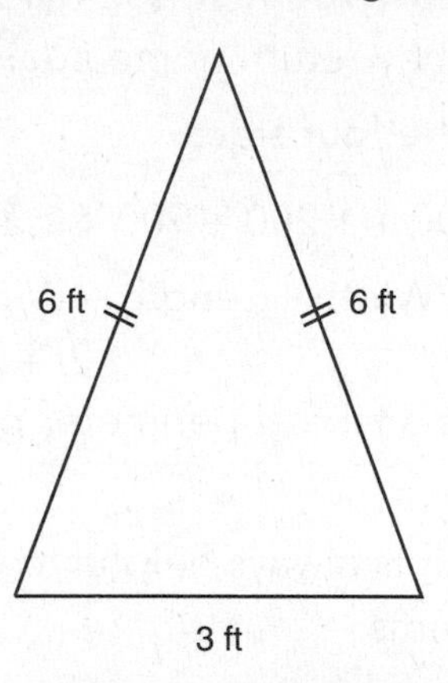


**Isosceles triangle**

This first triangle is an ***equilateral*** triangle.
All sides are the same length.
If one side = 4 ft, Perimeter = 4 + 4 + 4 = 12 ft
or Perimeter = 3(4) = 12 ft

This is an ***isosceles*** triangle.
Two sides are the same length.
If base = 3 ft and side $a$ = 6 ft,
$P = 3 + 6 + 6 = 15$ or
$P = 3 + 2(6) = 15$

**F.** Circles use different vocabulary. The perimeter of a circle is called its *circumference*. (See the *PARCC Grade 8 Mathematics Assessment Reference Sheet* on page 359.) The formula used to calculate the circumference is

*Circumference* = (π)(*diameter*)
$$C = \pi d$$

The approximation for π (pi) is 3.14 or 22/7. Find the approximate circumference of a circle whose radius is 4.5 feet. Round to nearest whole number. (*Remember*: We know the diameter is twice the radius, so the diameter is 9 feet.)

$C = \pi d$ or $(3.14)(d)$

$C = (3.14)(9)$

$C = 28.26$ feet

$C = 28$ feet (rounded to nearest whole number)

If you use a calculator, press [π] [×] [9] [=] and see [28.27433388]. Remember that the calculator uses a more accurate estimate for π; it uses 36,141592654 not just 3.14, so the calculator answer will be a little higher.

## PRACTICE: Perimeter

(For answers, see pages 245-246.)

**1.** David has a garage that is separate from his house. The garage measures $12\frac{1}{3}$ feet wide and $19\frac{1}{2}$ feet long. He is planning to decorate for a big family outdoor party and wants to buy some colorful lights to go around the garage. What is the total length he will need for these lights?

- ○ **A.** $63\frac{2}{3}$
- ○ **B.** $62\frac{2}{3}$
- ○ **C.** $31\frac{2}{3}$
- ○ **D.** $12\frac{1}{3} \times 19\frac{1}{2}$

**2.** If one side of a regular pentagon measures 8.4 inches, what is the perimeter of this pentagon?

Answer: ______ inches

**3.** Use the diagrams below to answers **Parts A, B**, and **C** below. (All lines are either horizontal or vertical.)

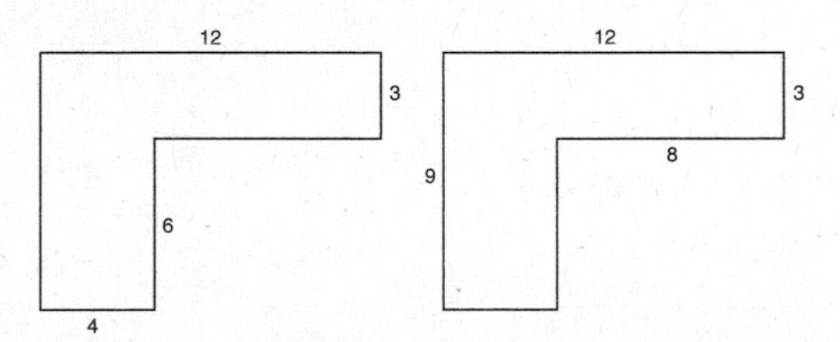


**Part A:** Brett is helping his grandmother prepare her vegetable garden. If he fences off either garden shown above, how much fencing will he need?

Answer: ______feet.

**Part B:** Brett also wants to find the area of the garden. To do this he needs to divide this irregular shape into rectangles. Draw on the shapes above to show him two different ways he could do this. Show your work and find the area of each section.

Answer: The areas of the two sections are ------ and ------ sq. feet.

**Part C:** What is the area of the garden?

Answer: ------ sq. feet

**4.** In a memorial site downtown, there is a statue placed in the middle of a triangular space. The space is actually an equilateral triangle. One side of it measures 9.5 feet long. The city landscaping crew is planning to put a border of stones around the space. Which expressions below will give them a correct length for the stone border? (Select all that are correct.)

- ☐ **A.** (9.5) (9.5)
- ☐ **B.** $9.5 \times 3$
- ☐ **C.** 27.5
- ☐ **D.** $(9.5)^2$
- ☐ **E.** $\frac{(9.5)(9.5)}{2}$
- ☐ **F.** $9.5 + 9.5 + 9.5$

**5.** If the base of an isosceles triangle measures 10.6 cm and one side measures 8.5 cm, what is the distance around the whole triangle?

- ○ **A.** 29.7 cm
- ○ **B.** 27.6 cm
- ○ **C.** 19.1 cm
- ○ **D.** 90.1 cm

**6.** What is the perimeter of the polygon below?

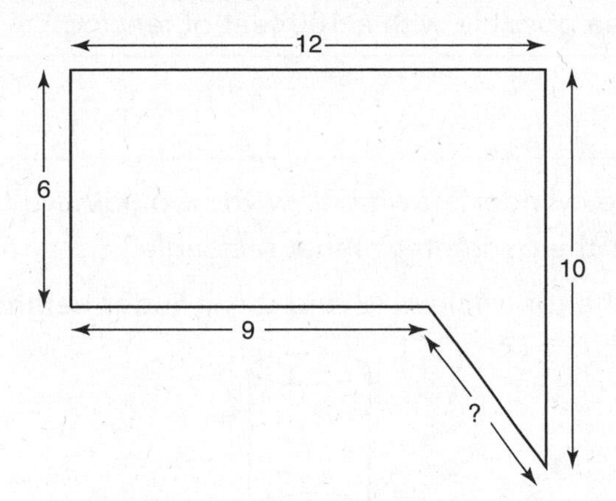


Answer: ______ units

**7.** The diameter of a circle is 12.6 feet. What is its circumference? (Use 3.14 as $\pi$ and round your answer to the nearest tenth.)

Answer: ______ feet

**8.** The diameter of a circle is 16 mm long. Mickaela says the circumference of this circle is $32\pi$ mm. Willie says that is wrong; the circumference of this circle is $16\pi$ mm. Suzie says they are both wrong; the circumference of this circle is $64\pi$ mm.

**Part A:** Who is correct?

Answer: ________________

**Part B:** Explain how you know.

**9.** Kevin has 100 feet of fencing. He wants to use it to make a rectangular play area for his dog. He wants to be sure his dog will have the largest area possible.

**Part A:** Should it be a long and narrow area or more like a square?

Answer: ________________

**Part B:** What should be the dimensions of this rectangle?

Answer: ______ × ______

**Part C:** Give examples to support your answer. (Explain why your choice was the largest area possible with a 100 feet of fencing.)

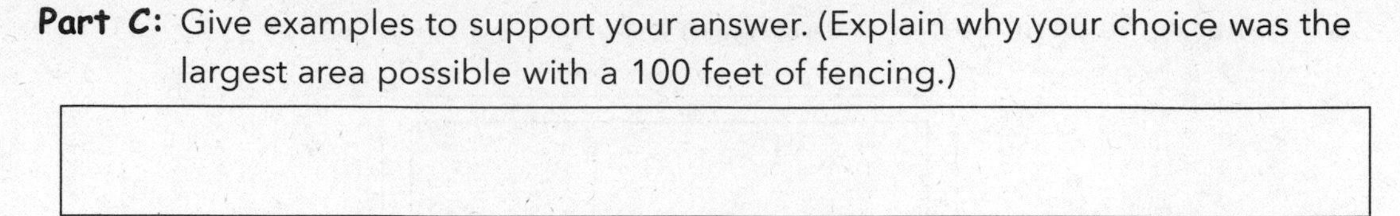

**10.** If you unrolled the cylinder drawn below you would have two circles and a rectangle. What is the perimeter of that rectangle?

The diameter of the top circle = 12 and the cylinder height = 32.

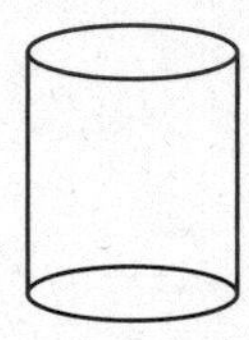

*(Diagram not drawn to scale.)*

**A.** 444

**B.** $444\pi$

**C.** 88

**D.** $64 + 24\pi$

## TRIANGLES AND OTHER POLYGONS

### Triangles

Triangles are three-sided polygons. Review the following chart to remind yourself of the different triangles you should be familiar with.

One way to study would be to make your own table, draw each triangle (or name each triangle), and then fill in the other columns without looking at the chart.

| Drawing of Triangle | Name of triangle | Angles | Sides* |
|---|---|---|---|
| | Scalene triangle | No angles are = | No sides are ≅ |
| | Isosceles triangle | Base angles are = | Opposite sides are ≅ |
| | Equilateral triangle | All angles are = | All sides are ≅ |
| | Obtuse triangle | One angle is greater than 90° (> 90°) | |
| | Acute triangle | All angles are less than 90° (< 90°) | |
| | Right triangle | One angle is 90° (= 90°) | The *hypotenuse* (the side opposite the right angle) is always the longest side. |

*The symbol ≅ means "are congruent," same size and shape.

Remember:

- The *sum* of the three angles in any triangle is always 180°.
- The *shortest* side of a triangle is always opposite the *smallest* angle.
- The *longest* side of a triangle is always opposite the *largest* angle.

## Examples

**A.** If one angle of a triangle is 20°, and another is 50°, then the third angle = ?

$$20 + 50 = 70, \quad 180 - 70 = 110°$$

**B.** If one angle of a triangle is 120°, and another is 35°, then the third angle = ?

$$120 + 35 = 155, \quad 180 - 155 = 25°$$

**C.** If you have a right triangle, and one angle = 40° then the third angle = ?

$$90 \text{ (right angle)} + 40 = 130, \quad 180 - 130 = 50°$$

**D.** In the triangle below, which angle is the smallest? (Do not just guess because it looks the smallest. Use what you know about the relationship of angles to sides in a triangle.)

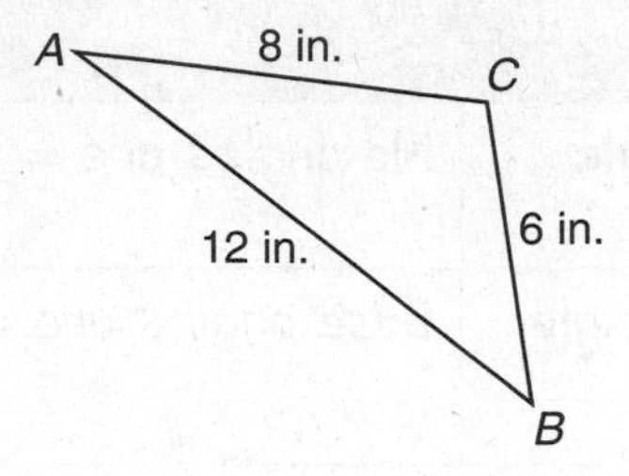


**A.** $\angle A$

**B.** $\angle B$

**C.** $\angle C$

**D.** not enough information

The answer is $\angle A$ because $\angle A$ is opposite the shortest side, which measures 6 inches.

**E.** Mark says that a triangle can have two obtuse angles; Aisha says that it cannot. Who is correct? Explain your answer.

If a triangle had two obtuse angles, then it would have two angles greater than 90°. They would add up to more than 180°; however, we know that the three angles in a triangle must add up to 180°, not just two. Aisha is correct.

## PRACTICE: Triangles

(For answers, see pages 246–247.)

**1.** Given: A triangle $ABC$, $m\angle A = 40°$, $m\angle B = 50°$, then $m\angle C =$ ______.

**2.** Given: An isosceles triangle $DEF$; each base angle = 30°, then the vertex angle = ______.

**3.** Given: A right triangle, one angle = 20°, then the third angle must = ______.

**4.** See the diagram below. The arrow is pointing to an angle that measures ______

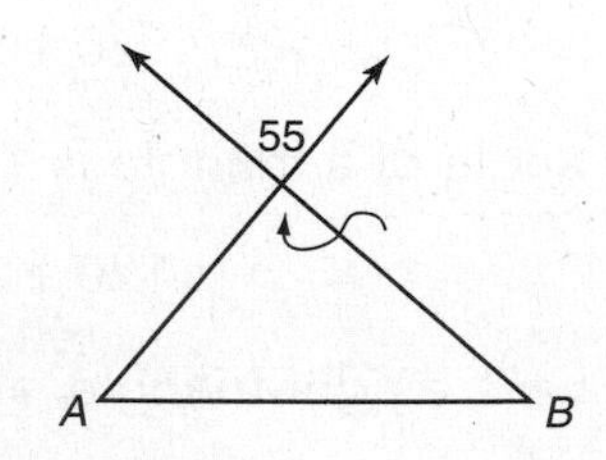

5. Using the same diagram as in question 4, you can determine that if the measure of angle $A = 65°$, then the measure of angle $B =$ ------.

For questions 6-8 you are given information about the sides of a triangle; classify each of the following triangles as equilateral, isosceles, or scalene.

6. Triangle *XYX*

**Part A:** $XY = 12$

$YZ = 12$

$XZ = 10$

--------------------

**Part B:** Triangle *UVW*

$UV = 15$

$VW = 15$

$UW = 15$

--------------------

**Part C:** Triangle *DEF*

$DE = 9$

$EF = 7$

$FD = 12$

--------------------

For questions 7-9 you are given information about the angles of each triangle; classify each of the following triangles as acute, right, or obtuse.

7. Triangle *ABC*

**Part A:** If $m\angle A = 82°$, $m\angle B = 56°$, then this is ------ triangle.

**Part B:** Triangle *MNP*

If $m\angle M = 102°$, then this is ------ triangle.

**Part C:** Triangle *QRS*

If $m\angle Q = 50°$ and $m\angle R = 40°$, then this is ------ triangle.

8. In the triangle below, which angle is the largest angle?

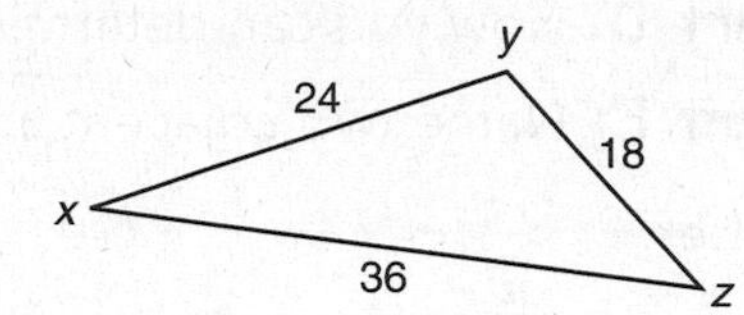


9. Area of triangle *A* is ------ sq. units; area of triangle *B* is ------ sq. units.

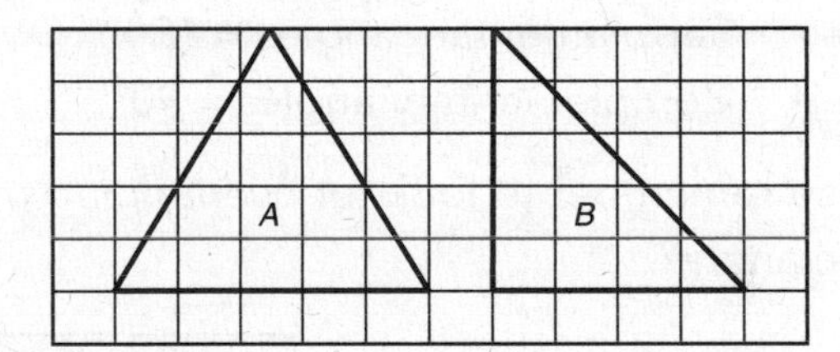

**10.** Use the diagram below. Copy the diagram below onto your own sheet of paper. Label the diagram. As you find each measurement, write it on the diagram.

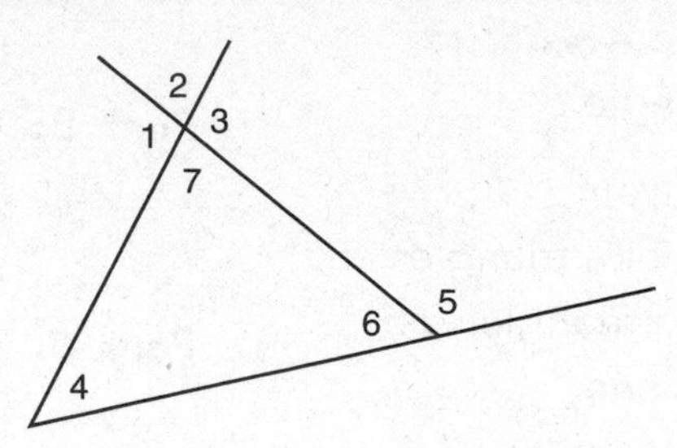


**Part A:** If m∠1 = 80°, then m∠3 = ______ because they are ______ angles.

**Part B:** If m∠1 = 80°, then m∠2 = ______ because together they add up to ______.

**Part C:** If m∠4 = 30°, then m∠6 = ______ because the three angles in a triangle add up to ______. (*Hint:* Find m∠7 first.)

**Part D:** Now you can determine that m∠5 = ______.

**Part E:** Name two adjacent angles. ______ and ______.

Remember:

- A straight line = 180°
- Vertical angles are equal
- The sum of all angles in a triangle = 180°
- Supplementary angles = 180°
- Complementary angles = 90°

Also remember to look for *overlapping* triangles. Do you see 8 triangles in this rectangle?

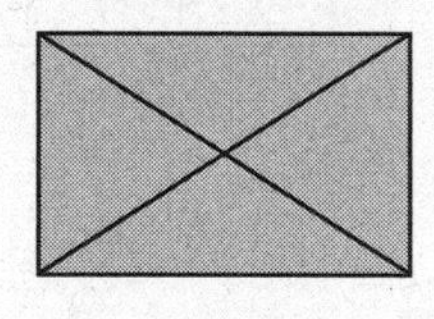

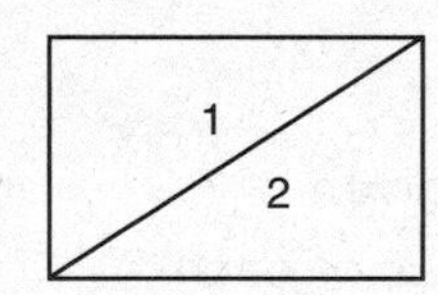


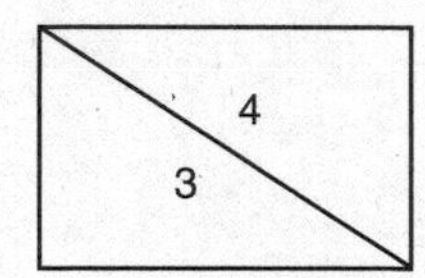


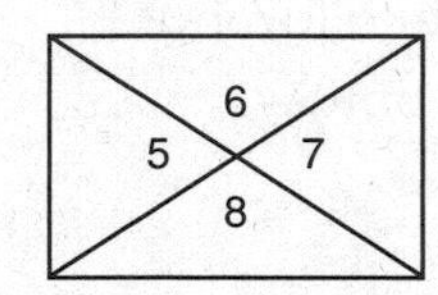

# Other Polygons

## Find the Sum of the Interior Angles of a Polygon

Because we know that the sum of the interior angles of a triangle = 180°, we can now find the sum of the interior angles of other polygons.

### Example

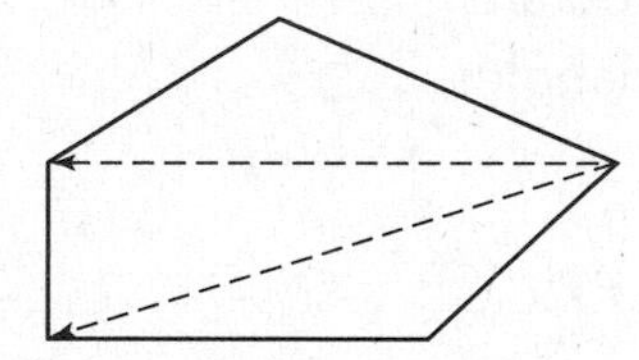

Use one of the basic polygons shown on page 80 and follow these steps.

1. Start at one vertex and draw a straight line to another vertex.
2. Continue from that same vertex and draw as many lines as you can to other vertices.
3. Now, count the number of triangles you have.
4. Notice that if you had chosen a five-sided polygon you would have been able to draw three triangles. Since one triangle has 180°, then three triangles would have $180 \times 3 = 540°$. This is the *sum of the interior angles* of that five-sided polygon.

## PRACTICE: Other Polygons

(For answers, see page 247.)

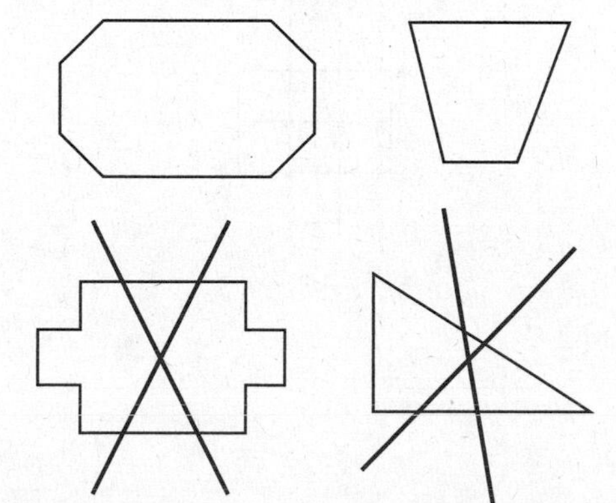

1. Draw a polygon with more than three sides. Use a ruler.

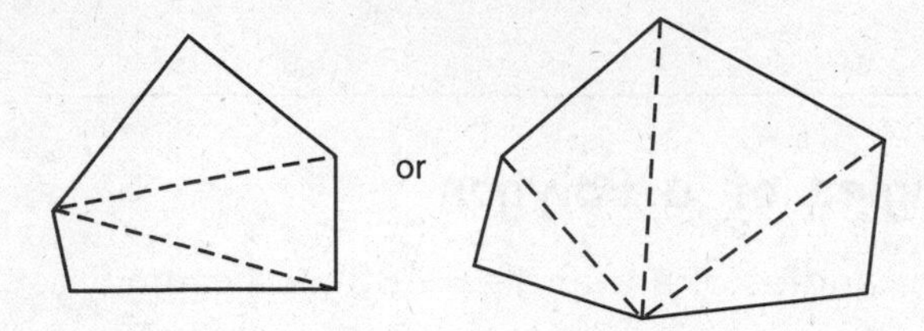


2. Use your drawing and start at one *vertex* (one point) and draw a line to every other vertex. Notice you are dividing the polygon into triangles. Use a ruler please.

3. How many sides does your polygon have?

4. How many triangles were you able to draw?

5. How many degrees in the sum of the interior angles of one triangle? ------

6. How many degrees in the sum of the interior angles of your polygon? ------

7. What is the area of the following shape that is made up of different rectangles? (*Hint:* Divide the shape into rectangles first.)

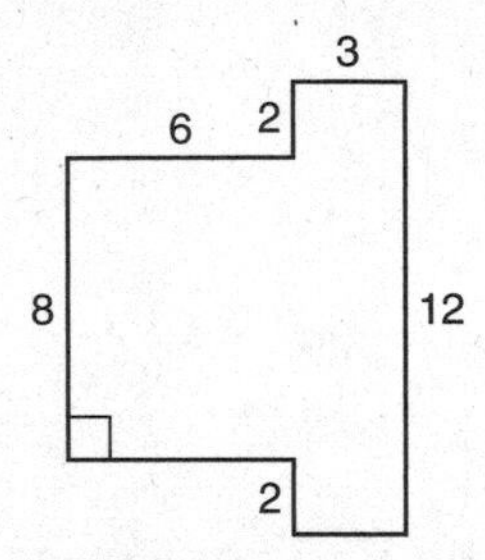


- A. 42 sq. units
- B. 78 sq. units
- C. 84 sq. units
- D. 92 sq. units

A *net* is another term you should know. A net is just a plan (a *pattern*) on a flat plane that could be folded to make a three-dimensional solid form.

## Examples

**A.** Here is a net that would fold up to be a box without a cover. The shaded area is the bottom of the box.

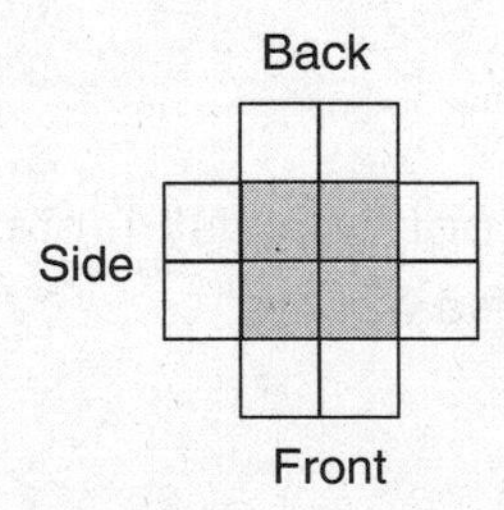


**B.** This net would make a box WITH a cover. Both bottom and cover are shaded.

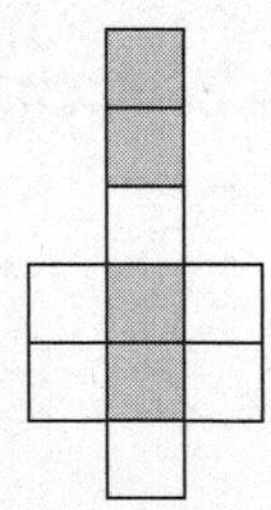

**8.** Which net will make a rectangular covered box (without overlapping)? The bottoms of the boxes have been shaded in to help you see the form.

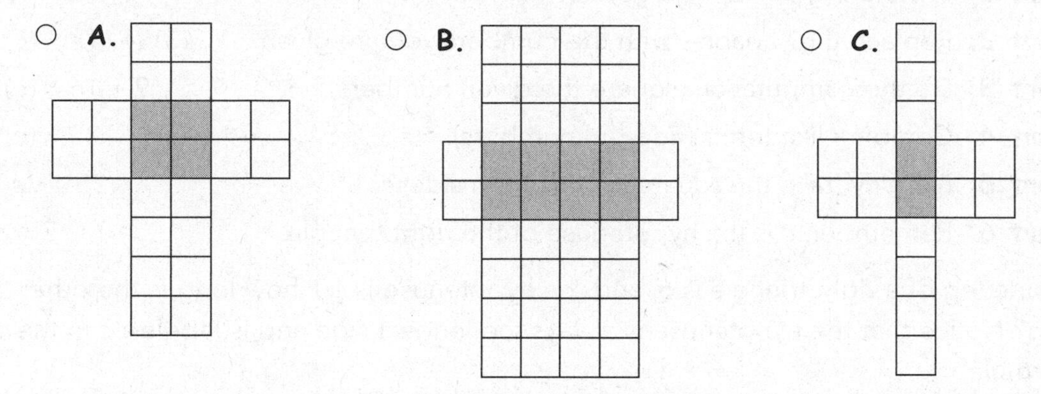


## RIGHT TRIANGLES

When we study right triangles, we study the relationships between their angles and their sides. This study is called *trigonometry*. The word comes from a Greek word that means *triangle measurement*.

### Pythagorean Theorem (Pythagorean Formula) (8.G.6–8)

First, let's look at a right triangle and label its parts. It has two sides with a right angle between them. These sides are called *legs*. The third and longest side is called the *hypotenuse*. We always label the legs $a$ and $b$, and the hypotenuse $c$.

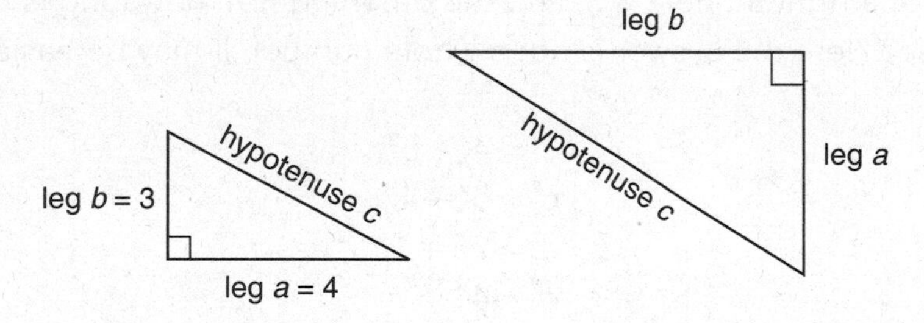


The theorem says that

$$(\text{Leg } a)^2 + (\text{Leg } b)^2 = (\text{Hypotenuse})^2$$

$$a^2 \quad + \quad b^2 \quad = c^2$$

## Examples

**A. Part 1:** Write the general Pythagorean formula. $a^2 + b^2 = c^2$

**Part 2:** Replace the variables with the numbers you are given. $(3)^2 + (4)^2 = (c)^2$

**Part 3:** Do the computation (square the given numbers). $9 + 16 = (c)^2$

**Part 4:** Combine like terms (add the numbers). $25 = (c)^2$

**Part 5:** Simplify (take the square root of both sides). $\sqrt{25} = \sqrt{c^2}$

**Part 6:** Remember, *c* is the hypotenuse of the right triangle. $5 = c$

**B.** If one leg of a right triangle is 6, and the hypotenuse is 10, how long is the other leg? Notice that the hypotenuse is always the longest side and is labeled *c* in the formula.

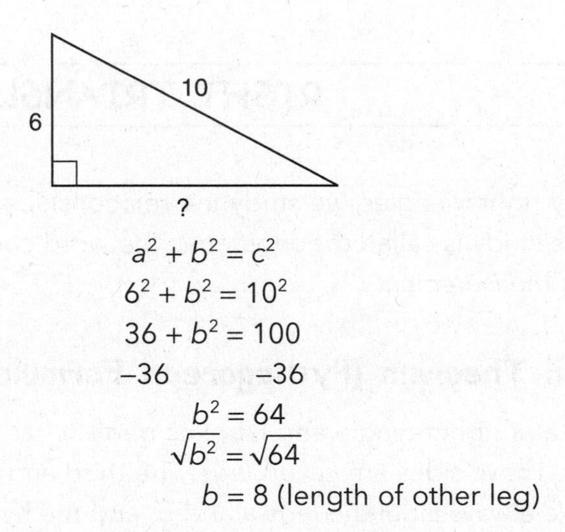


$$a^2 + b^2 = c^2$$
$$6^2 + b^2 = 10^2$$
$$36 + b^2 = 100$$
$$-36 \qquad -36$$
$$b^2 = 64$$
$$\sqrt{b^2} = \sqrt{64}$$
$$b = 8 \text{ (length of other leg)}$$

**C.** If one leg of a right triangle is 5, and the other leg is 4, how long is the hypotenuse? Here the answer is not a whole number. It may be easiest to use your calculator.

$$c^2 = a^2 + b^2$$
$$c^2 = 5^2 + 4^2$$
$$c^2 = 25 + 16$$
$$c^2 = 41$$

You should first *estimate* the answer.

Since $6^2 = 36$, and $7^2 = 49$, we know that the square root of 41 is between 6 and 7. To get an exact answer we can use a calculator.

Press [41] [second] [$\sqrt{x}$] [=]. You will see [6.4], the length of *c* (the hypotenuse).

## PRACTICE: Right Triangles

(For answers, see pages 247–248.)

**1.** If one leg of a right triangle is 12 cm long, and the other leg is 9 cm long, how long is its third side (the hypotenuse)? (*Hint*: Label the diagram, write the Pythagorean formula, show all steps, and use your calculator when appropriate.)

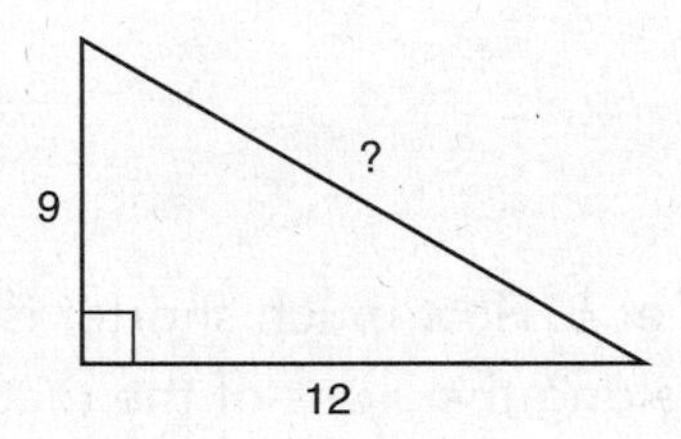


- ○ **A.** 10 cm long
- ○ **B.** 54 cm long
- ○ **C.** 15 cm long
- ○ **D.** 21 cm long

**2.** If leg $a$ in a right triangle measures 6 inches, and leg $b$ measures 5 inches, how long is the third side (the hypotenuse)? (*Hint*: Label the diagram, write the Pythagorean formula, show all steps, and use your calculator when appropriate.)

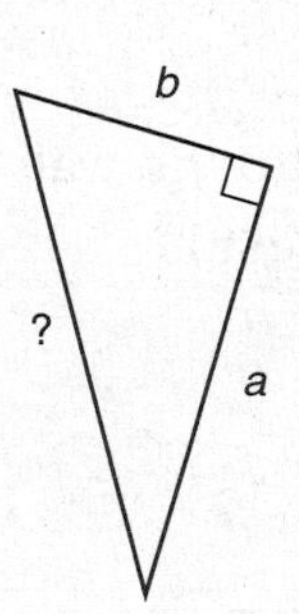


- ○ **A.** 8 inches long
- ○ **B.** 5 inches long
- ○ **C.** 11 inches long
- ○ **D.** 7.8 inches long

**3.** *Be careful, this one is a little different*. Here you are given the length of the hypotenuse and have to find the length of one of the legs.

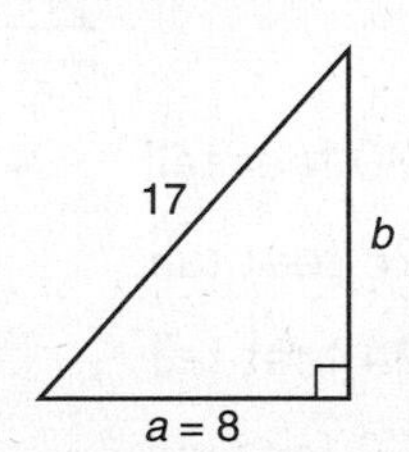


Given: a right triangle, leg $a = 8$ cm and the hypotenuse = 17 cm, how long is leg $b$? (*Hint*: Label the diagram, write the Pythagorean formula, show all steps, and use your calculator when appropriate.)

- ○ **A.** 15 cm
- ○ **B.** 12 cm
- ○ **C.** 9 cm
- ○ **D.** 6 cm

**4.** The window of a burning building is 24 feet above the ground. The base of the ladder is leaning against the wall and is 10 feet from the bottom of the wall. How tall must the ladder be to reach the window?

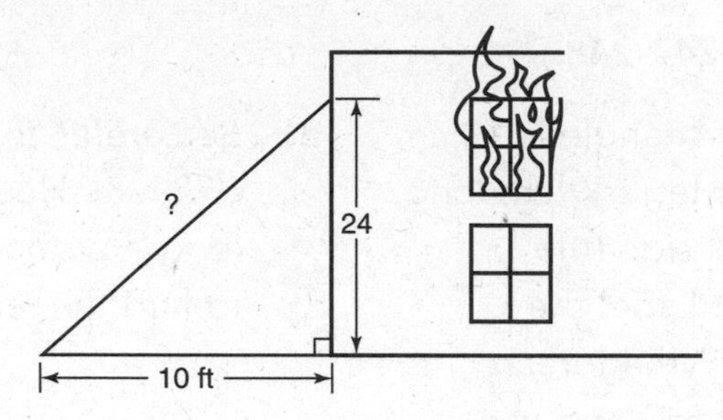


- ○ **A.** 30 feet tall
- ○ **B.** 26 feet tall
- ○ **C.** 24 feet tall
- ○ **D.** 20 feet tall

**5.** Jessica is in her backyard at *A* and her house is at *B*. How much shorter is her path along the diagonal compared to the walk along the sides of the rectangle?

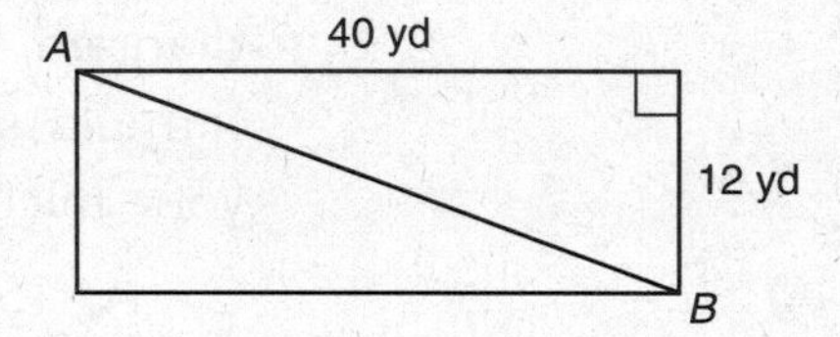


**6.** An isosceles right triangle has sides of length 5 meters each. Find the length of the hypotenuse to the nearest whole number. Show your work. (*Hint*: Use your calculator to find the square root.)

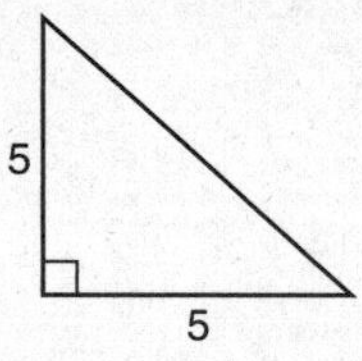


**7.** A 15-foot ladder is 6 feet away from the wall. How far up the wall does the ladder go? Show your work. (*Hint*: Label the diagram first.)

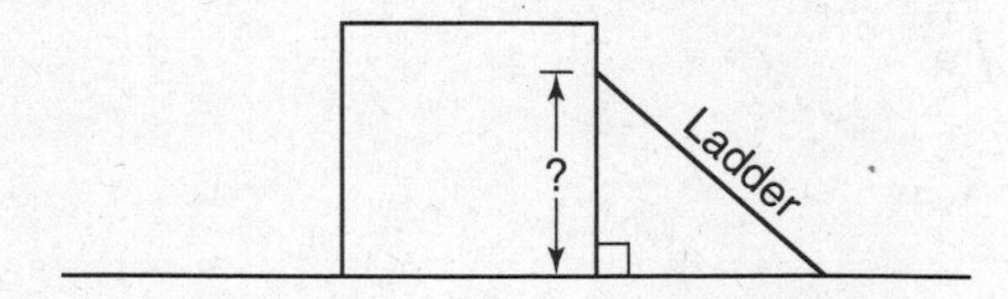

**8.** A square's sides are 8 cm long. Approximately, how long is its diagonal (the dotted line)? (*Hint*: Use your calculator to find the square root, or estimate.)

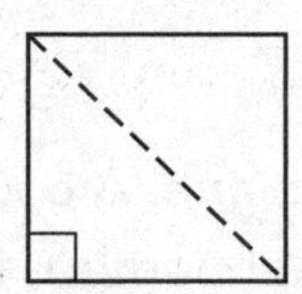

- ○ **A.** 9
- ○ **B.** 10
- ○ **C.** 11
- ○ **D.** 15

**9.** Below is a drawing of a right circular cone. The cone is 12 inches tall and has a diameter of 18 inches.

What is the distance from the point B (at the vertex) to point C (on the circumference of the base)?

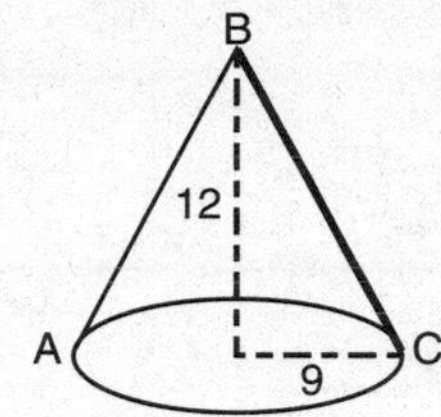


*(figure not drawn to scale)*

Answer: The length from B to C = ______ inches.

**10.** The entrance to the Native American Museum is in the shape of a large right circular cone.

Below is a drawing of that cone; it is 10 feet tall and has a diameter of 30 feet. What is the distance from the point B (at the vertex) to point D (on the circumference of the base)? Round your final answer to the nearest whole number.

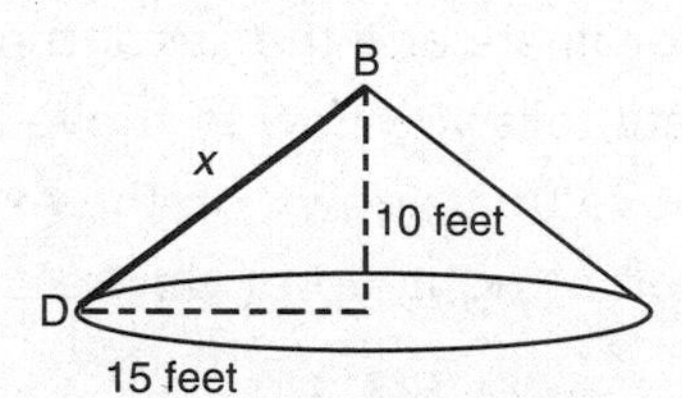


*(figure not drawn to scale)*

Answer: The length from B to D = __________ feet.

# COORDINATE GEOMETRY

## Plotting Points

The grid below is called a *coordinate plane*. A coordinate plane is formed by two number lines called *axes*. The horizontal number line is called the *x-axis* and the vertical number line is called the *y-axis*. The point where the number lines meet is called the *origin*. Each section of the coordinate plane is called a *quadrant*. To plot a point or to locate a point on this grid, you need to give its *x-coordinate* and its *y-coordinate*.

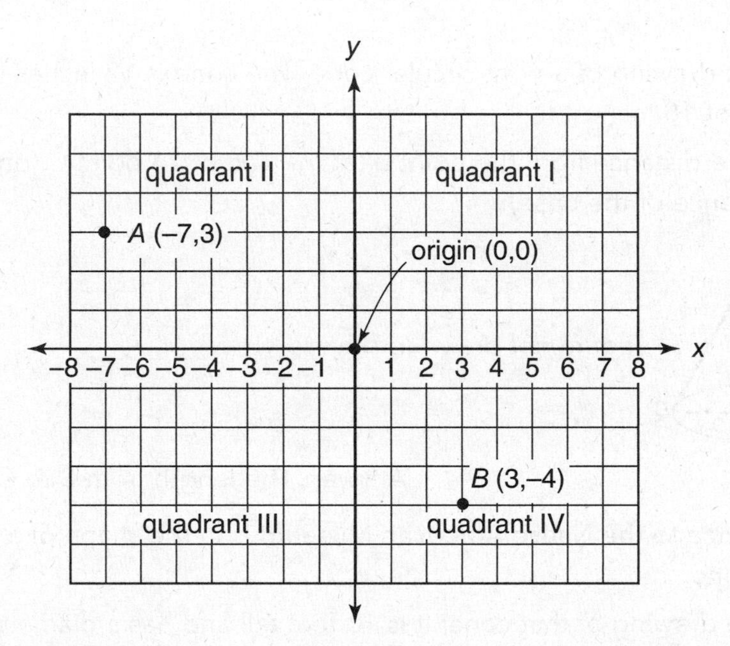


## Example

To help us locate a point, we give it an *ordered pair* of numbers. The first number gives you the *x*-coordinate and the second number gives you the *y*-coordinate. Notice that the *x*-coordinate tells you how to move left or right on the axis; the *y*-coordinate tells you how to move up or down on the axis. Always begin at the origin (0, 0) when you are counting left or right, up or down.

| $B(3, -4)$ | $A(-7, 3)$ |
|---|---|
| $(x, y)$ | $(x, y)$ |

## Steps to Graphing Any Point (x, y)

1. Start at the origin (0, 0).
2. Move $x$ units right or left along the horizontal $x$-axis.
3. Then move $y$ units up or down along the vertical $y$-axis.
4. Draw the point and label it with a capital letter.

## AREA AND PERIMETER ON THE COORDINATE PLANE

Now that we have reviewed how to plot points on the coordinate plane, it will be very easy to connect the points to form different geometric shapes. It also will be easy to find the area or perimeter of these geometric shapes.

### Examples

**A.** Find the area of triangle *ABC*. See the triangle on the coordinate plane below.

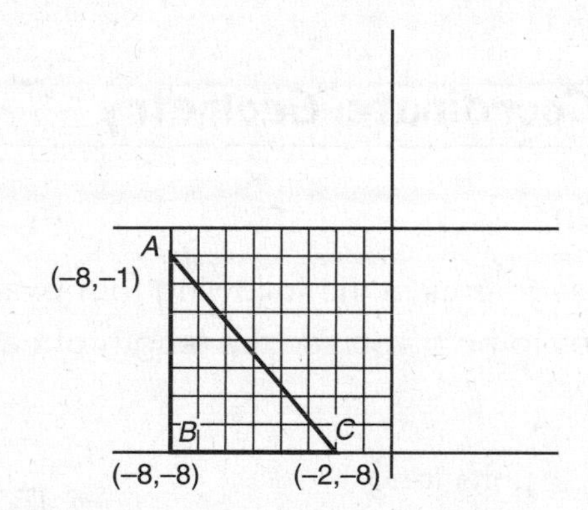


$A(-8, -1)$ $B(-8, -8)$ $C(-2, -8)$

*BC* (base) = 6 units long and *AB* (height) = 6 units long,

$$\text{Area triangle} = \frac{bh}{2} = \frac{(6)(6)}{2} = \frac{36}{2} = 18 \text{ square units}$$

**B.** Find the area and perimeter of the rectangle. Draw rectangle at *EFGH* on the grid below. Start with *E* at (0, 0), *F* at (0, 4), *G* at (7, 4), and *H* at (7, 0). Label these points. Now you can see that $EF = 4$ units and $EH = 7$ units.

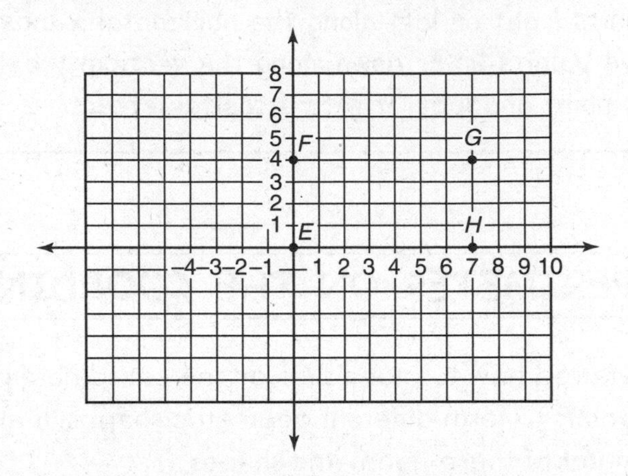


$$\text{Area} = bh = (7)(4) = 28 \text{ sq. units}$$

$$\text{Perimeter} = 2(w) + 2(l) = 2(4) + 2(7) = 8 + 14 = 22 \text{ units}$$

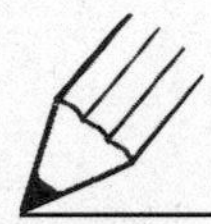

## PRACTICE: Coordinate Geometry

(For answers, see page 249.)

Find the length of the line segments in the following four examples. Plot the points and connect the coordinates to help you see the length of each line.

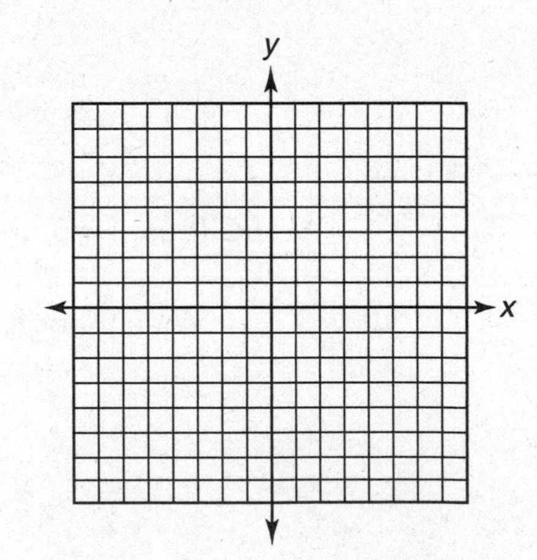


**1.** *A* (0, 0), *B* (6, 0)

Line segment *AB* is ------ units long.

**2.** *F* (5, 5), *G* (5, 0)

Line segment *FG* is ------ units long.

**3.** *C* (–4, 2), *D* (6, 2)

Line segment *CD* is ------ units long.

**4.** *R* (4, –3), *S* (4, 6)

Line segment *RS* is ------ units long.

**5.** Find the perimeter of a rectangle with vertices

*E* (–2, 0) *F* (–2, 8)

*G* (2, 8) *H* (2, 0)

(*Hint*: Plot the rectangle on the grid first.)

○ **A.** 24

○ **B.** 22

○ **C.** 12

○ **D.** 32

**6.** Find the area of a rectangle with vertices *P* (–5, –2), *Q* (–5, 8), *R* (4, 8), and *S* (4, –2).

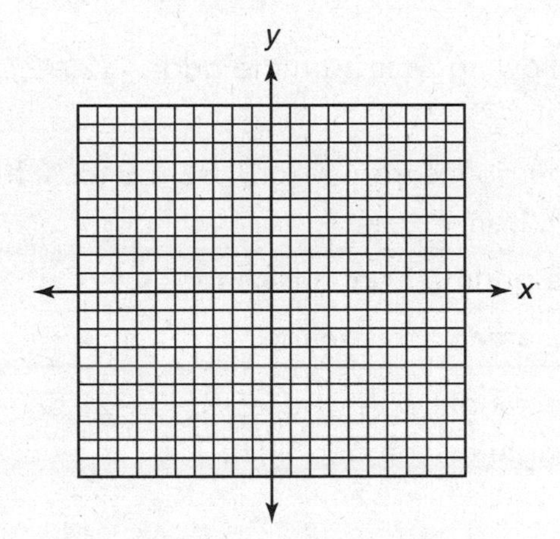


○ **A.** 90

○ **B.** 36

○ **C.** 38

○ **D.** 19

7. Remember to use a ruler to draw lines, and to show all work.

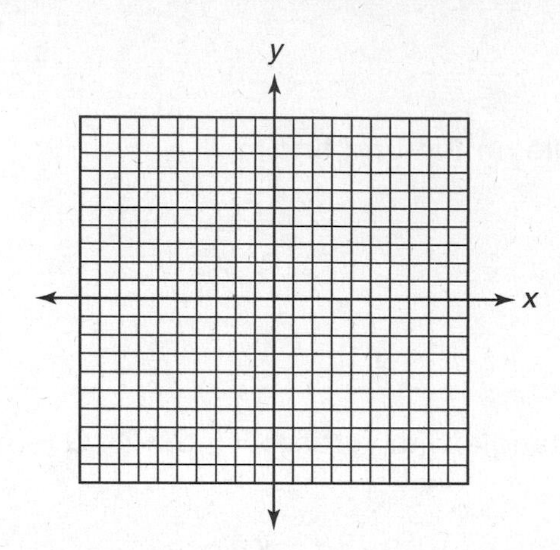


**Part A:** Plot the following points on the grid: *A* (2, –2), *B* (2, 4), *C* (6, 4), *D* (6, 1), *E* (4, 1), *F* (4, –2)

**Part B:** Connect the points from *A* to *B*, from *B* to *C*, from *C* to *D*, from *D* to *E*, from *E* to *F*, from *F* to *A*.

**Part C:** What is the perimeter of this shape?

**Part D:** What is the area of this shape?

8. Find the area of a rectangle with vertices *P* (–5, –2), *Q* (–5, 6), *R* (4, 6), and *S* (4, –2). (*Hint*: Plot the rectangle on the grid first.)

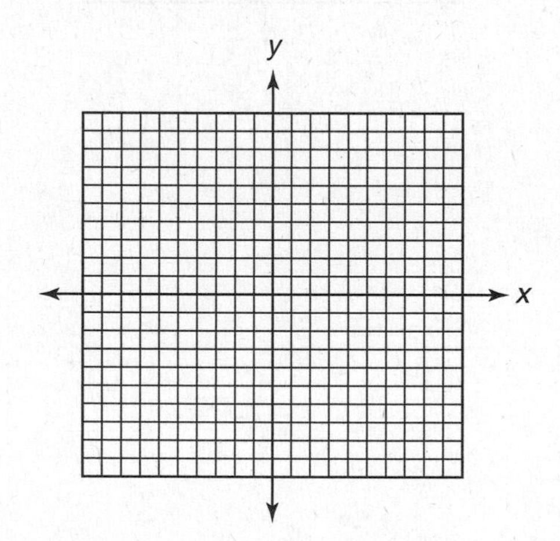


○ **A.** 34 sq. units

○ **B.** 72 sq. units

○ **C.** 56 sq. units

○ **D.** 90 sq. units

**9.** Find the perimeter of the triangle with vertices *A* (0, 0), *B* (0, 6), and *C* (–8, 0). (*Hint*: Plot the triangle on the coordinate plane first and remember, if necessary, you can use the Pythagorean theorem to find the length of the hypotenuse.)

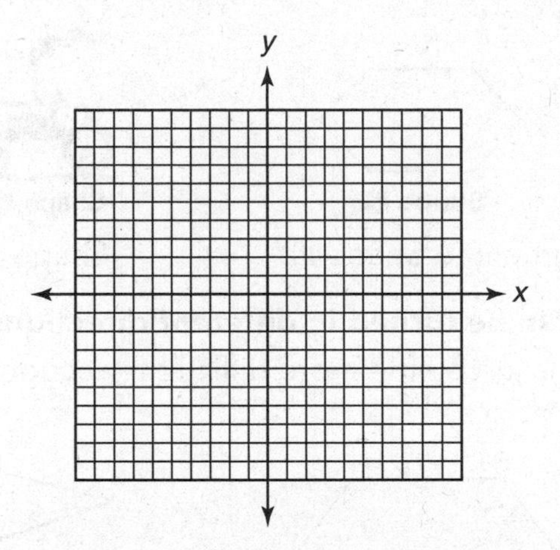


- ○ **A.** 24 units
- ○ **B.** 12 units
- ○ **C.** 48 units
- ○ **D.** 60 units

---

## CONGRUENCY

---

When you take a geometry course in high school, you will learn a great deal about lines and shapes and how to determine if they are congruent. For now, we'll just review some main ideas and work with some simple shapes and lines.

What does congruent mean? *Congruent* really means the SAME SIZE and SHAPE.

### Examples

**A. Lines can be congruent.** If line segments have the same length, then they are congruent. We use the symbol ≅ to replace the words *is congruent to.*

*A* ———————— *B*

*D* ———————— *E*

If line segment *AB* = 15 ft, and line segment *DE* = 15 ft, then line segment *AB* ≅ line segment *DE* (*AB* ≅ *DE*).

**B. Angles can be congruent.** If angles have the same measure, they are congruent.

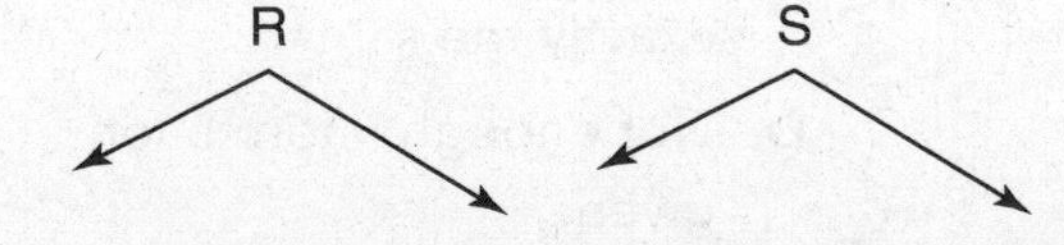


Angle *R* measures 120°.
Angle *S* also measures 120°.
Angle *R* is congruent to angle *S*.

**C. Shapes can be congruent.** For shapes to be congruent, their corresponding sides must be the same length, and their corresponding angles must be the same measure.

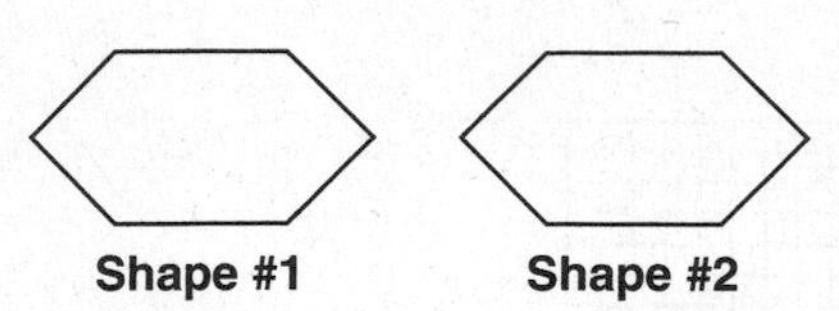


Shape #1 is congruent to Shape #2

Shape #3 ≅ Shape #4

**D. Congruent shapes can be turned in different directions,** but they are still congruent shapes. It is just a little more difficult to recognize them sometimes.

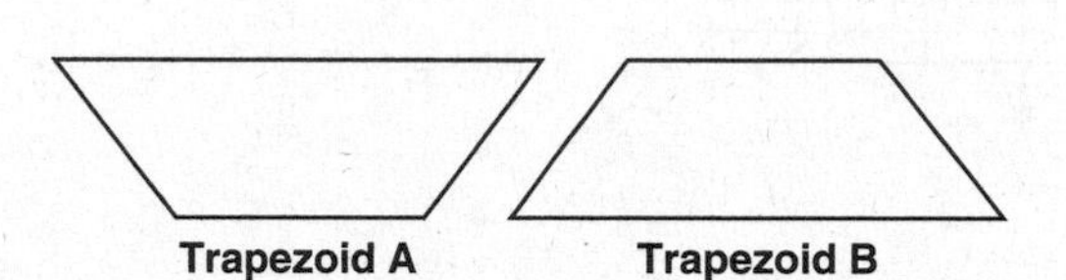


Trapezoid A ≅ Trapezoid B

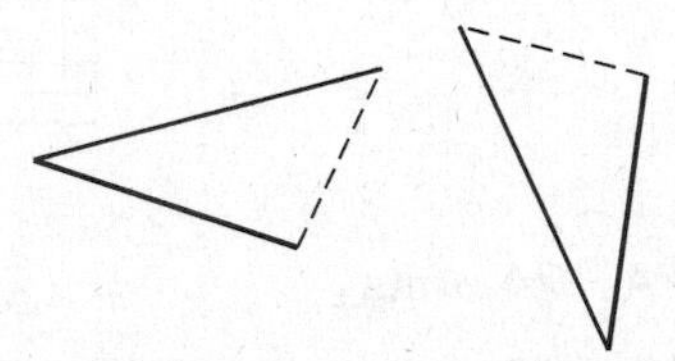

Triangle on the left ≅ Triangle on the right

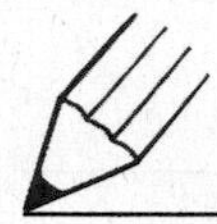

## PRACTICE: Congruency

(For answers, see pages 249–250.)

*For this section it is a good idea to work on graph paper and use a ruler.*

1. In triangle $ABC$, m$\angle A = 45°$, and m$\angle B = 90°$. In triangle $PQR$, m$\angle P = 45°$, and m$\angle Q = 90°$. Are these triangles congruent?

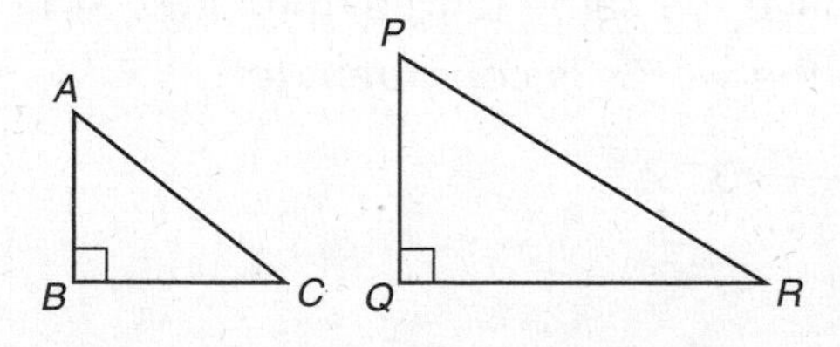


○ **A.** Yes

○ **B.** No

○ **C.** Not enough information

2. You are told that the corresponding angles of these two rectangles are *equal*, and the lengths of their sides are shown on the diagrams below. Are these rectangles congruent?

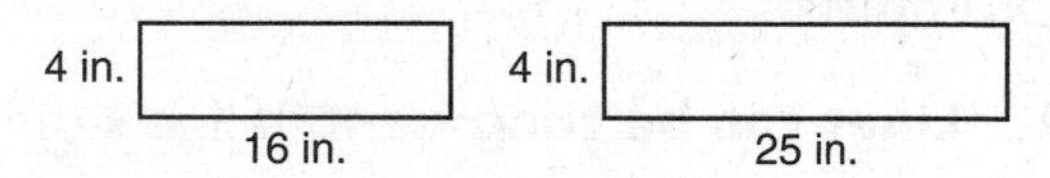


○ **A.** Yes, because their corresponding angles are equal.

○ **B.** No, because their corresponding sides are not equal.

○ **C.** Yes, because they look exactly the same.

○ **D.** Not enough information given.

3. If two quadrilaterals have the same perimeter, are they always congruent? Show all work and draw diagrams to support your answer.

4. If two squares have the same perimeter, are they always congruent? Show all work and draw diagrams to support your answer.

5. What additional information do you need to determine that the rectangles below are congruent?

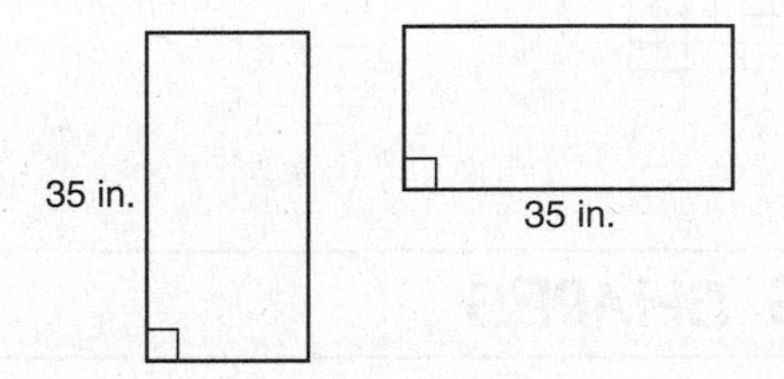


- ○ **A.** The measure of their angles
- ○ **B.** The length of the other long sides
- ○ **C.** The length of the short sides
- ○ **D.** No additional information is needed

6. Are these right triangles congruent? Show your work and explain your answer.

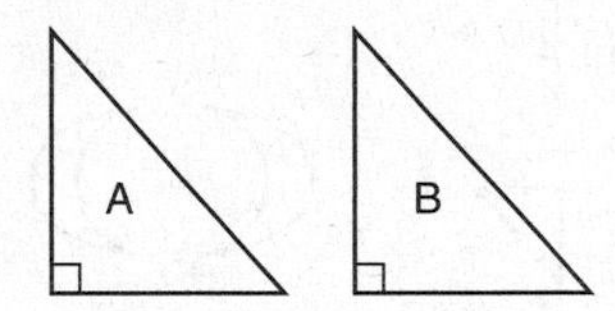


In triangle *A*, one leg = 3 in., and the other leg = 4 in.

In triangle *B*, the hypotenuse = 5 in., and one leg = 3 in.

7. Are these right triangles congruent? Show your work and explain your answer.

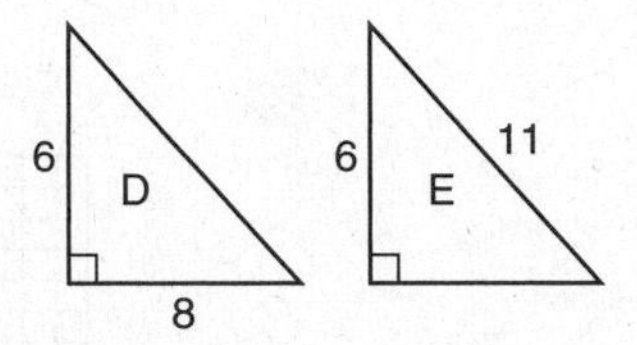


In triangle *D*, one leg = 6 in., and the other leg = 8 in.

In triangle *E*, the hypotenuse = 11 in., and one leg = 6 in.

8. If two rectangles each have a perimeter of 100, will they always be congruent rectangles? Give an example and explain your answer.

9. Are all circles congruent? Give an example and explain your answer.

10. Draw two congruent rectangles and explain why they are congruent.

11. Draw two squares that are *not* congruent and explain why they are not.

**12.** Plot the points (0, 0), (0, 4), (5, 0), and (5, 4), and connect them to make a rectangle. Now, plot the points (1, –1), (6, –1), and (6, –5). What should the other coordinate be if this is a rectangle *congruent* to the first one?

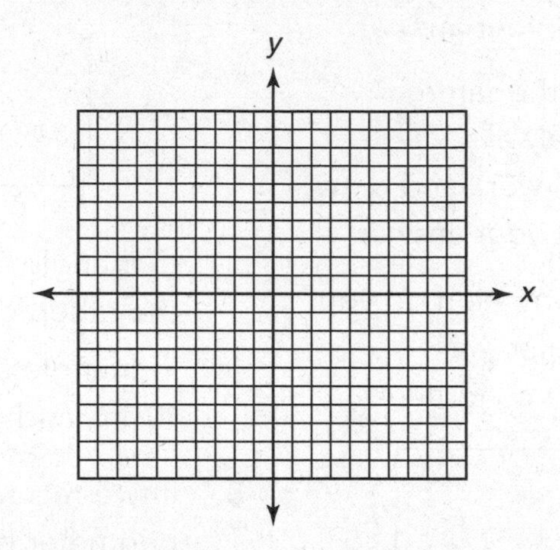


## TRANSFORMING SHAPES

*Transforming shapes* means we are taking a plane figure and all of its points and moving them to another location according to specific rules. We might move a rectangle 3 units to the left, or move a triangle 5 units down. There are different ways to move a shape.

The new figure is usually congruent to (an exact copy of) the original. The exception is an image of a figure under *dilation*. *Translations*, *reflections*, *rotations*, and *dilations* are different types of transformations. Many computer games use transformations.

### Examples

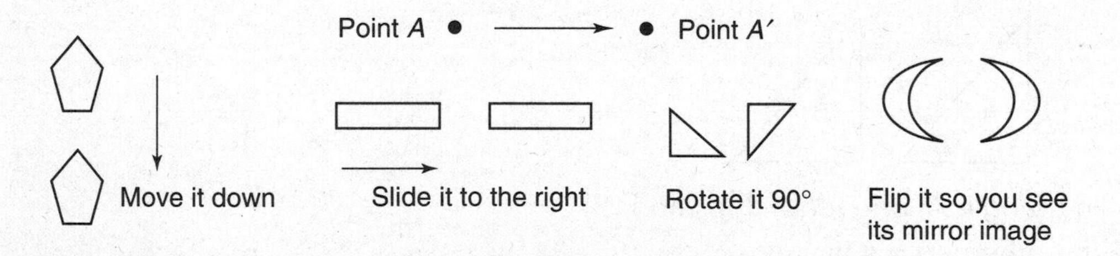


### Translation

The result of a movement in one direction is a transformation called *translation*. A *translation* moves all points the same distance and in the same direction. Think of a train speeding along a track to the station. (When the front of the engine moves 10 yards ahead, the back of the engine moves 10 yards ahead, too.)

To move shapes a particular distance we need to first place them on a grid. See the figure to the right for a few reminders and vocabulary terms to remember.

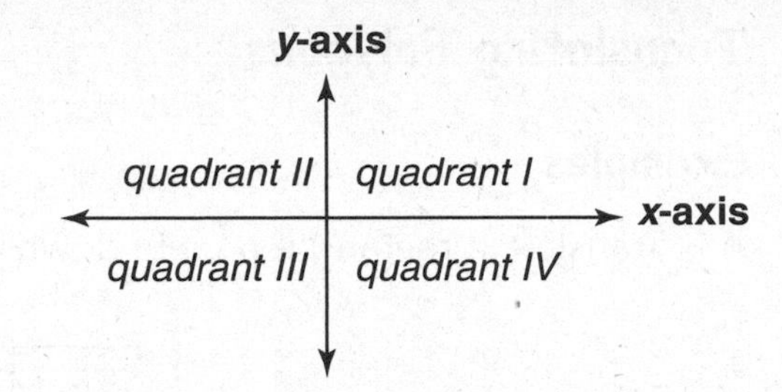


## Translating Coordinate Points

Remember, a translation is the same as sliding a figure to a new location.

### Example

Translate a coordinate (a point on the grid): Coordinate $A$ (5, 1) has been translated down 2 units to coordinate $A'$ (5, –1).

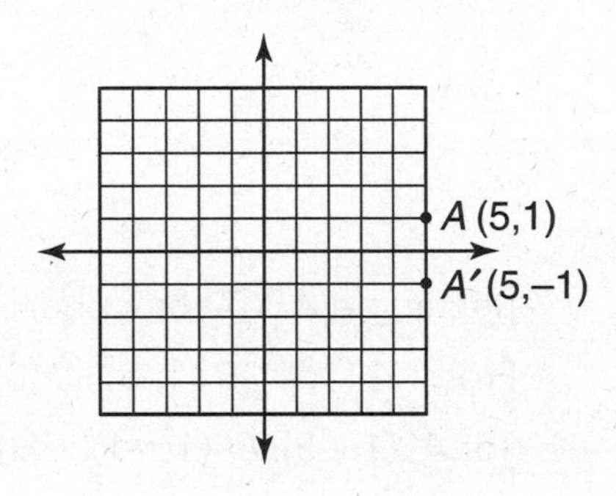

## PRACTICE: Translating Points

(For answers, see page 250.)

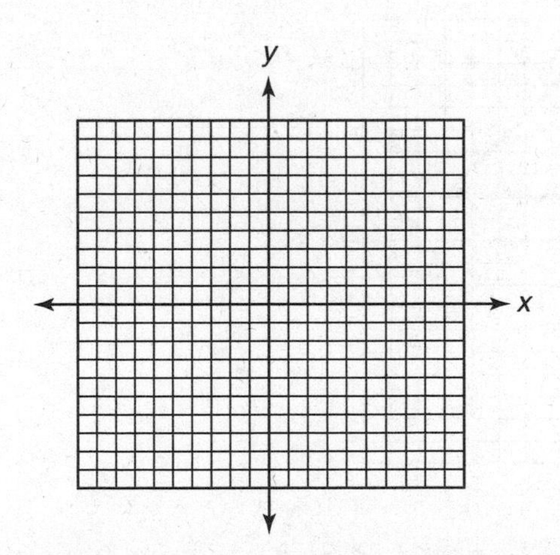


1. Translate a coordinate horizontally. Translate coordinate $B$ (2, 4) to coordinate $B'$ (–2, 4.) Draw and label both points on the grid above.
2. Translate coordinate $C$ (1, 3) horizontally, then vertically over the $x$-axis to $C'$ (–1, –3). Draw and label both points on the grid above.
3. Translate coordinate $D$ (–2, –1) to $D'$ (2, 1). Draw and label both points on the grid above.
4. Translate point $D$ four units to the right and label the new coordinate $D'$ on the number line.

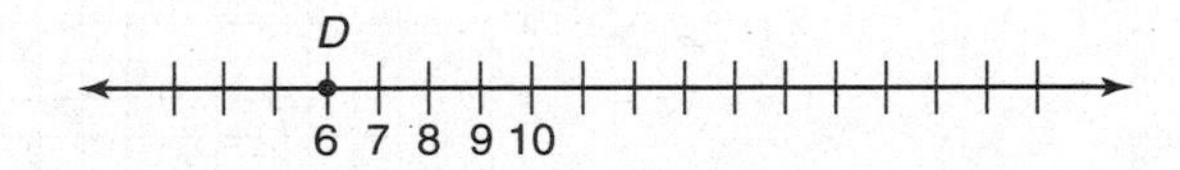


5. Translate point $A$ five units to the left and label the new coordinate $A'$ on the number line.

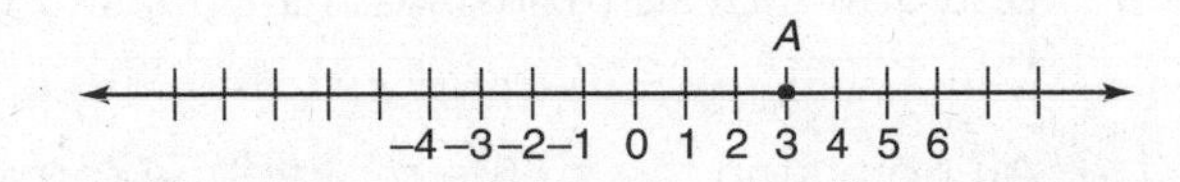

## Translating Polygons

### Examples

**A.** Translate a rectangle 6 units down.

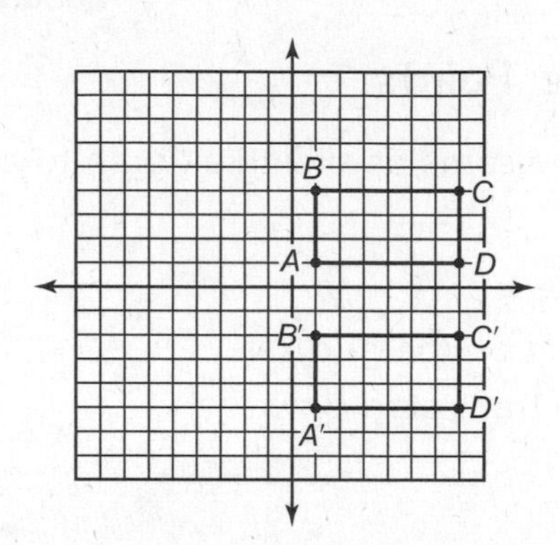


Rectangle *ABCD* has been translated down 6 units to rectangle *A′B′C′D′*.

From: *A* (1, 1), *B* (1, 5), *C* (8, 5), *D* (8, 1)

To: *A′* (1, –5), *B′* (1, –1), *C′* (8, –1), *D′* (8, –5)

Notice that there is no change in the *x*-values of the points. The difference between the *y*-values of the original point at its image is 6 for all four points.

**B.** Translate a triangle 11 units to the right.

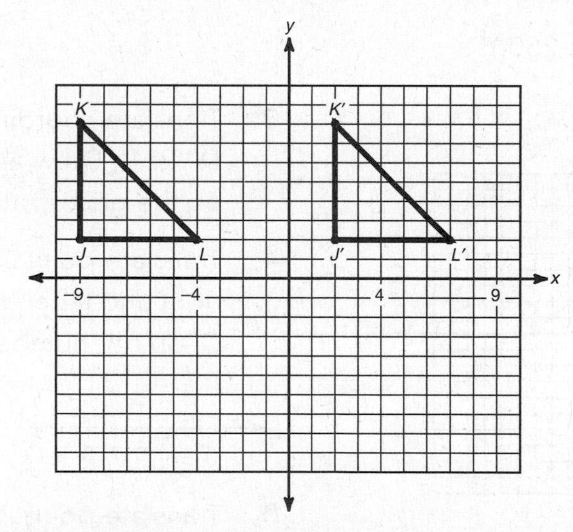


Triangle *JKL* has coordinates *J* (–9, 2), *L* (–4, 2), and *K* (–9, 8). It has been translated to *J′* (2, 2), *L′* (7, 2), and *K′* (2, 8), triangle *J′K′L′*. Notice that all points on the polygon moved (translated) 11 units to the right: *J* to *J′* (11 units to the right), *K* to *K′* (11 units to the right), and *L* to *L′* (11 units to the right).

All points on the shape moved the same distance (11 units to the right). Translating a figure horizontally will show that there is no change in the *y*-axis coordinates.

**C.** Translate from quadrant II to quadrant III; then translate the same shape to quadrant IV. Here we'll take rectangle *PQRS* that begins in quadrant II and slide it down into quadrant III, and then we'll slide it right into quadrant IV.

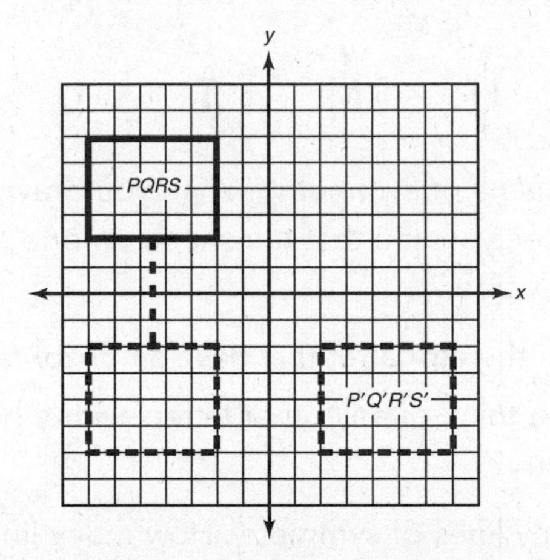


**D.** Triangle *ABC* has coordinates *A* (1, 4), *B* (1, 0), and *C* (4, 2).

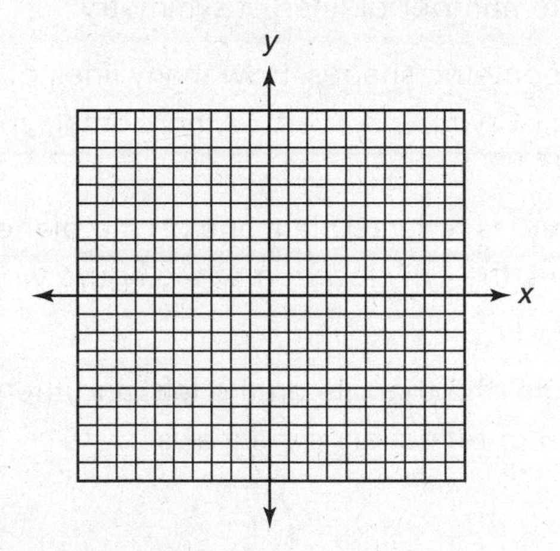


In the coordinate grid above, do the following:

1. Draw triangle *ABC*.
2. Draw a reflection of triangle *ABC*, and label the new triangle *A'B'C'*.
3. Draw a rotation of triangle *ABC*.

## Reflection

A *reflection* can be seen in water, in a mirror, or in a shiny surface. An object and its reflection have the **same shape and size**, but the **figures face in opposite directions.** In a mirror, for example, right and left are reversed.

In mathematics, the *reflection* of an object is called its **image**. If the original object (called the **preimage**) was labeled with letters, such as polygon *ABCD*, the image may be labeled with the same letters followed by a prime symbol, *A'B'C'D'*.

## Examples

**A.** To easily understand *reflection* think back to what you remember about symmetry. What do the following letters have in common?

A H M T U V

They all have a *vertical line of symmetry*. When you draw a vertical line through the center of each letter, you see that the shape on one side is a *mirror image* of the shape on the other side.

**B.** Now, think of letters in the alphabet that have a *horizontal line of symmetry*.

**B E** Can you think of any other letters with a horizontal line of symmetry?

**C.** Some letters have many lines of symmetry. How many lines of symmetry does the letter **X** or the letter **O** have? The letter **X** has two lines of symmetry, but the letter **O** has an infinite number of lines of symmetry.

**D.** Now think of some geometric shapes. How many lines of symmetry does a square have? It has four lines of symmetry–vertical, horizontal, and along the two diagonals.

In a *reflection*, the figure is reflected in a line on the plane. If you traced the figure on paper and cut it out, the new location of the figure would be "flipped" over the line of reflection.

**E.** See the original triangle (the triangle on the left) and the new reflected shape to the right. Here the line of reflection is the *y*-axis.

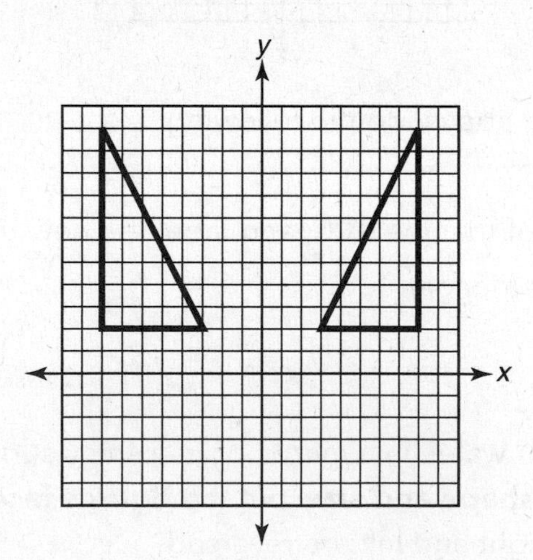

**F.** See how the trapezoid is reflected over the $x$-axis.

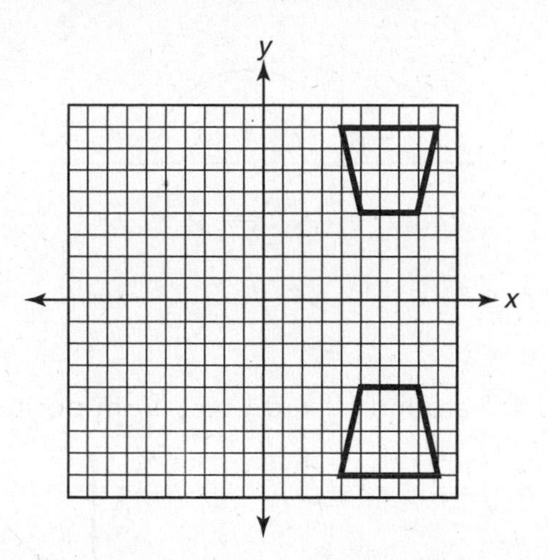


## Rotation

### Examples

***A.*** When a shape is *rotated* the figure is turned about a fixed point on the plane.

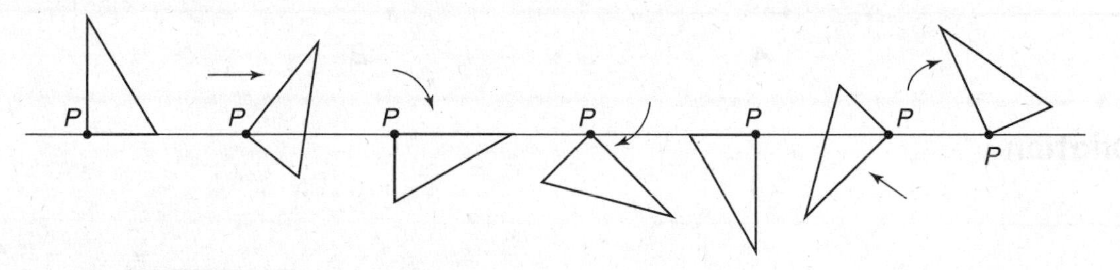


**B.** Think about the hour hand on a clock. The hand is pivoting from the center and rotates 360° throughout the day.

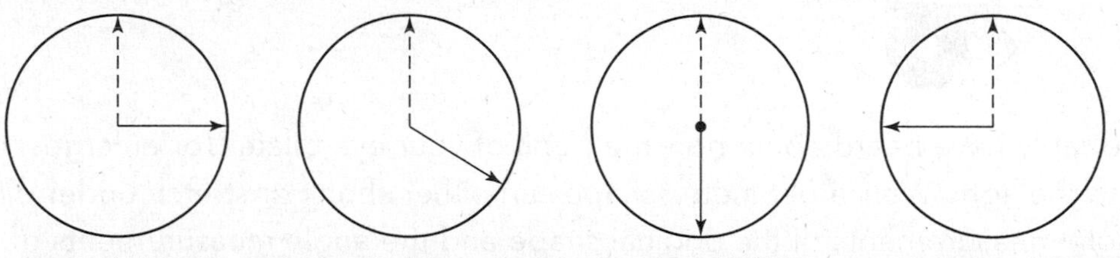

**Rotated 90°** **Rotated about 150°** **Rotated 180°** **Rotated 270°**

***C.*** Below you see a shape rotated 180° around the fixed point $P$.

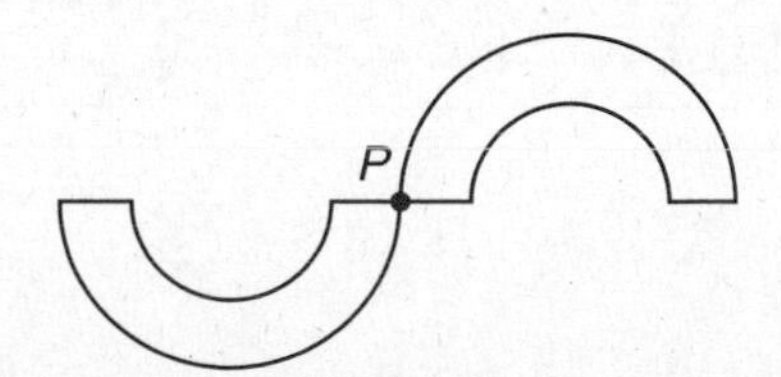

**D.** Here the shape is rotated only 90° around a fixed point *B*.

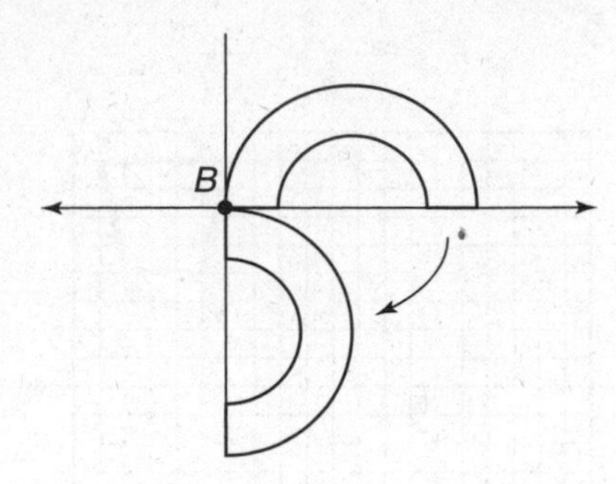


**E.** This shape is rotated 45° around a fixed point, from position ***A*** to ***B***.

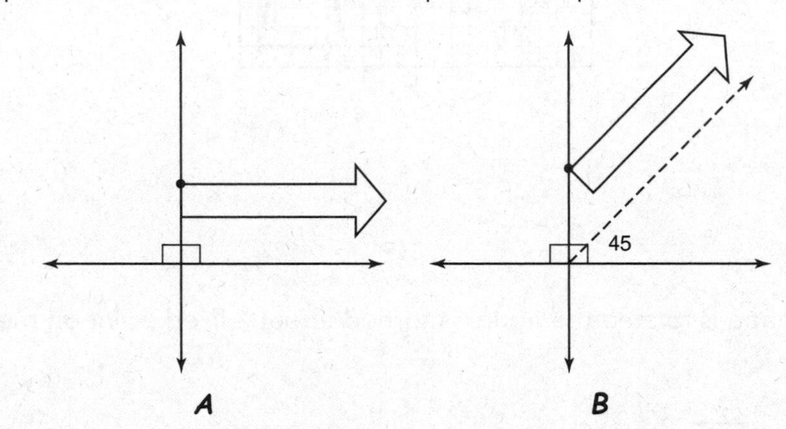


## Dilation

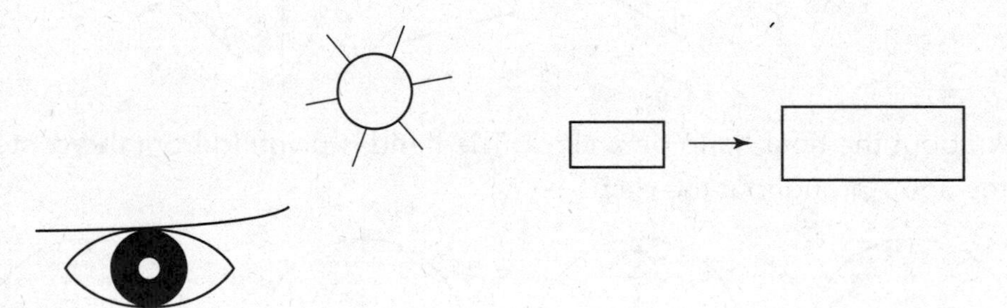

You probably have heard about how the pupil of your eye dilates (or enlarges) to adjust to the light. Well, a geometric shape can either shrink or stretch under *dilation*. The angle measurements of the original shape and the angle measurements of the new shape remain the same. The sides of the new shape are in the same proportion as the sides of the original shape. Dilating a shape creates a similar shape.

(See the examples on page 131.)

## Examples

**A.** The larger triangle $ABC$ is *similar* to the smaller triangle $A'B'C'$.

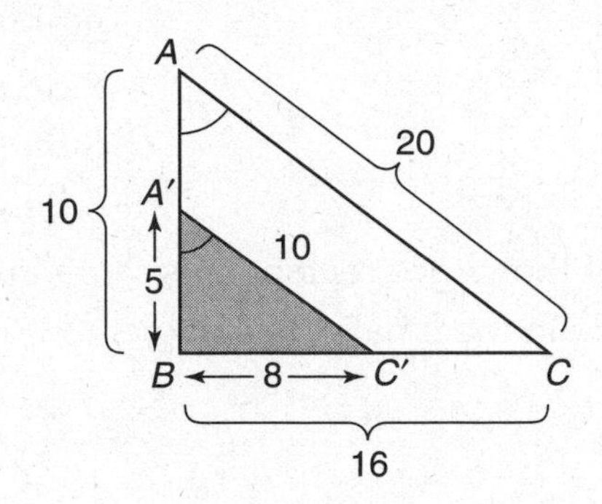


Corresponding angles are equal

$m\angle A = m\angle A'$ $m\angle B = m\angle B'$ $m\angle C = m\angle C'$

Corresponding sides are in proportion

| **Small triangle** | **Larger triangle** |
|---|---|
| Segment $A'B = 5$ | Segment $AB = 10$ |
| Segment $BC' = 8$ | Segment $BC = 16$ |
| Segment $AC' = 10$ | Segment $AC = 20$ |

Notice how each side of the larger triangle is *double* the length of its corresponding side in the smaller triangle.

**B.** The larger rectangle $DEFG$ is similar to the smaller rectangle $D'E'F'G'$.

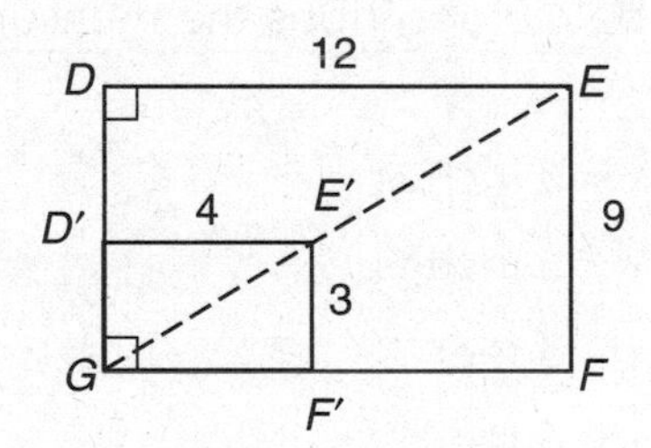


Corresponding angles are equal

$m\angle D = m\angle D'$ $m\angle E = m\angle E'$ $m\angle F = m\angle F'$

Corresponding sides are in proportion

| **Small rectangle** | **Larger rectangle** |
|---|---|
| Side $D'E' = 4$ | Side $DE = 12$ |
| Side $E'F' = 3$ | Side $EF = 9$ |
| Side $F'G = 4$ | Side $FG = 12$ |
| Side $GD' = 3$ | Side $GD = 9$ |

Notice how each side of the larger rectangle is *three times larger* than the length of its corresponding side in the smaller rectangle.

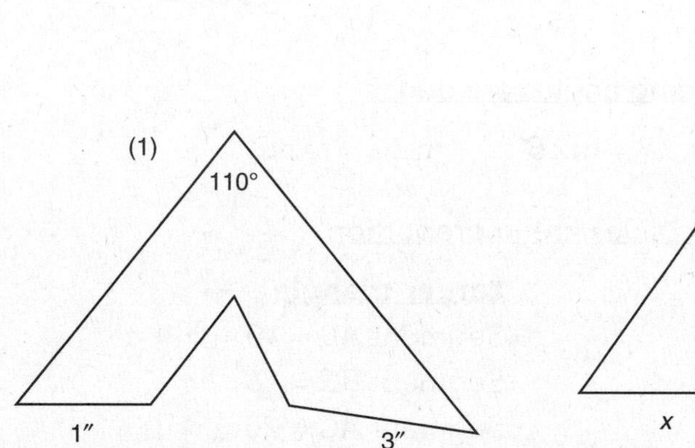


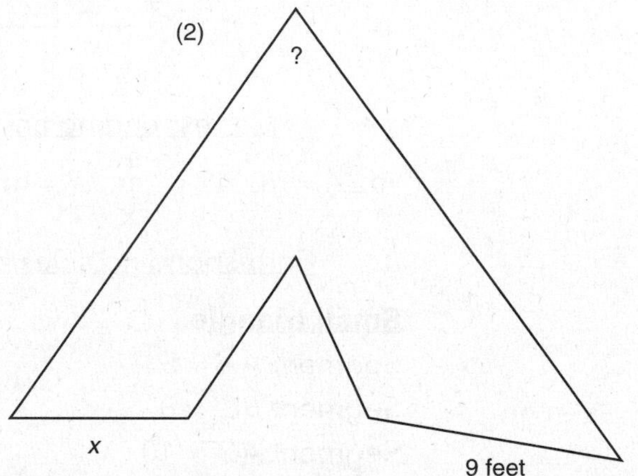


**C.** Our middle school's Art Club was assigned to design scenery for the school play. The drawing of a teepee showed the measurements in (1) above. The second figure (2) shows the measurements of the real-life teepee used on the stage. Which information below describes the measurements of the real-life teepee?

**A.** Top angle = 110° and $x = 3$ inches

**B.** Top angle = 130° and $x = 3$ inches

**C.** Top angle = 110° and $x = 3$ feet

**D.** Top angle = 130° and $x = 6$ feet

The answer is ***C***.

In similar shapes the angle measures remain the same, and the sides remain in proportion. $\frac{3"}{9"} = \frac{1"}{3"}$

## PRACTICE: Translating Polygons

(For answers, see pages 250–251.)

**1.** Which pair of figures shows a 180° rotation?

○ **A.** 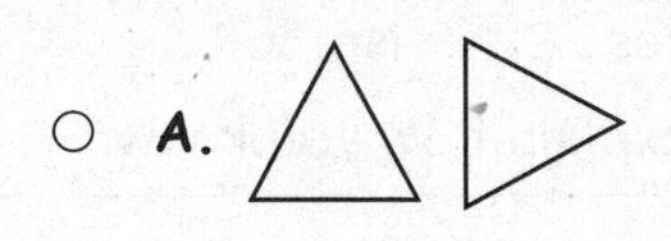

○ **B.** 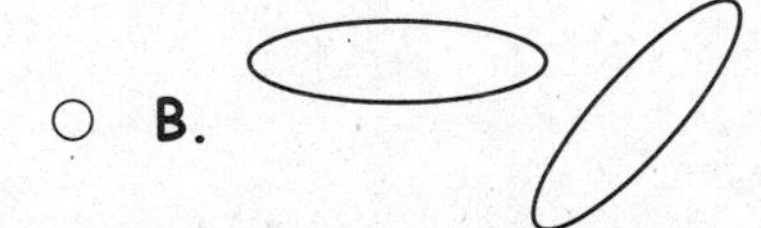

○ **C.** 

○ **D.** 

**2.** Which type of transformation is the following?

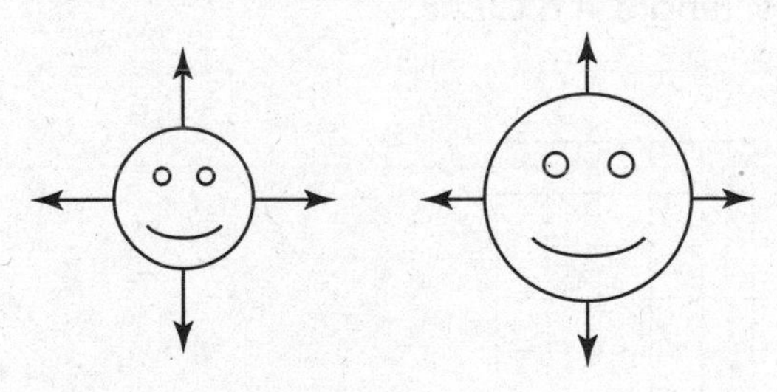

○ **A.** Translation
○ **B.** Reflection
○ **C.** Rotation
○ **D.** Dilation

**3.** Use the figure below for **Parts A–C**.

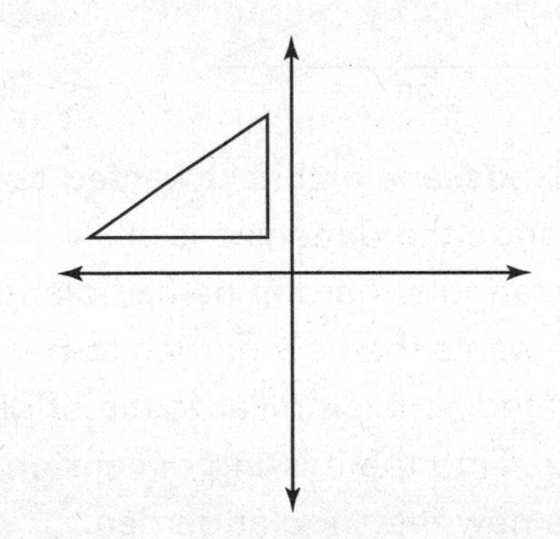

**Part A:** Sketch this shape so the *x*-axis is the line of reflection. Reflect triangle *A* over the *x*-axis.

**Part B:** If this shape was reflected over the *x*-axis and then over the *y*-axis what quadrant would it be in?

**Part C:** If we wanted to keep the figure above the same shape, but just slide it down 5 units, this transformation would be called a

○ **A.** translation
○ **B.** reflection
○ **C.** rotation
○ **D.** dilation

**4.** What is the missing length of the base of the smaller triangle if you know these two triangles are similar?

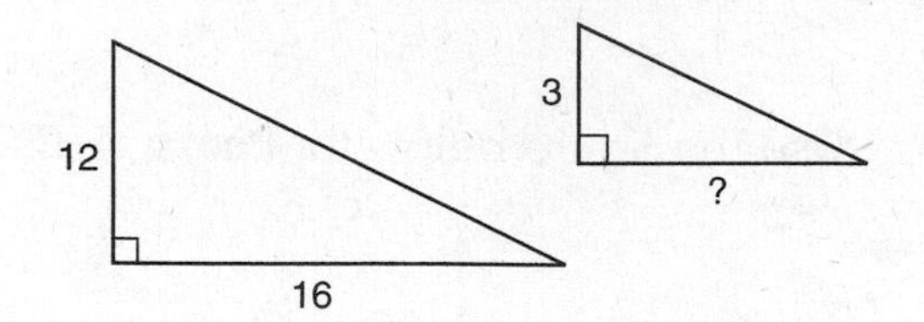


**5.** A landscape architect wanted to change the dimensions of a rectangular garden he had planned. He wants the new garden to be dilated so it is similar to the original one. Find the missing dimension of the new rectangular garden.

20 ft
50 ft
30 ft
?

**6.**

1.5 in.
62°
6 in.
28°
3.0 in.
58°
32°
12 in.

**Part A:** Are the two triangles drawn above similar?

Yes ______ No ______

**Part B:** Explain how you know.

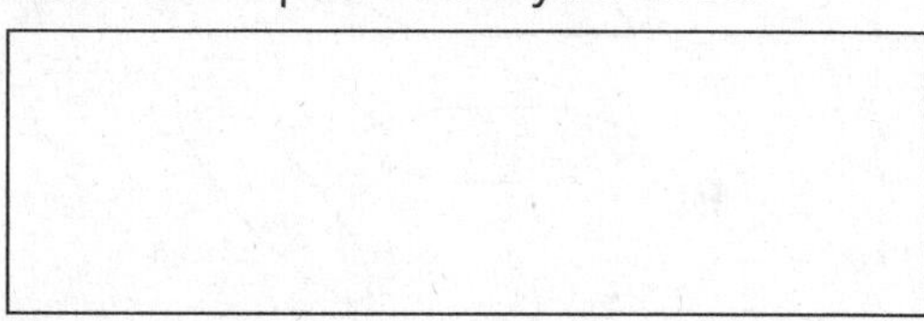

**7.** Use a scale factor of 2 to dilate the rectangle *ABCD* drawn below. What will be the coordinates of the new rectangle *A′B′C′D′*?

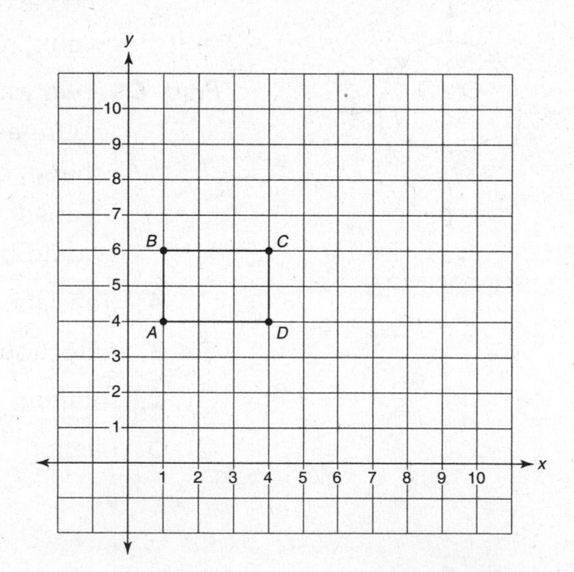


○ **A.** A′ (2, 8) B′ (2, 12) C′ (8, 12) D′ (8, 8)
○ **B.** A′ (1, –4) B′ (1, –6) C′ (4, –6) D′ (4, –4)
○ **C.** A′ (1, 8) B′ (1, 12) C′ (4, 12) D′ (4, 8)
○ **D.** A′ (2, 4) B′ (2, 6) C′ (8, 6) D′ (8, 4)

## PRACTICE: Non-calculator Questions

Each question is worth 1 point. No partial credit is given.

(For answers, see pages 251–253.)

1. If triangle *ABC* is translated over the *x*-axis and then over the *y*-axis, what quadrant would it be in?

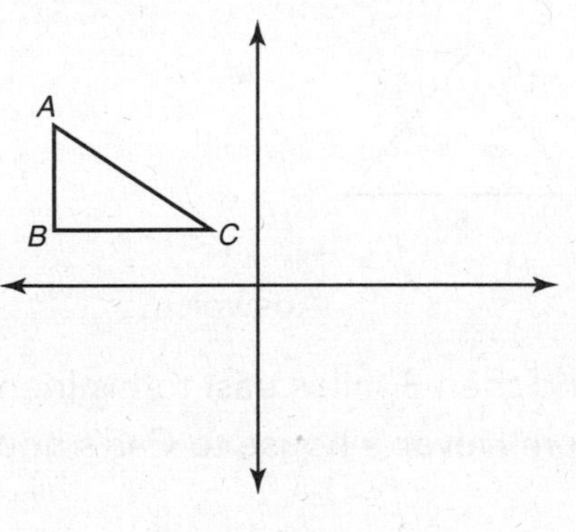


Answer: ________________

2. What are the possible coordinates of points *C* and *D* that would create a rectangle *ABCD* that is 3 units tall?

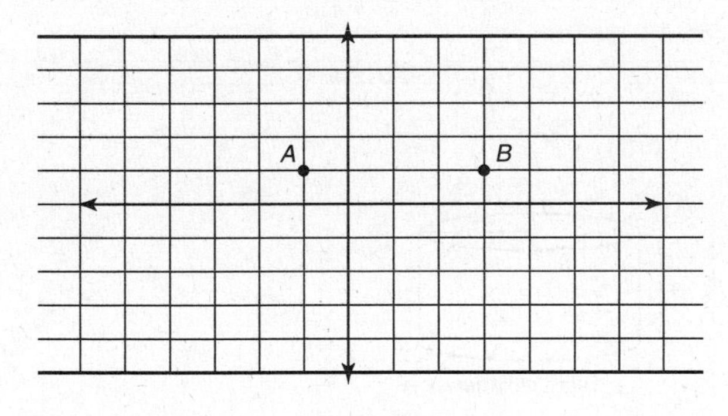


Answer: ________________

3. Which of the following polygons has a different area from the others?

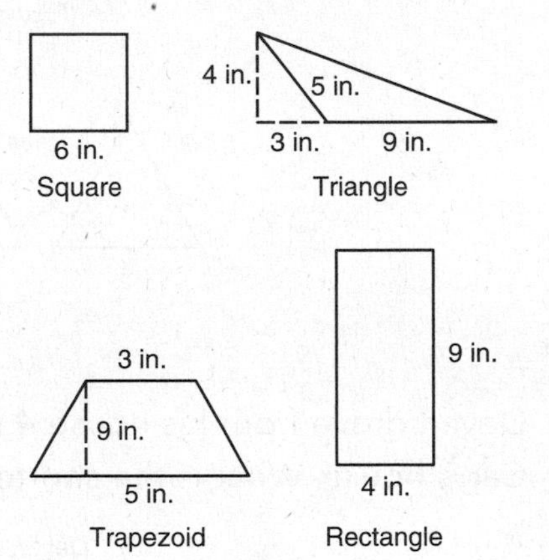


Answer: ________________

4. These polygons all have the same what?

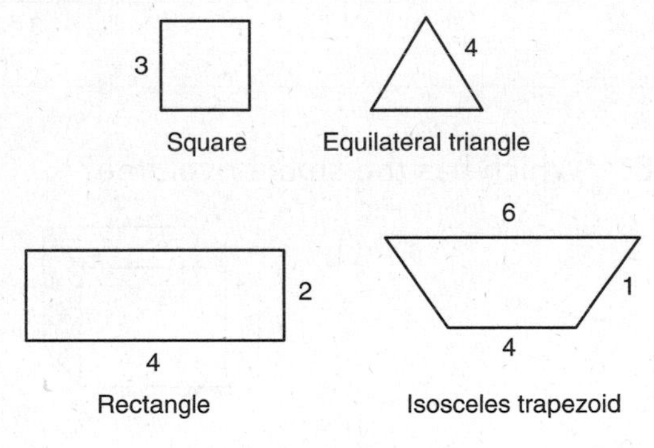


Answer: ________________

5. What is the value of $x$ in the triangle shown?

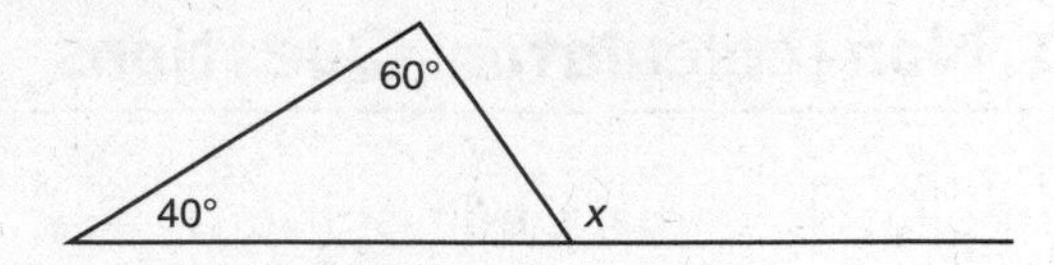


Answer: ________________

6. Which two triangles are similar?

(Figures not drawn to scale.)

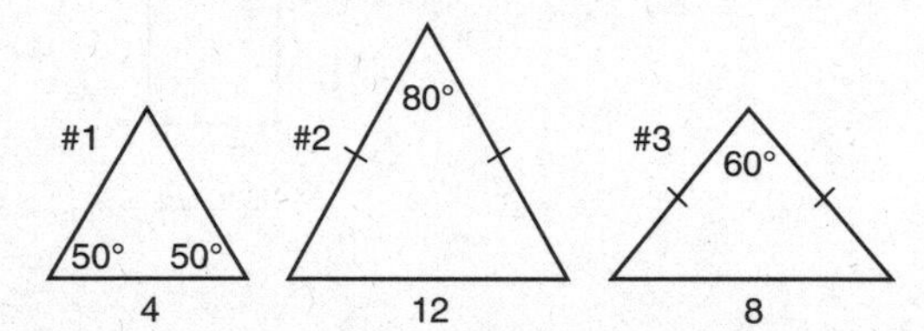


Answer: ________________

7. Devan drove from his house 4 miles south and then 3 miles east to his friend Carl's house. What is the shortest distance from Devan's house to Carl's house?

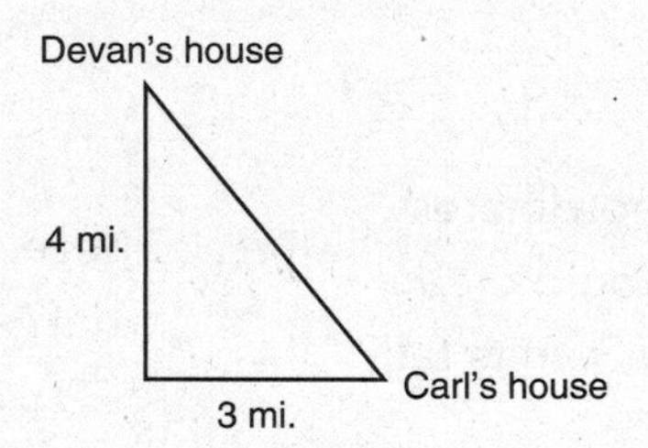


Answer: ________________

8. Which has the smaller volume?

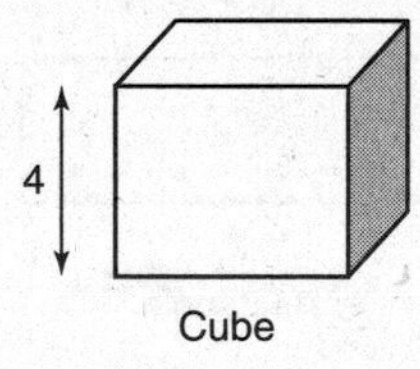


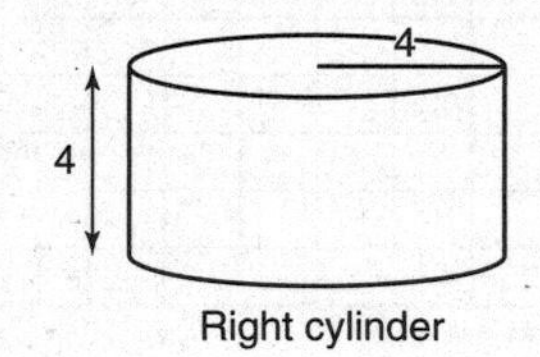


Answer: ________________

9. What is the perimeter of a square that has an area of 25?

Answer: ________________

**10.** What is the perimeter of the irregular figure drawn below? (All angles are right angles.)

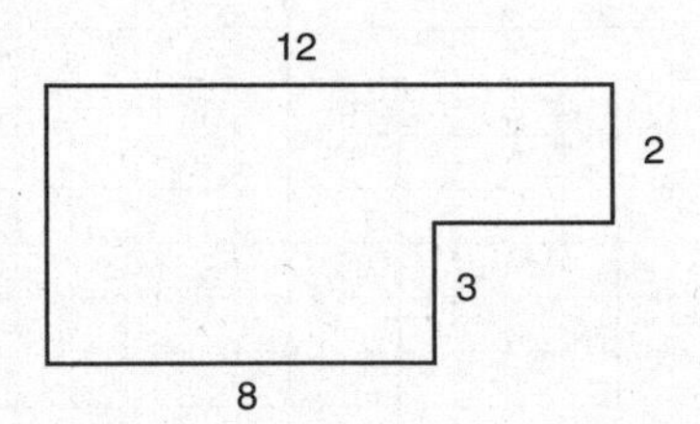


Answer: ________________

**11.** The *radius* of a sphere is the same as the length of the base of the right triangle shown below. Use 3.14 for $\pi$, and round your answer to the nearest whole number.

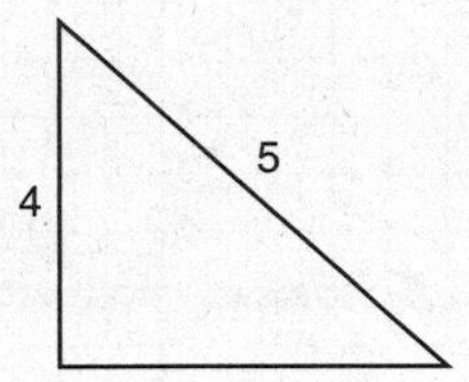


What is the *volume* of that sphere?

Answer: ________________

**12.** Find the *volume of a sphere* that has a *radius* the same length as the side **X** in the rectangle shown below. Use 3.14 for $\pi$, and round your answer to the nearest whole number.

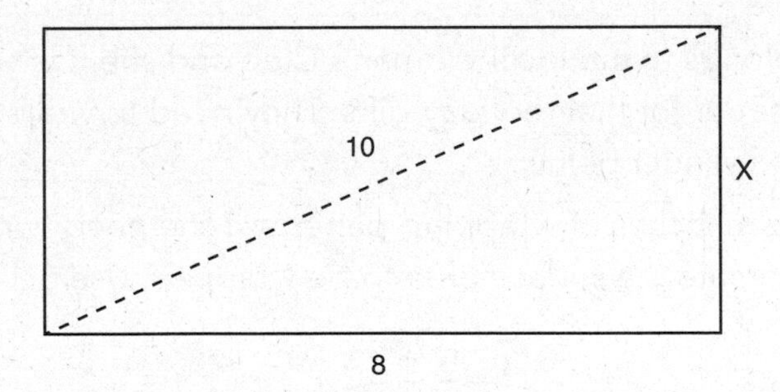


Answer: ________________

Which of the following are correct expressions for the volume of that sphere?

- ☐ **A.** $(8)(36)\pi$
- ☐ **B.** $288\pi$
- ☐ **C.** $\frac{500}{3}\pi$
- ☐ **D.** $\frac{864}{3}\pi$
- ☐ **E.** $864\pi$

**13.** Use a scale factor of 2 to dilate the rectangle *MNOP*.

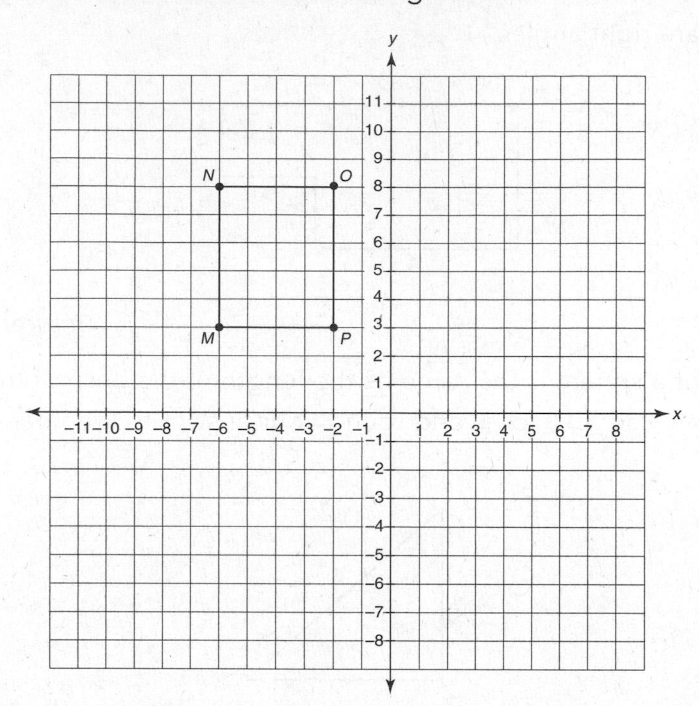


The new coordinates *M′N′O′P′* will be:

○ **A.** *M′* (–6, 9) *N′* (–6, 24) *O′* (–2, 24) *P′* (–2, 9)

○ **B.** *M′* (–18, 3) *N′* (–18, 8) *O′* (–6, 8) *P′* (–6, 3)

○ **C.** *M′* (–12, 6) *N′* (–12, 16) *O′* (–4, 16) *P′* (–4, 6)

○ **D.** *M′* (–18, 9) *N′* (–18, 24) *O′* (–6, 24) *P′* (–6, 9)

**14.** Sally's mom belongs to the local Woman's Club and she has volunteered to buy the wrapping paper for their holiday gifts. They need to wrap 25 containers shaped like the cylinder below.

How many square inches of wrapping paper will she need (without overlapping any paper)? Estimate the *surface area* to be wrapped. Use 3.14 for $\pi$.

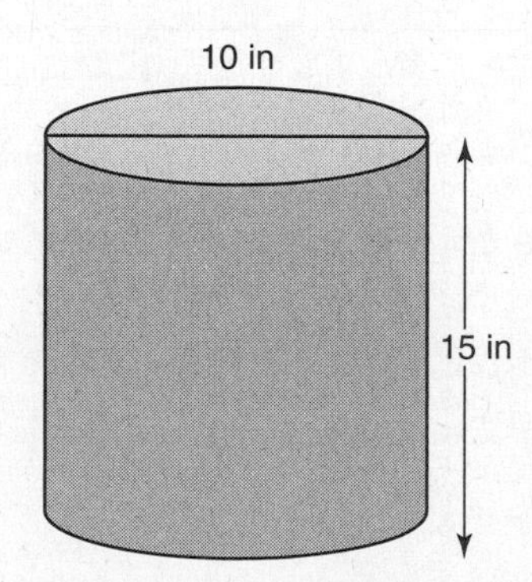

## PRACTICE: Performance-Based Assessment Questions

(For answers, see pages 253–256.)

**1.** In the figure below, you see a drawing of an A-frame tent that Nancy and Rod are taking on their camping trip.

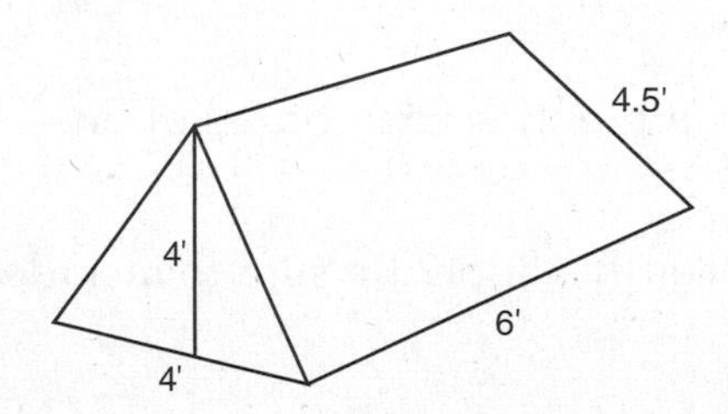


How much canvas was used to make this A-frame tent? Show all work. (Remember to include the floor of the tent, too.)

Area of front and back:

Area of sides:

Area of floor:

Total surface area:

**2.** Below are three different dartboard games. The object of each game is to throw a dart so it hits the shaded area of the dartboard. If you had your choice of any of the three boards below, which board would give you the best chance of getting the most points? Show all work and explain your answer.

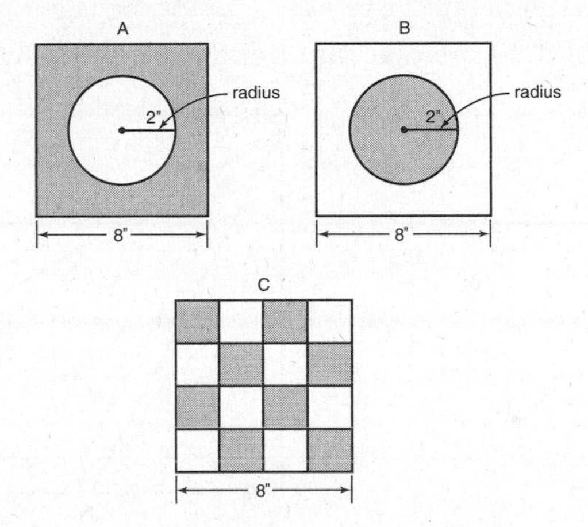


**3.** Holly wants to bake a large cake for her mom's birthday and needs to use the largest baking pan she can find. She has three choices:

A rectangular pan that is 12" long, 16" wide, and 5" high.

A round pan that has a diameter of 16" and is 5" high.

A square pan that measures 15" on each side and is only 4" high.

**Part A:** What is the maximum volume of each pan? Use 3.14 for $\pi$.

**Part B:** Which would hold the most batter?

**Part C:** If she made a second cake and chose the smallest pan and filled it only 3" deep, what would be the maximum volume of that batter?

**4.** Mr. and Mrs. B. are putting a new tile floor down in their entrance foyer. See figure below.

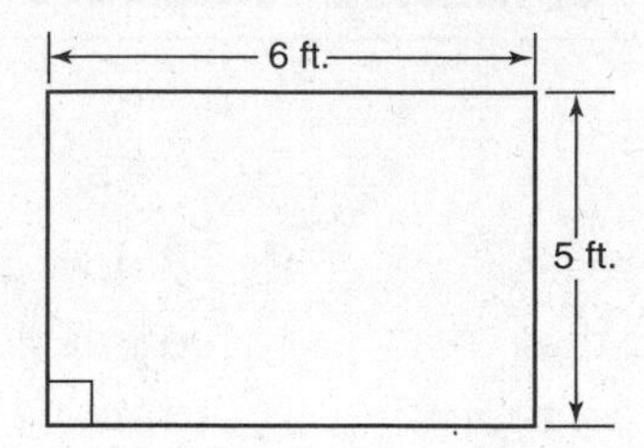


**Part A:** What is the maximum number of 6" × 6" square tiles that would fit on this floor?

**Part B:** The tiles are not sold separately. They are only sold in boxes of one dozen tiles for $21.60.

What will it cost to buy the tiles to cover this floor? Be sure to include a 6% sales tax in your total cost.

**5.** Use the diagram below to help answer the following questions.

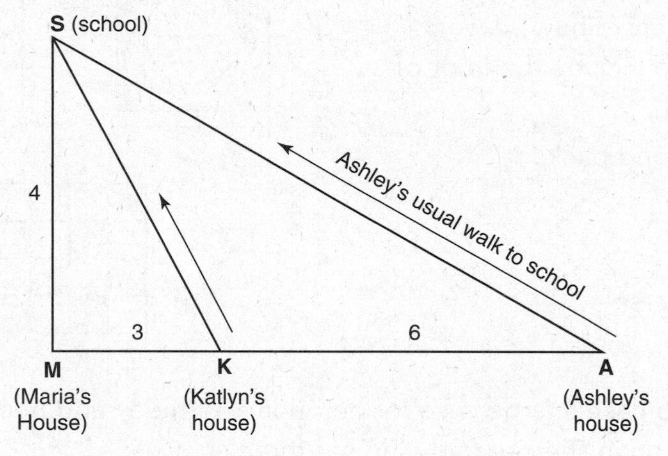


**Part A:** How does the distance of Ashley's usual walk to school compare to the length of Katlyn's walk to school?

○ **A.** < **twice as long** as Katlyn's walk

○ **B.** = Katlyn's walk (the same distance)

○ **C.** > **twice as long** as Katlyn's walk

○ **D.** **exactly twice as long** as Katlyn's walk

**Part B:** Explain or show how you know. Be specific.

**Part C:** What if Ashley walked to Katlyn's house first and then to school; would her walk be longer or shorter than if she walked her usual way, directly from her house to school? Explain or show how you know.

**6.** Devan drove with his dad to pick up his mother and sister at the airport. His mother and sister were flying back from visiting their grandmother in Orlando, Florida. Devan and his dad left from Clarke, New Jersey, point "C" (see diagram below). They drove south for 12 miles, then east for 9 miles to the airport. Devan's dad said he wished the new road had been completed; it would have gone from point "C" directly to the airport.

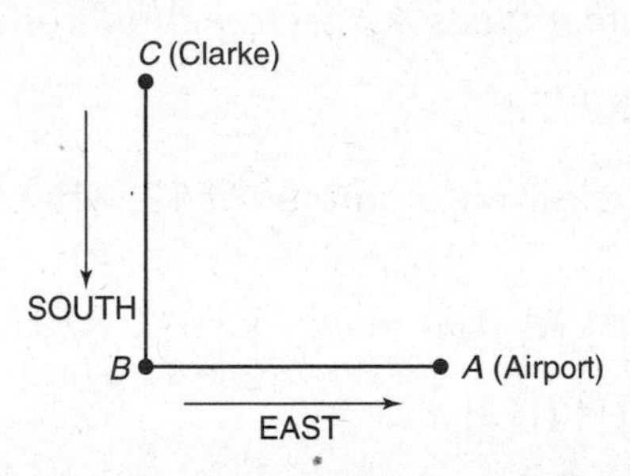


**Part A:** How much shorter would the car ride have been if they could have driven from point "C" straight to point "A" instead of from C to B to A?

**Part B:** The plane fare for Devan's mom and sister costs them each $325.00 round-trip. Considering the other expenses listed in the table below, what would have been the cost if his mother and sister had driven to Orlando?

**Part C:** How much money would they have saved by driving instead of taking a plane?

| Travel Information | Your work space to determine total cost |
|---|---|
| • 1,072 miles from Clarke, New Jersey to Orlando, Florida.<br>• 1,072 miles from Orlando back home to New Jersey<br>Mom's car would have an average of 24 miles per gallon, and we'll estimate gasoline costs $2.40 per gallon. | Gasoline Cost: |
| • Tolls round-trip would be approximately $16.00. | Tolls: |
| • $68.50 hotel for one night while driving down to Florida<br>• $59.00 hotel for one night on the drive home | Hotel Cost: |
| • Food purchased on trip from NJ to Florida ($15.00 + $28.00)<br>• Food purchased on trip from Florida back to NJ ($12.00 + $6.00 + $18.00) | Food Cost: |
| | **Total cost by car:** ______ |

Name: ________________ Date: ____________

## Chapter 2 Test: Geometry

35 minutes

(Use the *PARCC Grade 8 Mathematics Assessment Reference Sheet* on page 359.)

(For answers, see pages 256–257.)

**1.** To find the area of a circle with diameter of 12, what buttons should I press on my calculator?

- ○ **A.** [3] [.] [1] [4] [×] [6] [=]
- ○ **B.** [3] [.] [1] [4] [×] [12] [=]
- ○ **C.** [3] [.] [1] [4] [×] [1] [4] [4] [=]
- ○ **D.** [3] [.] [1] [4] [×] [6] [×] [6] [=]

**2.** Use the diagram of the square below. What is the area of the shaded region?

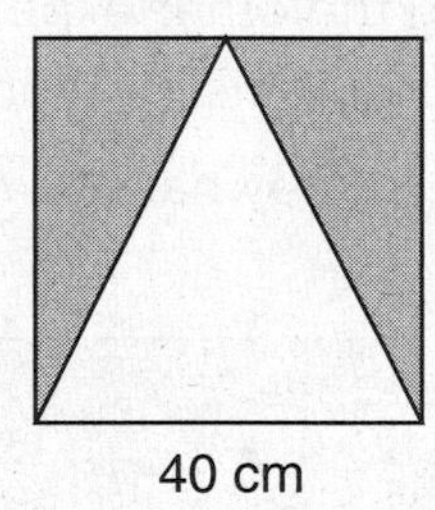


- ○ **A.** 80 sq. cm
- ○ **B.** 400 sq. cm
- ○ **C.** 800 sq. cm
- ○ **D.** 1,000 sq. cm

**3.** The following two right triangles are congruent. Angle *A* measures 60°. What is the measure of angle *F*?

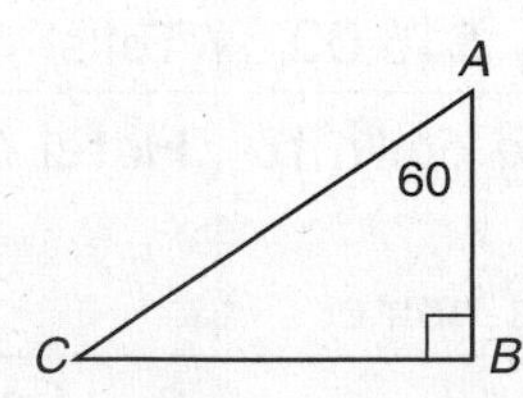


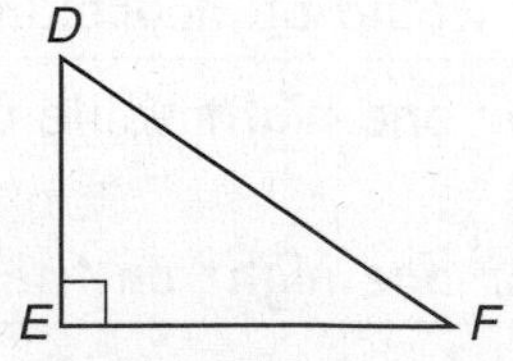


- ○ **A.** 20°
- ○ **B.** 30°
- ○ **C.** 40°
- ○ **D.** 60°

**4.** The perimeter of this figure is 50 yards and all angles are right angles. What is the measure of line segment *FG*?

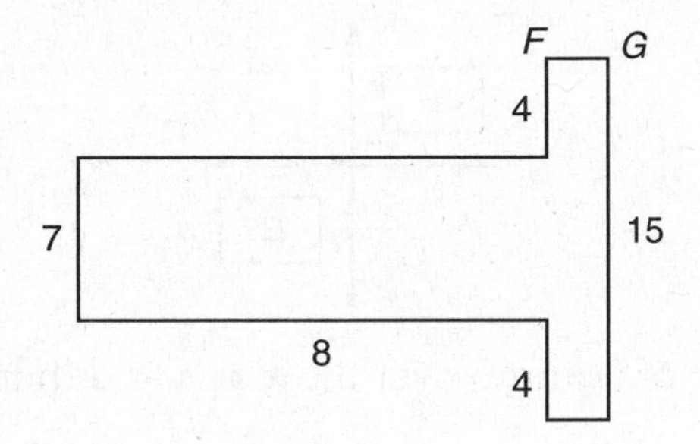


- **A.** 2 yd.
- **B.** 4 yd.
- **C.** 7 yd.
- **D.** 8 yd.

**5.** Point Q has the coordinates (–2, 4). What are the coordinates of its image point if it is translated 3 units to the right and then reflected over the *x*-axis?

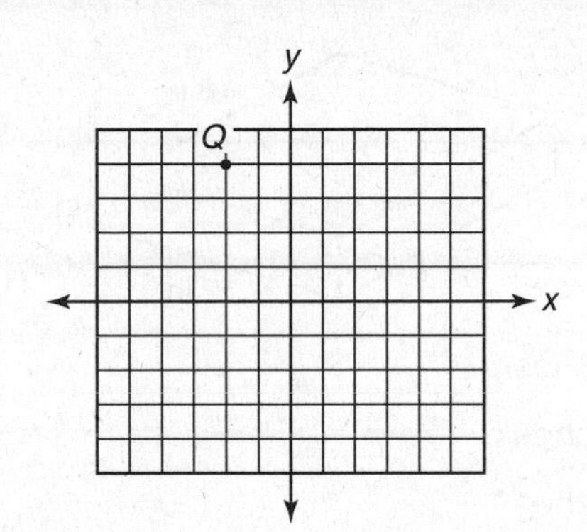


- **A.** (3, 4)
- **B.** (–3, 4)
- **C.** (1, –4)
- **D.** (1, 4)

6. Figure $B'$ is the result of a sequence of transformations of Figure $B$. Which of the following does *not* describe a correct possible sequence of transformations?

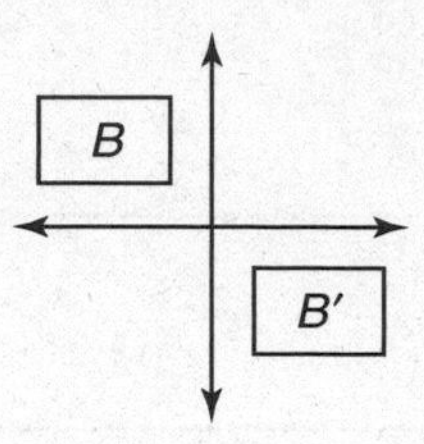


○ **A.** A translation of figure B over the $x$-axis and then a translation over the $y$-axis?

○ **B.** A reflection over the $y$-axis and a second reflection over the $x$-axis?

○ **C.** A translation to quadrant I and then a translation over the $x$-axis?

○ **D.** A translation over two axes and then a dilation.

7. What is the surface area of the box drawn below? (What is the area we would need if we were to cover this box with self-stick wrapping paper without overlapping?)

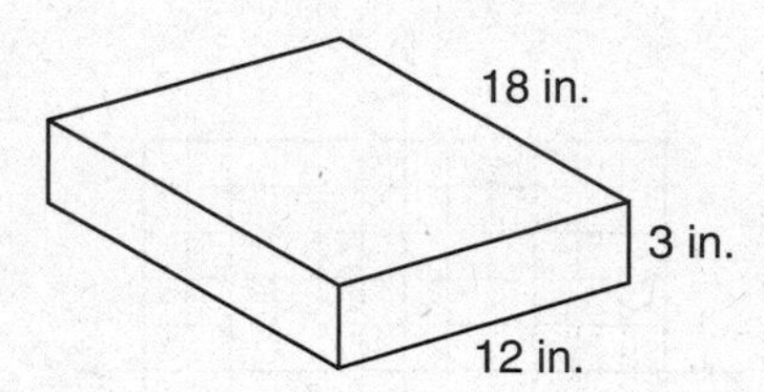


○ **A.** 648 square inches

○ **B.** 612 square inches

○ **C.** 306 square inches

○ **D.** 252 square inches

**8.** Bill is an artist who works for an advertising company that helps design food containers. He is asked to compare the size and shape of two cylindrical containers drawn below.

Container A

Radius = 3 in.
Height = 6 in.

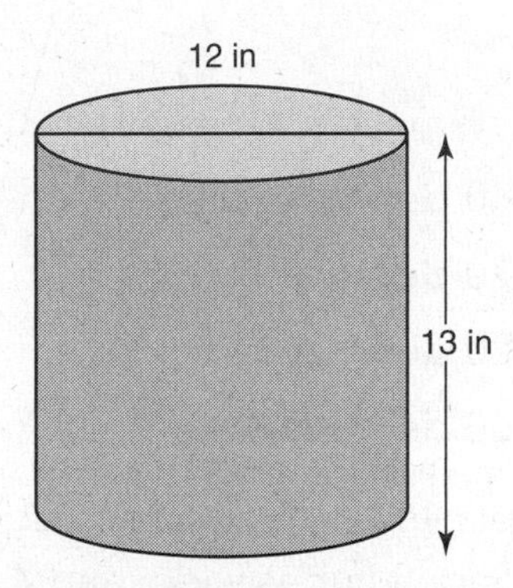


**Part A:** Will the two cylinders hold the same amount of product? Yes ------ No ------

**Part B:** If not, which cylinder will hold more? A ------ B ------

**Part C:** Explain and show details to support your answer. (Write your answer in the box below.)

|   |
|---|
|   |

**9.** Find the *volume of a sphere* that has a *radius* the same length as the side **"X"** in the rectangle shown below. Leave your answer in $\pi$ form.

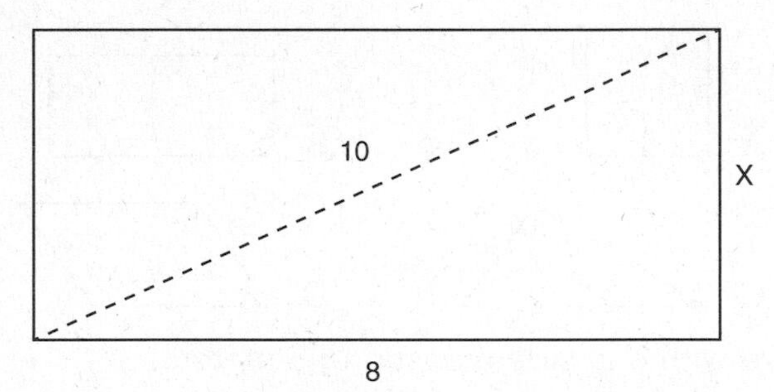


Answer: ------------------

**10.** What is the sum of the interior angles of the pentagon drawn below?

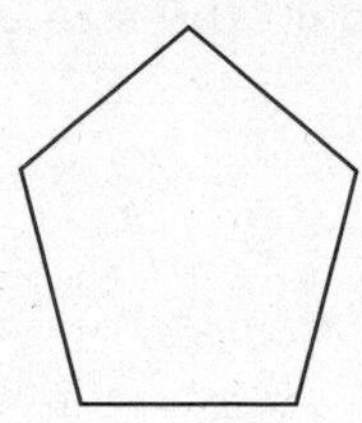

- ○ **A.** 540 degrees
- ○ **B.** 270 degrees
- ○ **C.** 720 degrees
- ○ **D.** 360 degrees

**11.** Nicole is playing with wooden cubes. She just built a box-shaped structure that is 7 cubes long, 5 cubes wide, and 4 cubes high. How many wooden cubes did Nicole use in all?

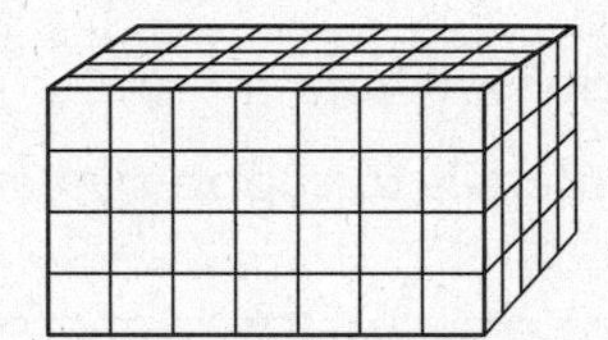

- ○ **A.** 70 cubes
- ○ **B.** 228 cubes
- ○ **C.** 140 cubes
- ○ **D.** 221 cubes

**12.** Which figure has an area of 64 square meters?

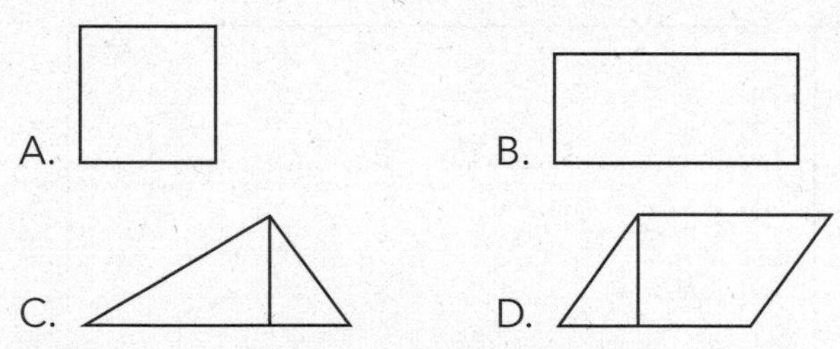


- ○ **A.** A square with one side measuring 8 meters.
- ○ **B.** A rectangle that is 20 meters wide and 12 meters high.
- ○ **C.** A triangle that has a base of 16 meters and is 4 meters high.
- ○ **D.** A parallelogram that has a base of 16 meters and is 2 meters high.

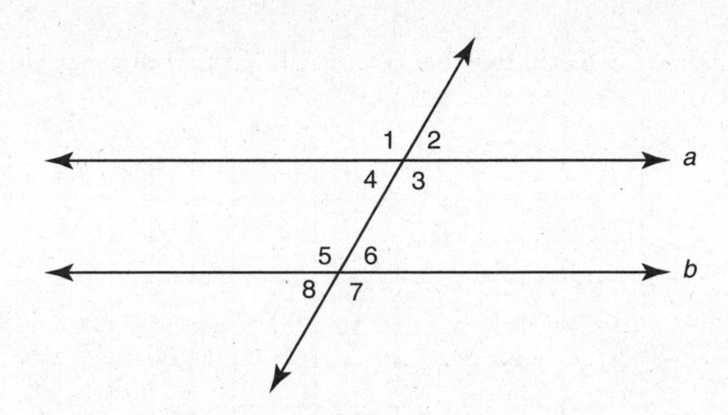


**13.** Answer the following questions or statements based on the diagram above. Lines ***a*** and ***b*** are parallel lines. Select Yes ..✔... or No ..✔... .

**A.** Are $\angle 2$ and $\angle 6$ vertical angles? Yes ...... No ......

**B.** Are $\angle 3$ and $\angle 4$ complementary angles? Yes ...... No ......

**C.** The measure of $\angle 2$ = the measure of $\angle 4$. Yes ...... No ......

**D.** If $\angle 6$ measures 80° then $\angle 5$ measures 100°. Yes ...... No ......

**E.** If $\angle 7$ measures 110° then $\angle 3$ measures 110°. Yes ...... No ......

**F.** $\angle 6 = \angle 8$. Yes ...... No ......

**14.** Find the length of the base of the right triangle drawn below, then find the area of the triangle.

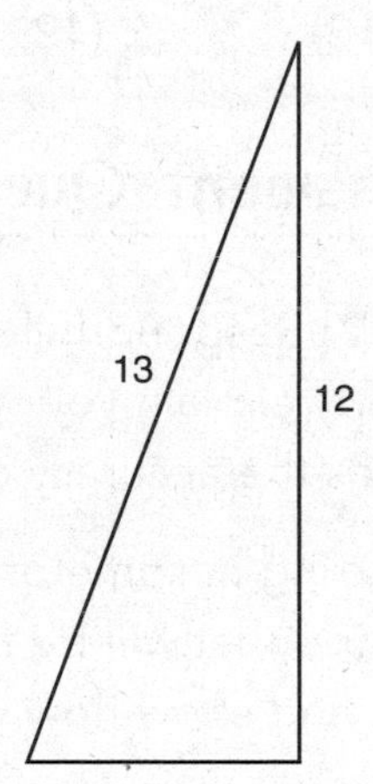


**Part A:** The length of the base of the triangle is ...... units long.

**Part B:** The area of the triangle shown is ...... square units.

**15.** Use a scale factor of 3 to dilate the rectangle *EFGH* shown below.

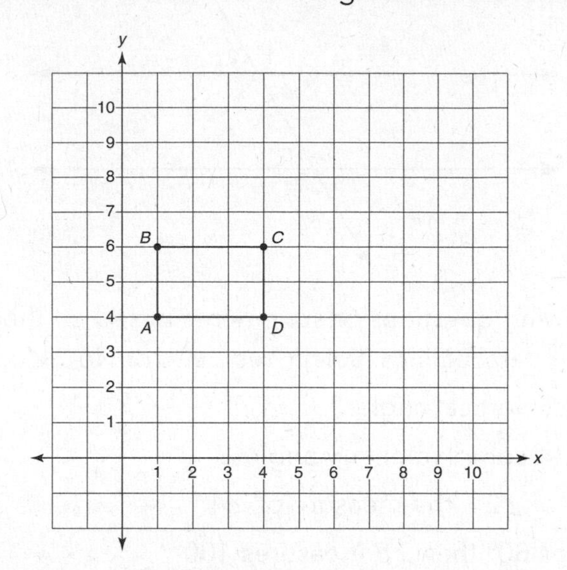


The new coordinates *E′F′G′H′* will be:

○ **A.** *E′* (−6, 9) *F′* (−6, 24) *G′* (−2, 24) *H′* (−2, 9)

○ **B.** *E′* (−18, 3) *F′* (−18, 8) *G′* (−6, 8) *H′* (−6, 3)

○ **C.** *E′* (−12, 6) *F′* (−12, 16) *G′* (−4, 16) *H′* (−4, 6)

○ **D.** *E′* (3, 12) *F′* (3, 18) *G′* (12, 18) *H′* (12, 12)

## Performance-Based Assessment Questions

**Directions for Questions 16 and 17:** Respond fully to the PBA questions that follow. Show your work separately and clearly explain your answer. You will be graded on the correctness of your method as well as the accuracy of your answer.

**16.** The school is repairing the flooring in some of the older science rooms. Find the number of square feet of flooring that will be needed for this irregularly shaped room. Use the diagram below, and show how you found the area of the floor.

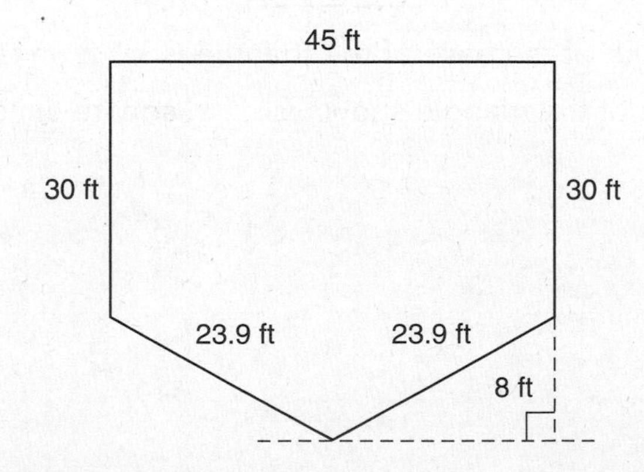

**17.** A triangle *ABC* is made by connecting the points *A* (0, 0), *B* (5, 0), and *C* (5, 6).

**1.** Plot and label the points on the coordinate plane provided below.

**2.** Connect the points to make the triangle *ABC*.

**3.** Classify the triangle as right, isosceles, equilateral, or obtuse.

**4.** Use the Pythagorean theorem (Pythagorean formula) to find the length of the side *AC*. Round your answer to the nearest tenth

**5.** Find the perimeter of triangle *ABC*.

**6.** Find the area of triangle *ABC*.

# Expressions, Equations, and Functions

CHAPTER 3

## WHAT DO PARCC 8 EXPRESSION, EQUATION, AND FUNCTION QUESTIONS LOOK LIKE?

## Multiple-Choice Questions (MC)

### Example 1: Patterns

What is the next number in the sequence?

$$4 \quad -8 \quad 16 \quad -32 \quad \text{------}$$

○ **A.** –49

○ **B.** 46

○ **C.** 64

○ **D.** –64

**Strategies and Solutions**

Look at the sequence; notice if the "numbers" are getting larger or smaller. If they are getting larger, you probably need to add or multiply to get from one number to the next. If the sign changes every other number it is probably because each number is multiplied by a negative number. Remember, when multiplying

$$(-)(+) = - \quad \text{and} \quad (+)(-) = -$$

$$(-)(-) = + \quad \text{and} \quad (+)(+) = +$$

$$(-32)(-2) = 64$$

Correct choice is **C**.

### Example 2:

Is this an *arithmetic or geometric* sequence?

$$6 \quad 2 \quad -2 \quad -6 \quad -10 \quad -14$$

○ **A.** arithmetic

○ **B.** geometric

○ **C.** both

○ **D.** neither

**Strategies and Solutions**

Here you see the numbers begin as positive and then all become negative. By trial and error, you see that the pattern is to add a –4 to each number to get the next number. When you *add* to continue a sequence this is called an *arithmetic* sequence.

Correct choice is **A**.

## What Do New Multiple-Choice Expressions and Equations Questions Look Like?

### Example 3a:

Use the equations listed below. Select Yes if true, No if it is not true.

- $y = 12 + 8 \div 4 = \mathbf{5}$ Yes ------ No ------
- $y = 3 + 2(-2) = \mathbf{-1}$ Yes ------ No ------
- $x = 4\,(-2)^2 = \mathbf{-16}$ Yes ------ No ------
- $\frac{1}{2}x = 10,\ x = 20$ Yes ------ No ------
- if $y - 2 = 4,\ y = \mathbf{6}$ Yes ------ No ------
- $M = \frac{4}{3}(9);\ M = \frac{32}{3}$ Yes ------ No ------

**Strategies and Solutions**

Solutions: Remember to follow the correct algebraic order of operations (PEMDAS).

- $8 \div 4 = 2;\ 12 + 2 = \mathbf{14}$ Yes ------ **No** --✔--
- $+2(-2) = -4;\ 3 - 4 = \mathbf{-1}$ **Yes** --✔-- No ------
- $x = (-2)^2 = 2;\ 4(2) = \mathbf{+16}$ Yes ------ **No** --✔--
- *multiply both sides by 2* **Yes** --✔-- No ------
- *add 2 to both sides* **Yes** --✔-- No ------
- $M = \frac{4}{3}(9) = \frac{36}{3} = 12$ Yes ------ **No** --✔--

### Example 3b:

Refer to the expression shown below and check ALL that are true.

$$2(6 - 4) + (10 \div 2)^2 - 2(-2)^3$$

☐ **A.** A correct first step would be 2(6).

☐ **B.** A first step could be $10 \div 2$.

☐ **C.** $(-2)^3 = 8$.

☐ **D.** When this expression is simplified the solution is a negative integer.

☐ **E.** When simplified the expression = 45.

**Strategies and Solutions**

$$2(6-4)+(10\div 2)^2-2(-2)^3$$

☐ **A.** No; work inside parentheses first.

☑ **B.** Yes; work inside parentheses first.

☐ **C.** No; $(-2)(-2)(-2)=(4)(-2)=\mathbf{-8}$.

☐ **D.** No; see solution below.

☑ **E.** Yes $2(2)+(5)^2-2(-8)=4+25+16=20+25=\mathbf{45}$.

## Example 4a:

Use the functions listed below. Check Yes if it is linear and No if it is not linear.

| | | |
|---|---|---|
| • $y=-3x^2+2$ | Yes ------ | No ------ |
| • $y=5x$ | Yes ------ | No ------ |
| • $A=\pi r^2$ | Yes ------ | No ------ |
| • $y=0.45+0.2(x-2)$ | Yes ------ | No ------ |
| • $y=3$ | Yes ------ | No ------ |
| • $V=\frac{4}{3}\pi r^3$ | Yes ------ | No ------ |

**Strategies and Solutions**

| | | |
|---|---|---|
| • Nonlinear (See $x^2$) | Yes ------ | **No** ✔ |
| • Yes, linear. | **Yes** ✔ | No ------ |
| • No, nonlinear (See $r^2$) | Yes ------ | **No** ✔ |
| • Yes, linear. | **Yes** ✔ | No ------ |
| • Yes, linear. (This is a horizontal line.) | **Yes** ✔ | No ------ |
| • Nonlinear (See $r^3$. This is the formula for the volume of a sphere.) | Yes ------ | **No** ✔ |

## Example 4b:

Refer to the function $y=-\frac{2}{3}x+9$ and select all that are true.

☐ **A.** This represents a line that rises from left to right.

☐ **B.** This represents a line that crosses the *y-axis* at (0, 9).

☐ **C.** The line $y=-\frac{2}{3}x-6$ is parallel.

☐ **D.** The rate of change is $\frac{2}{3}$.

☐ **E.** This represents a line that crosses the *x-axis* at (4, 0).

**Strategies and Solutions**

☐ **A.** This represents a line that rises from left to right.

☑ **B.** This represents a line that crosses the *y-axis* at (0, 9).

☑ **C.** The line $y = -\frac{2}{3}x - 6$ is parallel.

☐ **D.** The rate of change is $\frac{2}{3}$.

☐ **E.** This represents a line that crosses the *x-axis* at (4, 0).

## Short Constructed-Response Question (SCR)

### Example 5:

(No calculator permitted.)

What is the slope of the line represented by the following equation?

$$4y = -2x + 16$$

Answer: ________________

**Strategy and Solution**

Explanation: First, you need to divide all terms by 4 so it is in the correct form to easily see the slope.

$$4y = -2x + 16$$

$$\frac{4y}{4} = \frac{-2x}{4} + \frac{16}{4} \text{ or } y = -\frac{1}{2}x + 4$$

Therefore, the slope is $\frac{-1}{2}$.

Answer: $\frac{-1}{2}$

Remember, no partial credit is given for short constructed-response answers and usually no calculators are permitted.

### Example 6:

Solve the equation for ***a***. Show all steps.

$$\frac{4}{5}(2a - 5) = 3a - (4a + 1)$$

**Solution:**

$a=\frac{15}{13}$ or $1\frac{2}{13}$

$\frac{8}{5}a-\frac{20}{5}=3a-4a-1$ Distribute on the left and on the right sides.

$\frac{8}{5}a-4=-a-1$ Combine like terms $3a - 4a$.

Simplify the fraction $-\frac{20}{5}$ to $-4$.

$\frac{8}{5}a=-a+3$ Add **4** to both sides.

$\frac{8}{5}a+a=3$ Add ***a*** to both sides.

$\frac{13}{5}a=3$ Combine $8/5a + a$; use $8/5a$ + $\mathbf{5/5a}$ = $13/5a$.

$\left(\frac{5}{13}\right)\frac{13}{5}a=3\left(\frac{5}{13}\right)$; $a=\frac{15}{13}$ *or* $1\frac{2}{13}$ Multiply both sides by the reciprocal (5/13).

## MONOMIALS, TERMS, AND EXPRESSIONS (From Grade 7.EE.4)

In algebra you work with *numbers* and *variables* and different combinations of them. Here are some terms to remember and recognize:

- **Monomial:** A number, variable, or product of numbers, variables, or numbers and variables.

  Numbers: $5$ $-4.5$ $\frac{1}{2}$ $-0.008$

  Variables: $a$ $b$ $x$ $m$

  Different ways to write the *product* of monomials, variables, or numbers and variables:

  $(5)(9)$ $\left(\frac{1}{2}\right)(-4)(5)$ $5(n)$ $-16x$ $20ab$ $xyz$ $25abx^2$

- **Expressions:** Combinations of monomials (numbers and variables)

  $3x + 2.5$ $16 - 19a$ $24a+\left(3b-\frac{3}{4}\right)$ $-12+\frac{1}{2}x-6ab$ $xy + 2a - 3(4b)^2$

## SIMPLIFY AND EVALUATE EXPRESSIONS
## (From Grade 7.EE.1)

Evaluate each of the following expressions if $a = 2$, $b = 4$, and $x = -2$.

| The Expression | Work Shown (replace variables with numbers and do the computation) | The Evaluation |
|---|---|---|
| $3x + 2.5$ | $3(-2) + 2.5 = -6 + 2.5 =$ | $-3.5$ |
| $16 - 19a$ | $16 - 19(2) = 16 - 38 =$ | $-22$ |
| $24a + 3b - 4$ | $24(2) + 3(4) - 4 = 48 + 12 - 4 = 60 - 4 =$ | 56 |
| $-12 + \frac{1}{2}b(8)$ | $-12 + \frac{1}{2}(4)(8) = -12 + 2(8) = -12 + 16 =$ | 4 |
| $3(x - 4)^2 + b$ | $3(-2 - 4)^2 + 4 = 3(-6)^2 + 4 = 3(36) + 4 = 108 + 4$ | 112 |
| | *Remember*: PEMDAS (work inside Parentheses first, then work with Exponents.) | |

Combine like terms. Sometimes you do not know the value of a variable. Often you are asked to just simplify an expression by combining like terms:

- 8 apples + 15 cherries + \$3 + 4 cherries – \$2;
  simplified: 8 apples, 19 cherries, and \$1
- $2a + 15b - 6 + 3a + 4b - 2$; put like terms next to each other:
  $\underline{2a + 3a} + \underline{15b + 4b} - \underline{6 - 2}$ and then combine like terms: $5a + 19b - 8$ (answer)
- $10a - 2b + 15a + 6b - 4$; put like terms next to each other:
  $\underline{10a + 15a} - \underline{2b + 6b} - \underline{4}$ and then combine like terms: $25a + 4b - 4$ (answer)

Use the distributive property:

- $2(4x + 5)$ $8x + 10$ This is the same as $2(4x) + 2(5) = 8x + 10$ (answer)
- $-6(3x - 3)$ $-18x + 18$ This is the same as $-6(3x) -6(-3) = -18x + 18$ (answer)

# EQUATIONS

After learning how to combine like terms to simplify expressions, you now can learn how to solve basic linear equations.

## Solving Linear Equations with One Variable on One Side

Remember, do the same thing to both sides of the equation to keep the equation balanced. Study the following examples and remember the correct order of operations (PEMDAS) and the rules about working with positive and negative integers.

### Solving One-Step Equations

Study each chart. The original equations are in bold, with steps and explanations below.

| $10 = x - 2$ | $b + 5 = 8$ | $3x = 15$ | $w + 12 = -10$ |
|---|---|---|---|
| $+2 \quad +2$ | $-5 \quad -5$ | $\frac{3x}{3} = \frac{15}{3}$ | $-12 \quad -12$ |
| Undo the subtraction; add 2 to both sides. | Undo the addition; subtract 5 from both sides. | Undo the multiplication; divide both sides by 3. | Undo the addition; subtract 12 from both sides. |
| $12 = x$ | $b = 3$ | $x = 5$ | $w = -22$ |
| $-15 = 4 + w$ | $\frac{1}{4}m = 5$ | $-5a = 15$ | $10 = x + 2.5$ |
| $-4 \quad -4$ | $\left(\frac{4}{1}\right)\frac{1}{4}m = 5\left(\frac{4}{1}\right)$ | $\frac{-5a}{-5} \quad \frac{15}{-5}$ | $-2.5 \quad -2.5$ |
| Undo the addition; subtract 4 from both sides. | Multiply both sides by the reciprocal of $\frac{1}{4}$; use $\frac{4}{1}$. $m = \frac{20}{1}$ | Undo the multiplication; divide both sides by −5. | Undo the addition; subtract 2.5 from both sides. |
| $-19 = w$ | $m = 20$ | $a = -3$ | $7.5 = x$ |

## Solving Two-Step Equations

Study each chart. The original equations are in bold, with steps and explanations below. The equations have been solved working one step at a time.

| $3x + 6 = 36$ | $100 - 2x = 60$ | $\frac{1}{3}x - 8 = -4$ | $16 = 11 + 2x$ |
|---|---|---|---|
| $-6 \quad -6$ | $-100 \qquad -100$ | $+8 \quad +8$ | $-11 \quad -11$ |
| Subtract 6 from both sides. | Subtract 100 from both sides. | Add 8 to both sides. | Subtract 11 from both sides. |
| $3x = 30$ | $-2x = -40$ | $\frac{1}{3}x = 4$ | $5 = 2x$ |
| Divide both sides by 3 | Divide both sides by −2 | Multiply both sides by 3 | Divide both sides by 2 |
| $x = 10$ | $x = 20$ | $x = 12$ | $\frac{5}{2} = x$ |

## Solving Linear Equations with Variable on Both Sides

| | | | |
|---|---|---|---|
| $5x = -75 + 2x$<br>$-2x \qquad -2x$<br>$3x = -75$ | $16 + 4w = -48 + 3w$<br>$-3w \qquad -3w$<br>$16 + w = -48$<br>$-16 \qquad -16$<br>$w = -64$ | $2x + 8 = 6x + 80$<br>$-2x \qquad -2x$<br>$8 = 4x + 80$<br>$-80 \qquad -80$<br>$-72 = 4x$<br>$-18 = x$ | $12 + 2(x - 2) = 3x$<br>$12 + 2x - 4 = 3x$<br>$8 + 2x = 3x$<br>$-2x \quad -2x$<br>$-8 = x$ |

## PRACTICE: Solving Equations

(For answers, see pages 258–259.)

**1.** $3w - (5)^2 = 20$

- ○ **A.** $w = -15$
- ○ **B.** $w = 15$
- ○ **C.** $w = 10$
- ○ **D.** $w = -10$

**2.** Which step would **not** be a correct first step for solving the equation below algebraically?

$$\frac{1}{4}(2x-3)-3\frac{1}{4}=8-\frac{1}{2}x$$

- ○ **A.** multiply every term in the equation by 4
- ○ **B.** multiply $-3$ by $\frac{1}{4}$
- ○ **C.** add $\frac{1}{2}x$ to $2x$
- ○ **D.** add $3\frac{1}{4}$ to 8

**3.** $2(a - 3) = 3(2a + 10)$

- ○ **A.** $a = -1$
- ○ **B.** $a = -4$
- ○ **C.** $-9 = a$
- ○ **D.** $-8 = x$

**4.** $\frac{1}{4}z = 20$

- ○ **A.** $z = 5$
- ○ **B.** $z = 16$
- ○ **C.** $z = 80$
- ○ **D.** $z = -16$

**5.** $x(5 - 3)^2 = -64$

- ○ **A.** $x = 4$
- ○ **B.** $x = -4$
- ○ **C.** $x = 16$
- ○ **D.** $x = -16$

**6.** $16 + \frac{1}{2}y = -24$

- ○ **A.** $y = -80$
- ○ **B.** $y = 80$
- ○ **C.** $y = 20$
- ○ **D.** $y = -20$

7. **Part A:** Solve the following equation for $x$.

$$4 - 2(3x - 5) = 3(-4) - 5(2)$$

Write your answer here: $x =$ ________________

**Part B:** Which of the equations or expressions below are equivalent to your answer for $x$? Select all that are correct.

- ☐ **A.** $-3x = -18$
- ☐ **B.** $3x = 3^3$
- ☐ **C.** $4^2 - 2(\sqrt{25})$
- ☐ **D.** $\sqrt{81} - \sqrt{9}$
- ☐ **E.** $2^3 + 3^2 - 7$

8. Solve the following equation. Show all steps.

$\frac{-3}{2}(2 + x) = 24$    Answer: $x =$ ________________

9. Solve the following equation. Show all steps.

$\frac{2}{3}(w - 9) = \frac{8}{9}$    Answer: $w =$ ________________

10. Solve the equation, and select the correct solution.

$-\frac{1}{3}(a + 1) = \frac{1}{6}$

- ○ **A.** $a = 3/2$
- ○ **B.** $a = -3/2$
- ○ **C.** $a = 1/2$
- ○ **D.** $a = -1/2$

11. Solve the equation. Show all steps.

$-.05\ (x - 10) = 6$    Answer: $x =$ ________________

12. Solve for $w$. Leave your answer in decimal form; round to the nearest tenth.

$0.02\ (w - 1.5) = 1.9$    Answer: $w =$ ________________

## FUNCTIONS AND RELATIONS

In this section, we will review graphing simple functions (seen as various lines) and discuss their general behavior (how they slant, where they are placed on the grid, and how they relate to other lines); we'll recognize if they are parallel, perpendicular, or intersecting lines.

- In Chapter 1, you reviewed graphing *integers* on a number line and sometimes connected them to see a range of points that created a *line segment*.

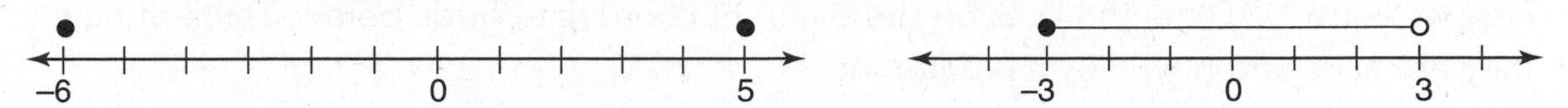


- In Chapter 2, you extended this knowledge and graphed *points* on a coordinate plane and sometimes connected the points to make a *line segment* and eventually, a *geometric shape*.

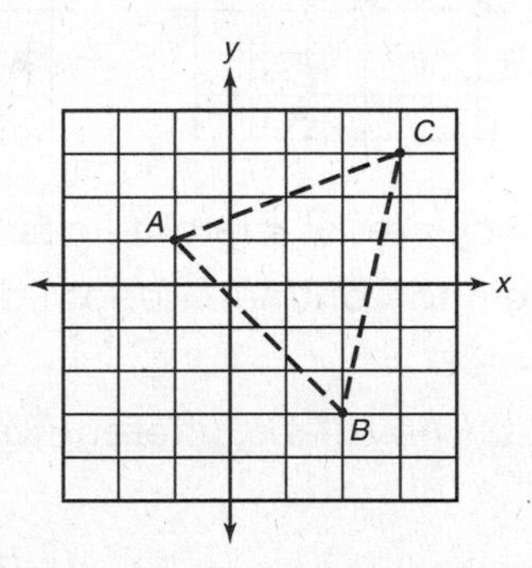


- In this section, you will review how to determine which $x$ and $y$ values match a particular *equation* of a line and what that means. You also will learn how to match equations to lines on the coordinate plane. How does the line described by $y = x$ compare to the line $y = -x$? How does the line $y = 4x$ compare to the line $y = \frac{1}{2}x$?

## Lines, Slopes, and Linear Equations

There are four basic areas you will review in this section:

- Lines with positive and negative slopes
- Intersecting, parallel, and perpendicular lines
- The slope-intercept form of a linear equation ($y = mx + b$)
- Finding the slope and y-intercept of a line if you are given
    - the slope-intercept form of the equation; for example, $y = \frac{-2}{3}x + b$
    - two points on the line; for example, (2, 3) and (4, –8)

## Slope

First we'll just LOOK at the lines on the different coordinate grids below. Think of how they are alike and how they are different.

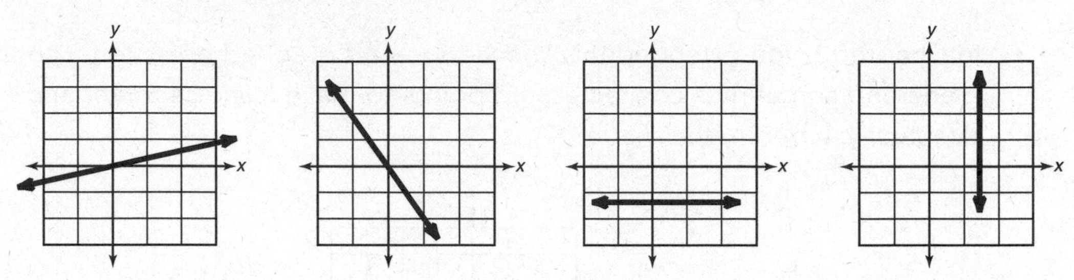


- How are they alike? They are all straight lines, two of them cross the origin (0, 0), and all lines seem to continue in both directions. (Notice the arrows at both ends.)
- How are they different? They *slant* different ways; two of them do not cross the origin (0, 0).

Now, think of each line as a hill or a mountain. Which line looks like a small hill that would be easy to walk up? Which line looks like a very steep hill, one that would be difficult to ride up with a bike? When we talk about the slant or steepness of lines, we describe this as the *slope* of the line. At other times, you measured a specific line segment, used a ruler to measure its length, and used a number line to describe its location.

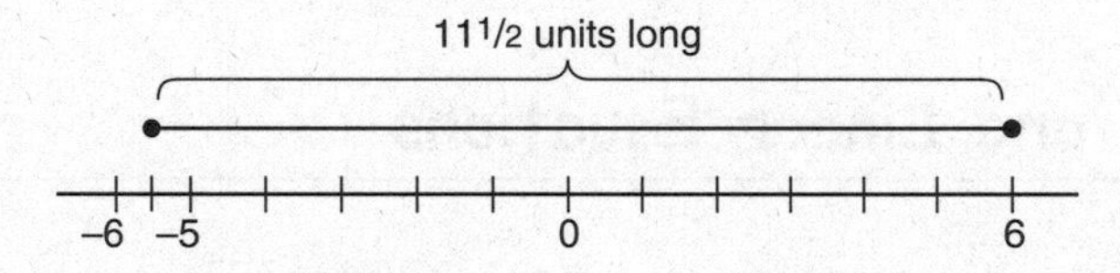


But when a line is on a coordinate plane it is not always a horizontal line and not usually a segment. We must know its slope to describe it accurately. We use a grid to help us be exact.

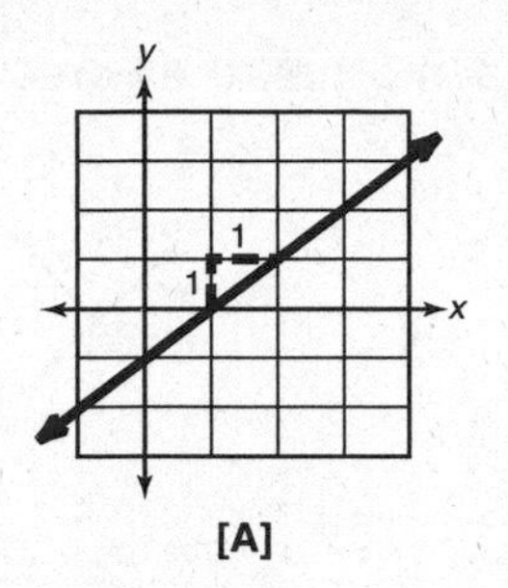


**[A]**

Slope here is 1/1 = 1.

The equation of this line is $y = 1x$ or $y = x$.

This is like a small hill.

**[B]**

Slope here is 1/2.

The equation of this line is $y = \frac{1}{2}x$.

This is a much smaller hill.

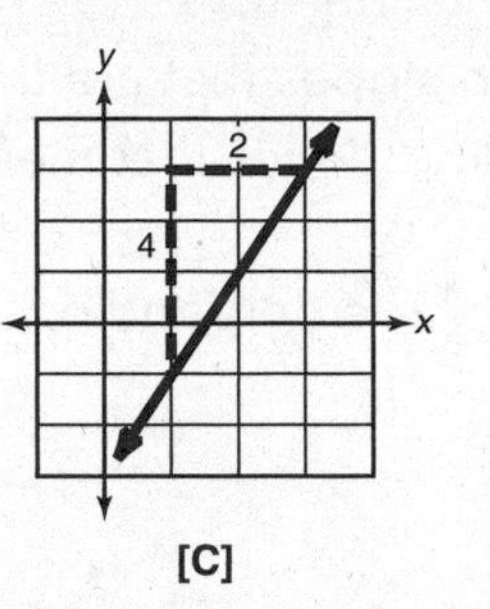


**[C]**

Here, the slope is 4/2, which reduces to 2/1.

The equation of this line is $y = 2x$.

This is a steeper hill.

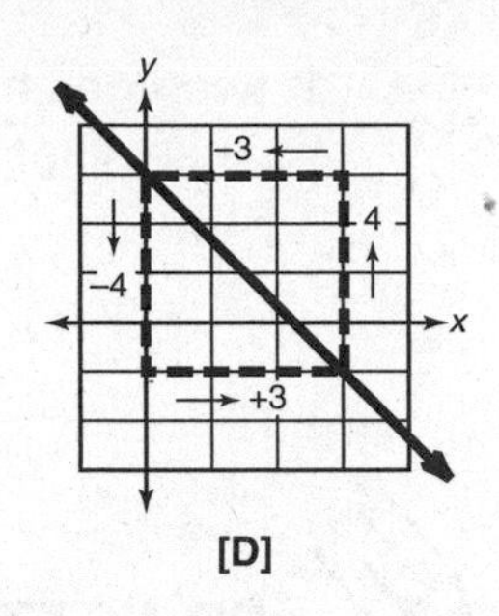


**[D]**

What is the slope of this line?

$\text{Slope} = \frac{-4}{3}$ or

$\text{slope} = \frac{4}{-3}$.

The equation of this line is $y = -\frac{4}{3}x + 3$

To determine the slope of any line, begin at *any* point on the line and move up or down (count the number of boxes you moved (a *y* movement); then move right or left (an *x* movement) until you reach the line (count the number of boxes you moved left or right). The slope is written as the fraction:

$$\text{Slope} = \frac{\text{The } y \text{ movement}}{\text{The } x \text{ movement}}$$

Remember $\frac{+}{+} = + \quad \frac{2}{5} = \frac{2}{5}$ Remember $\frac{-}{-} = + \quad \frac{-2}{-5} = \frac{2}{5}$

- For **A, B,** and **C** above, our movements were all "positive" movements (up and right) or (down and left). The slopes of these lines were all positive.
- For **D** above, we see a line that slants in a different direction. From a point on this line we moved down (a negative *y* movement), and then right (a positive *x* movement). The slope of this line is $\frac{-}{+}$ or – (negative).

Here are two facts to remember:

- If two lines have the same slope, they are parallel lines.

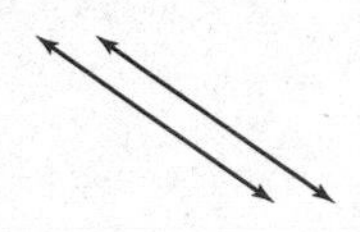

$$y = \frac{3}{4}x + 12 \text{ is parallel to the line } y = \frac{3}{4}x - 8$$

- If two lines have slopes that are the *negative reciprocals* of each other, they are perpendicular lines (lines that meet at right angles).

$$y=\frac{2}{5}x+12 \text{ is perpendicular to the line } y=\frac{-5}{2}x+16$$

## PRACTICE: Lines and Slope

(For answers, see pages 259–260.)

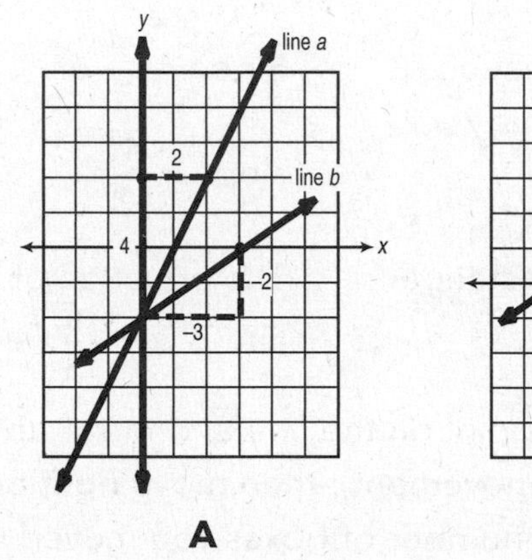

A

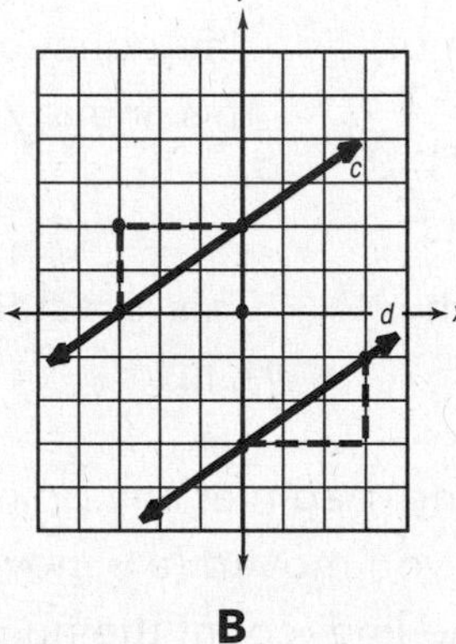

B

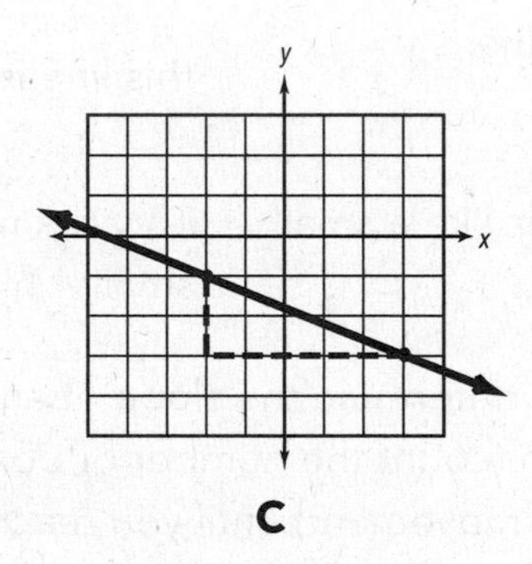

C

**1.** Look at Figure **A** above. What is the slope of the line labeled line *a*?

○ **A.** $\frac{1}{3}$ ○ **B.** $\frac{3}{1}$ ○ **C.** $\frac{2}{4}$ or $\frac{1}{2}$ ○ **D.** $\frac{4}{2}$ or 2

**2.** Look at Figure **A** above. What is the slope of the line labeled line *b*?

○ **A.** $\frac{2}{3}$ ○ **B.** $\frac{3}{2}$ ○ **C.** $\frac{2}{1}$ ○ **D.** $\frac{3}{1}$

**3.** Look at Figure **B** above. What do you notice about the two lines (*c* and *d*)? They seem to be

○ **A.** parallel ○ **C.** intersecting
○ **B.** perpendicular ○ **D.** the same line

**4.** Look at Figure **B** above. What is the slope of line *c*? What is the slope of line *d*?

○ **A.** slope of line $c=\frac{-3}{2}$; slope of line $d=\frac{-3}{2}$

○ **B.** slope of line $c=\frac{-2}{3}$; slope of line $d=\frac{-2}{3}$

○ **C.** slope of line $c=\frac{2}{3}$; slope of line $d=\frac{2}{3}$

○ **D.** slope of line $c=\frac{3}{2}$; slope of line $d=\frac{3}{2}$

**5.** From what you have observed complete this statement. If the slope of one line *equals* the slope of another line then

○ **A.** the two lines are intersecting lines

○ **B.** the two lines are perpendicular lines

○ **C.** the two lines are parallel lines

○ **D.** the two lines always have very steep slopes

**6.** Look at Figure **C** on page 164. What is the slope of the line here?

**7.** In the following two equations the coefficient of $x$ is the slope of each line.

$$\text{Line } m\text{: } y = \frac{-3}{2}x + 12 \qquad \text{Line } n\text{: } y = \frac{2}{3}x + 12$$

Just by looking at the slope of the two equations above, what can you tell about their lines?

○ **A.** They are parallel.

○ **B.** Line $n$ has a negative slope and line $m$ has a positive slope.

○ **C.** They are perpendicular.

○ **D.** They have the same slope.

**8.** Use the equation $y = -2x + 6$ to determine if the following statements are correct. Check Yes ..✔... or No ..✔... .

**A.** This equation represents a curved line. Yes ...... No ......

**B.** The $y$-intercept is at (0, 6). Yes ...... No ......

**C.** The slope is negative. Yes ...... No ......

**D.** $2y = 4x + 12$ is an equivalent equation. Yes ...... No ......

**E.** When $x = 2$, then $y = 2$. Yes ...... No ......

**F.** $y = ½x + 4$ is a line parallel to $y = -2x + 6$. Yes ...... No ......

There are four basic ways to get information about a line so that you can determine its slope. The table on page 166 outlines these four ways. Study each example and then do the practice examples on the following pages.

## Finding the Slope of a Line

**Remember:** It is always best to leave the slope as a fraction: $\frac{3}{2}$ not $1\frac{1}{2}$.

### Graph

**Example A**

Start at any point on the line; count up ($y$) and then count across ($x$) until you reach the line again.

$\frac{y}{x}$ = Slope

In this figure the slope is $\frac{1}{3}$.

**Example B**

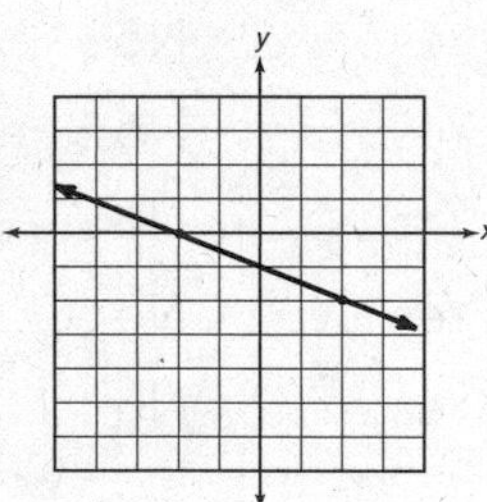


Start at any point on the line; count up or down ($y$) and then count across, left or right, ($x$) until you reach the line again.

$\frac{y}{x}$ = Slope

In this figure the slope is $= -\frac{1}{2}$.

### Table

**Example C**

This is a list of different coordinate points $(x, y)$

| $x$ | $y$ |
|---|---|
| 0 | 2 |
| 1 | 3 |
| 2 | 4 |
| 3 | 5 |
| 4 | 6 |

Look at the change in the y values.

$2 + 1 = 3$
$3 + 1 = 4$
$4 + 1 = 5$
$5 + 1 = 6$

The slope of this line is +1 or $\frac{1}{1}$.

**Example D**

| $x$ | $y$ |
|---|---|
| 0 | 0 |
| 1 | –2 |
| 2 | –4 |
| 3 | –6 |
| 4 | –8 |

Look at the change in the $y$ values.

$0 + -2 = -2$
$-2 + -2 = -4$
$-4 + -2 = -6$
$-6 + -2 = -8$

The slope of this line is –2 or $-\frac{2}{1}$.

### Equation in slope-intercept form*

**Example E**

$y = 4x + 6$

The slope of this line is 4 or $\frac{4}{1}$, and the $y$-intercept (where the line crosses the $y$-axis) is 6.

**Example F**

$4y = -2x + 16$

(Here we need to divide by 4 to get the equation in the correct form.)

$y = -\frac{2}{4}x + \frac{16}{4}$

$y = -\frac{1}{2}x + 4$

The slope of this line is $-\frac{1}{2}$ and the $y$-intercept is 4.

**Example G**

$y + 4x = 12$

(Here we need to add –4x to both sides.)

$-4x \quad -4x$
$y = -4x + 12$

The slope of this line is –4 or $-\frac{4}{1}$.

### Two Points $(x, y)$ and $(x_1, y_1)$

**Example H**

Use the formula

$\frac{y_1 - y}{x_1 - x}$ = Slope

(1, 0) and (5, 3)
$(x, y)$ $(x_1, y_1)$

$\frac{3-0}{5-1} = \frac{3}{4}$

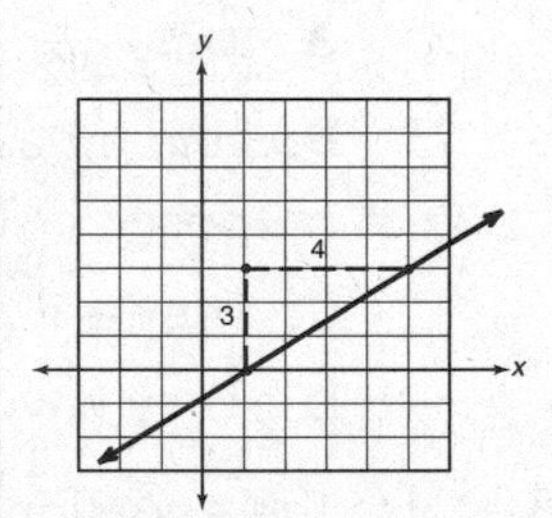


or

(5)(3) and (1)(0)
$(x, y)$ $(x_1, y_1)$

$\frac{0-3}{1-5} = \frac{-3}{-4} = \frac{3}{4}$

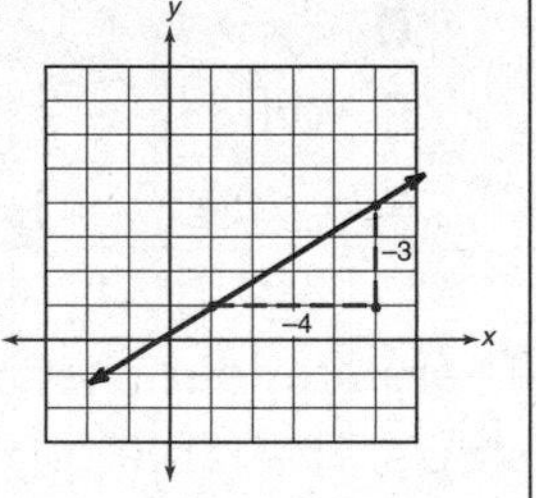


**Remember*: The slope-intercept form of an equation is $y = mx + b$. The $m$ is the slope of the line, and the $b$ is the point where the line intersects the $y$-axis.

## Sample Example:

Of the linear functions represented below, which has the greatest rate of change?

**A.** ------ a number, $y$ is four less than three times a number $x$.

**B.** ------

| $x$ | $f(x)$ |
|---|---|
| 6 | 12 |
| 4 | 8 |
| –2 | –4 |
| –4 | –8 |

**C.** ------

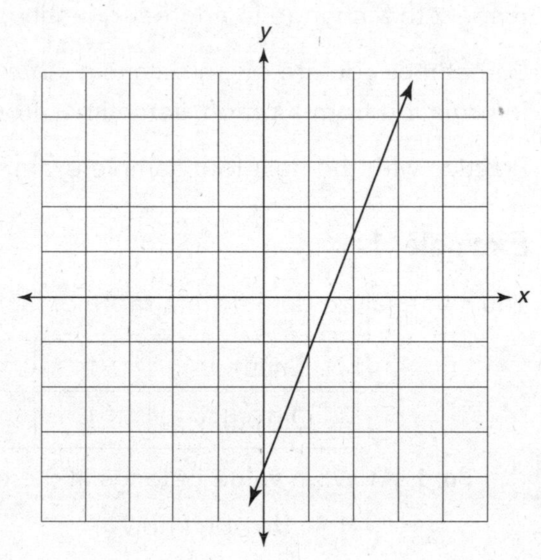


**D.** ------ a line that goes through (3, 0) and (0, –1)

**E.** ------ the line represented by $y - \frac{1}{3}x = -8$

### Solutions

In **A** the *rate of change* (the slope) is **3**. $y = \mathbf{3}x - 4$

In **B** the *rate of change* (the slope) is **2**. $\frac{12-8}{6-4} = \frac{4}{2} = 2$

In **C** the *rate of change* is $\frac{8}{3}$ which is $2\frac{2}{3}$ $\frac{4-(-4)}{3-0} = \frac{8}{3} = 2\frac{2}{3}$

In **D** the *rate of change* is $\frac{1}{3}$. Use the slope formula: $\frac{-1-0}{0-3} = \frac{-1}{-3} = \frac{1}{3}$

In **E** *the rate of change is also* $\frac{1}{3}$ $y = \frac{1}{3}x - 8$

## SLOPE AND RATE OF CHANGE

As discussed on pages 162-163, another way to think about *slope* is to think about *rate of change. Rate of change* is a *ratio* that describes how one quantity changes with respect to a change in another quantity.

Sometimes you are given data in a *table format*, sometimes you can find out the information from a *graph*, sometimes from an *equation* or from a *word problem*.

Practice with the next four sample examples.

### Example 1:

You can analyze data from a **table** of facts.

| Input $x$ | 1 | 4 | 9 | 16 | 25 | 36 |
|---|---|---|---|---|---|---|
| Output $y$ | 1 | 2 | 3 | 4 | 5 | |

**Part A:** What value belongs in the empty cell above? ______________

How do you know?

**Part B:** Does this show a positive or negative correlation between the input and output?

**Solutions**

**Part A:** The number 6.

I know because each $y$ is the square root of the $x$ value and $\sqrt{36} = 6$.

**Part B:** It shows a positive correlation because as the input ($x$) increases, the output ($y$) also increases.

## Example 2:

You also can analyze data from a **graph**.

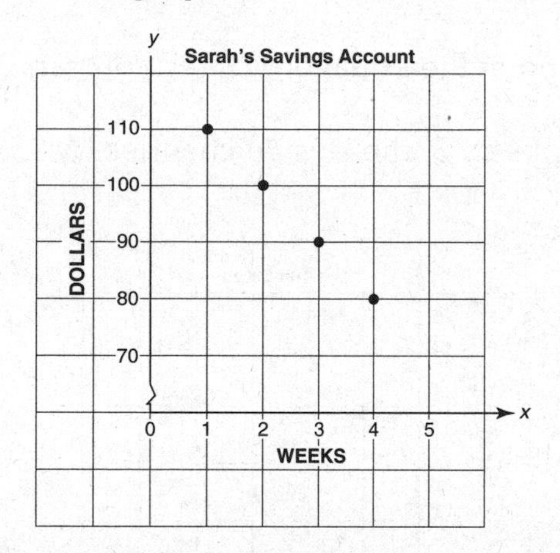


This graph gives you information about Sarah and her savings account.

**Part A:** Did Sarah add or withdraw money from her account each week?

------------------------------------------------------------

**Part B:** How much money would you expect to be in Sarah's account at week 5 if she continues with this same pattern? There would be ________________ in her account.

**Part C:** Does this graph show a positive or negative correlation between input and output? It shows a ________________ correlation because ________________________________________.

**Solutions**

**Part A:** Sarah withdrew money each week.

**Part B:** There would be $70 in her savings account at week 5.

**Part C:** This graph shows a *negative* correlation because as the weeks *increase*, the amount of money in Sarah's savings account *decreases*.

## Example 3:

You also can analyze the *rate of change* from an **equation.**

**Part A:** Just by looking at the equation $y = \frac{3}{2}x$ you can tell that each time the *y-value changes by 3,* the *x-value* changes by ________________ .

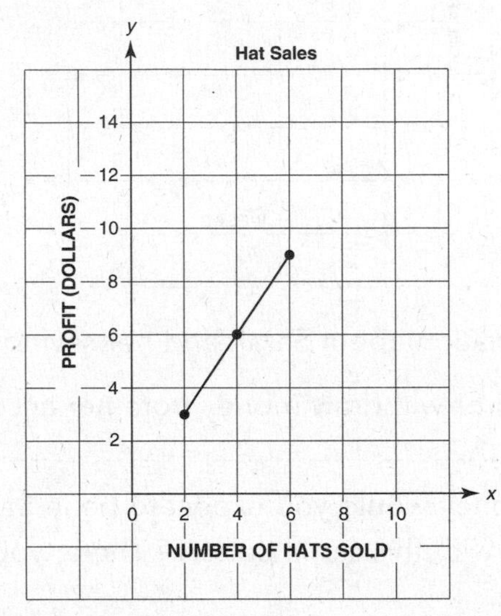


**Part B:** If the *numerator* of the slope (the change in *y*) represents the amount of profit, and the *denominator* (the change in *x*) represents the number of hats sold. How much profit should the store owner expect if she sold 8 hats? $ ___________

**Part C:** Does this slope show a *positive* or *negative* rate of change? ___________

**Part D:** What does the coordinate (0, 0) represent in this scenario?

______________________________________________________________________________ .

**Solutions**

**Part A:** Each time the *y-value* changes by a factor of **3** the *x-value* changes by a factor of **2**.

**Part B:** If the store owner sold **8** hats she should expect a profit of **$12.00.** The coordinate is (8,12). $\frac{\$3}{2\,hats} = \frac{\$x}{8\,hats} = \frac{3}{2} = \frac{x}{8}$ or $(2x) = (3)(8)$ $2x = 24$, $x = 12$

or think that for $x$ to change from 2 to 8 it was multiplied by **4**. Therefore, $y$ also should be multiplied by 4; $3 \times \mathbf{4} = 12$.

**Part C:** The (0, 0) coordinate means that if the store-owner sells no hats she makes no profit.

### Example 4:

Another way is to recognize *rate of change* in an everyday **word problem.**

**Part A:** A special pancake mix says that for every 2 cups of milk you should use 1.5 cups of flour. If the chef at the local diner uses 6 cups of milk, how many cups of flour should he add?

He should add __________________ cups of flour.

**Part B:** Each time the chef adds more milk he also should __________________ more flour. This is a __________________ rate of change. As the input increases the __________________ increases, too.

**Solutions**

**Part A:** 2 cups of milk was multiplied by **3** to = 6 cups of milk; therefore, 1.5 cups of flour should also be multiplied by **3**.

2**(3)** = 6 cups of milk ... and ... 1.5**(3)** = 4.5 cups of flour.

$$\frac{2\,milk}{1.5\,flour} = \frac{6\,milk}{?\,flour} \quad 2(?) = (6)(1.5) \quad 2? = 9 \quad ? = \frac{9}{2} = 4.5\,cups\ of\ flour$$

**Part B:** Each time the chef adds more milk he also should **add** more flour. This is a **positive** rate of change. As the input increases the **output** increases, too.

## EXTRA PRACTICE: Slope

(For answers, see pages 260–261.)

1. Look at the graph and determine the slope of the line.

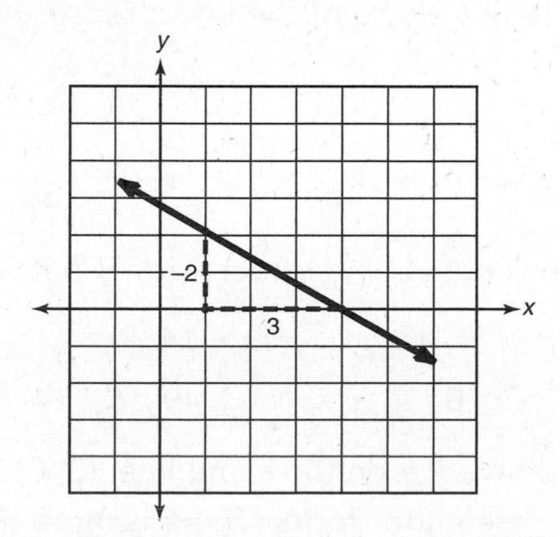


○ **A.** $\frac{2}{3}$ ○ **B.** $\frac{-2}{3}$ ○ **C.** $\frac{3}{2}$ ○ **D.** $\frac{-3}{2}$

**2.** Look at the graph and determine the slope of the line.

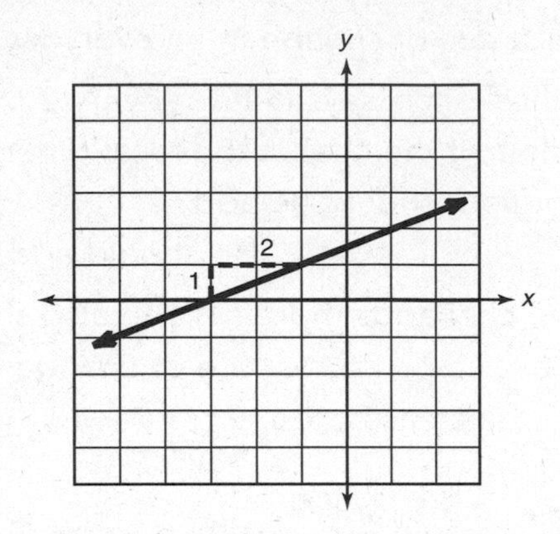


○ **A.** $\frac{1}{2}$ ○ **B.** $\frac{2}{1}$ ○ **C.** $\frac{-1}{2}$ ○ **D.** $\frac{-2}{1}$

**3.** Look at the following linear equation. It already is in the required form. What is the slope of the line represented by this equation?

$$y = \frac{2}{3}x + 5$$

○ **A.** 2 ○ **B.** 3 ○ **C.** $\frac{2}{3}$ ○ **D.** 5

**4.** Look at the following linear equation. It is not yet in the correct form. Put it in the correct form; then determine the slope of the line represented by this equation.

$$2y = 6x + 10$$

○ **A.** 6 ○ **B.** 3 ○ **C.** 10 ○ **D.** 5

**5.** Here is another linear equation that is not yet in correct form. Put it in correct form; then determine the slope of the line it represents.

$$y - 6 = 4x$$

○ **A.** –6 ○ **B.** 6 ○ **C.** $\frac{-2}{3}$ ○ **D.** 4

**6.** This linear equation is not yet in correct form. What is the slope of this line?

$$y - 2x = 14$$

○ **A.** –2 ○ **B.** 2 ○ **C.** 14 ○ **D.** 7

**7.** The following two points are on the same line: (3, 0) and (2, 4). Which of the following could represent the slope. (There is more than one correct answer.)

☐ **A.** $\frac{4}{-1}$ ☐ **B.** $\frac{-4}{1}$ ☐ **C.** 4 ☐ **D.** –4

☐ **E.** 2 ☐ **F.** $\frac{2}{1}$ ☐ **G.** $\frac{-1}{2}$ ☐ **H.** $\frac{-1}{-2}$

**8.** You are told that the slope of a line is $\frac{-1}{3}$. Which of the following pairs of coordinate points are on this line? (There are two correct answers.)

☐ **A.** (2, 4) (–5, –3)

☐ **B.** (2, 4) (5, 3)

☐ **C.** (–2, –4) (5, 3)

☐ **D.** (–4, 4) (5, 1)

☐ **E.** (5, 3) (–2, –4)

**9.** In both tables there is a rule that determines the output in each situation. In Table A, the rule is $2x + 1$ (take the input number, multiply it by 2 and add 1). In Table B, the rule is $x^2$ (take the input number and square it).

| Table A<br>Shows Linear Growth | |
|---|---|
| *Input* | *Output* |
| $x$ | $2x + 1$ |
| 0 | 1 |
| 1 | 3 |
| 2 | 5 |
| 3 | 7 |
| 4 | ? |
| 5 | 11 |
| 6 | 13 |
| 7 | ? |
| ↓ | ↓ |
| 100 | 201 |

| Table B<br>Shows Exponential Growth | |
|---|---|
| *Input* | *Output* |
| $x$ | $x^2$ |
| 0 | 0 |
| 1 | 2 |
| 2 | 4 |
| 3 | ? |
| 4 | 16 |
| 5 | ? |
| 6 | 36 |
| 7 | ? |
| ↓ | ↓ |
| 100 | 10,000 |

Let's say these tables represent the way two different banks determine the interest you would receive on money deposited at their bank.

**Part A:** Use the rule for each table and fill in the missing numbers (?) in the shaded cells.

**Part B:** According to Table A, how much money would you receive if you deposited $50?

**Part C:** According to Table B, how much money would you receive if you deposited $50?

**Part D:** If you deposit $50 in each bank explain why you receive so much more interest in one bank than in the other?

**10.** Use the equation $\mathbf{y = 3x + 4}$ to determine if the following statements are correct. Check Yes --✔--- or No --✔---.

**A.** This equation represents a curved line. Yes ------ No ------

**B.** The $y$-intercept is at (0, 4). Yes ------ No ------

**C.** The slope is negative. Yes ------ No ------

**D.** $3y = 9x + 12$ is an equivalent equation. Yes ------ No ------

**E.** When $x = 2$, then $y = 10$. Yes ------ No ------

**F.** $y - 3x = 6$ is a line parallel to $y = 3x + 4$. Yes ------ No ------

**11.** If a line contains the points on the table below, which equation below does **NOT** represent this line?

| $x$ | $y$ |
|---|---|
| –2 | –10 |
| 0 | –4 |
| 2 | 2 |
| 4 | 8 |

○ **A.** $y = 3x - 4$

○ **B.** $y = 3x - 4$

○ **C.** $2y - 6x = -8$

○ **D.** $f(x) = 3(x - 2) + 2$

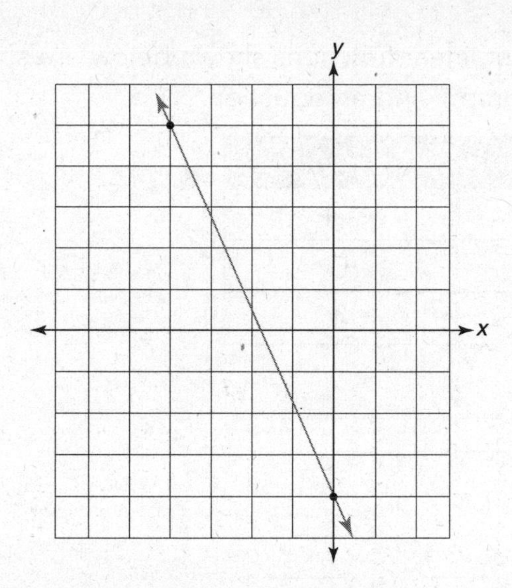


**12.** If a line passes through the two points on the graph above, what is the equation of that line?

○ **A.** $y=\frac{-9}{4}x-4$

○ **B.** $y=\frac{1}{4}x+b$

○ **C.** $y=\frac{9}{4}x-4$

○ **D.** $y=\frac{-4}{9}x+4$

**13.** The three different linear functions shown below are shown in three different ways: a table, a graph, and an equation.

**A.**

| $x$ | $f(x)$ |
|---|---|
| 2 | –4 |
| 0 | 0 |
| –2 | 4 |

**B.**

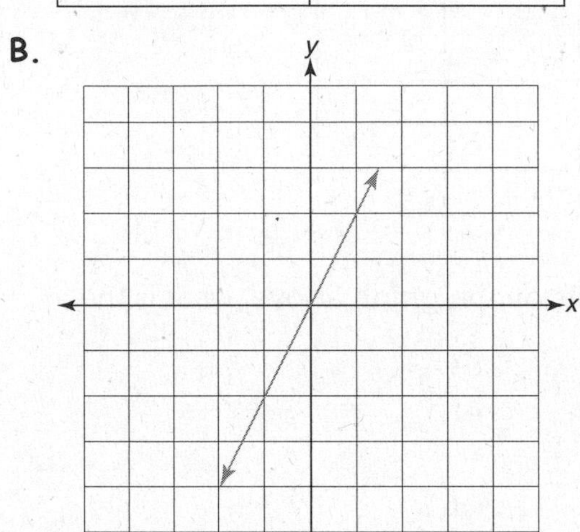


**C.** $3y + 6 = 2x$

**Part A:** Which function has the *smallest* rate of change?

**Part B:** Do any pair of functions have the same rate of change?

**Part C:** Justify your answer.

**14.** What is the relationship between the $x$ and $y$ in the graph below?

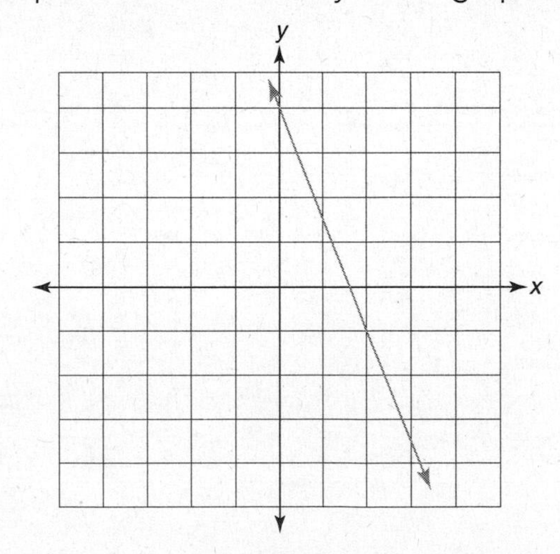


○ **A.** $y = \frac{-5}{2}x + 1.5$

○ **B.** $y = \frac{-5}{2}x + 4$

○ **C.** $y = \frac{-2}{5}x - 4$

○ **D.** $y = \frac{-3}{2}x - 4$

# SYSTEMS OF EQUATIONS

*A system of equations* is a set of equations that you deal with all together at once. We will work with pairs of linear equations (ones that graph as straight lines) to see if they intersect (if they meet) or if they don't intersect.

We will look at two possibilities:

1. **Parallel Lines:** If two lines (on the same plane) have the same slope, they are parallel lines and will never intersect (they will never meet). Their **slopes are the same** but they intersect the *y*- or *x*-axis at different points.

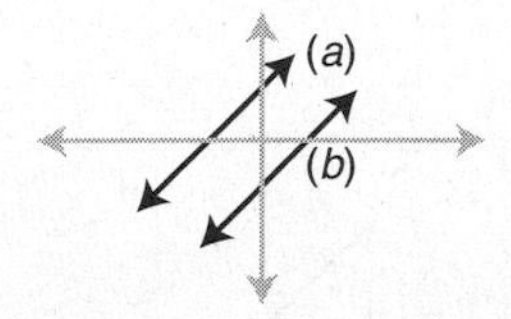


The equation of line *(a)* might be $y = \mathbf{2}x + 2$ — line *(c)* might be $y = 4$
The equation of line *(b)* might be $y = \mathbf{2}x - 1.5$ — line *(d)* might be $y = -1$

2. **Intersecting Lines:** If two lines (on the same plane) have **different slopes**, they will intersect somewhere; they will meet at one point. Finding this point can be very important when solving real-life situations.

Below is an example of a *system of equations* showing two *intersecting lines*.

The equation for line *(e)* is $y = x$

The equation for line *(f)* is $y = -3x + 4$

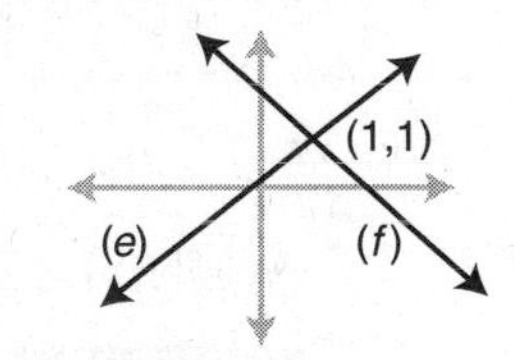


These lines ***intersect*** at the point (1,1); they *meet* at the point (1,1).

Look at the two graphed lines to the right.

*Line a:* $\mathbf{y} = \mathbf{x} - \boldsymbol{l}$ and
*line b:* $\mathbf{y} = -\,\mathbf{x} + \mathbf{3}$

The two equations above are in a system; we consider them together at the same time.

We graph them together on the same *x*/*y*-axis.

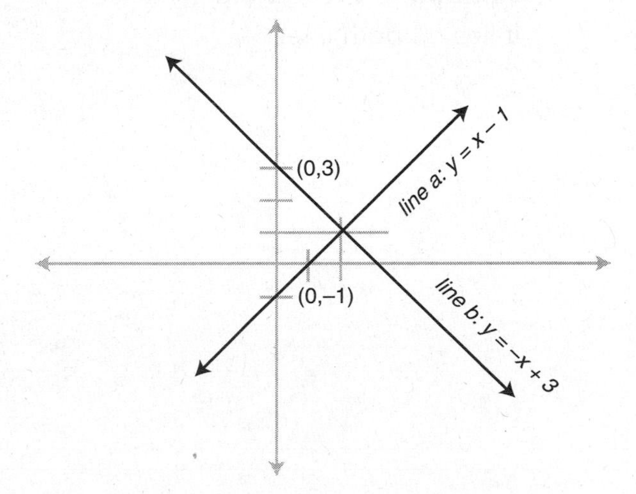

Notice that the point (–**2,3**) does not lie on either line. It is **not** a solution to either equation.

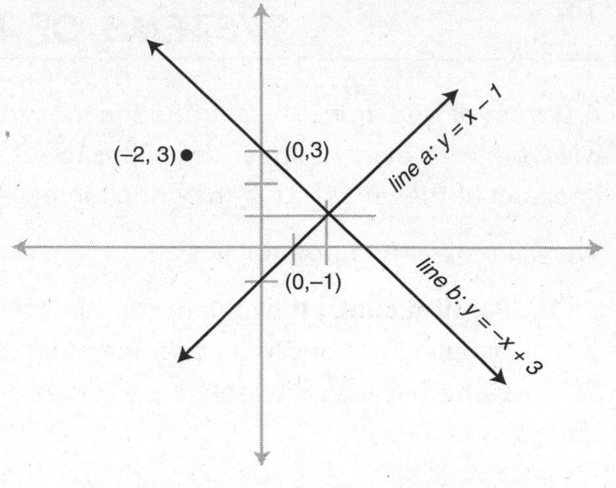


Notice that the point (**0,3**) lies on *line b*, but not on *line a*. It is **not** a solution to the system of equations.

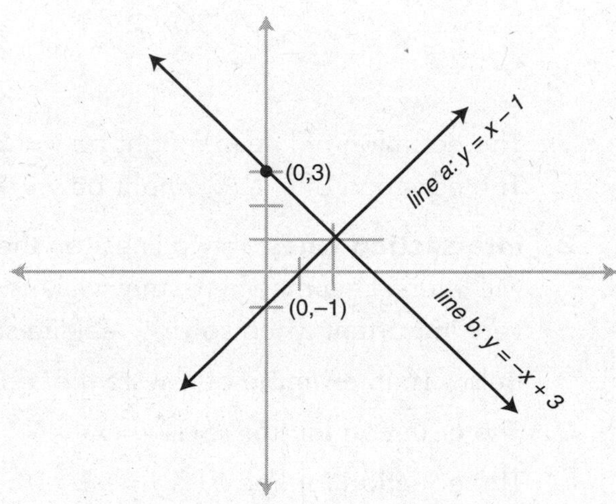


Finally, the point (**2,1**) **is a solution to the system**, because it lies on both lines.

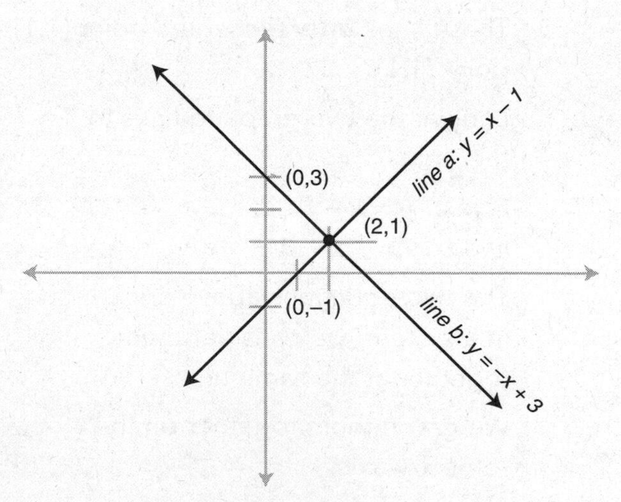

## PRACTICE: Systems of Equations

(For answers, see pages 262–263.)

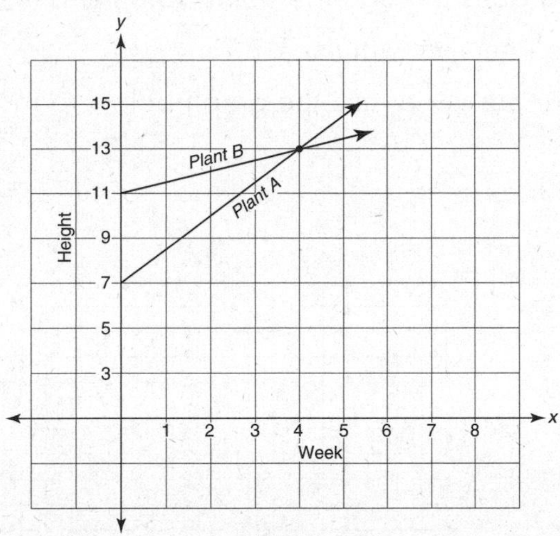


**1.** The graph above shows how two different plants grew over a number of weeks.

**Part A:** Write an equation to represent the rate of Plant A.

Let "*h*" = height, and "*w*" = weeks ____________________

**Part B:** Write an equation to represent the rate of Plant B.

Let "*h*" = height, and "*w*" = weeks ____________________

**Part C:** When will the two plants be the same height?

They will be the same height at week ____________________

**2.** You are given two lines.

One line, *line p* goes through the points (0, 1) and (4, 5).

The other line, *line q* goes through the points (2, 3) and (4, 0).

**Part A:** Does *line p* intersect with *line q*? Yes ______ No ______

**Part B:** Explain your answer or use the graph below to model your answer. (Show your work below.)

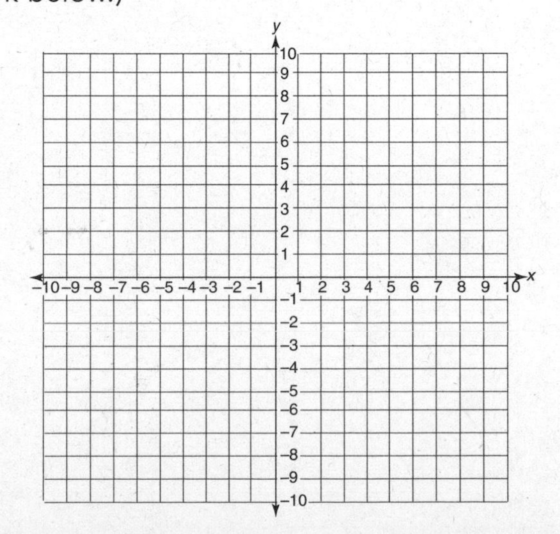

**3.** You are given two lines.

One line, *line p* goes through the points (4, 8) and (2, 4).

The other line, *line q* goes through the points (5, 7) and (4, 5).

**Part A:** Does *line p* intersect with *line q*?

**Part B:** Explain your answer or use the graph below to model your answer. (Show all work.)

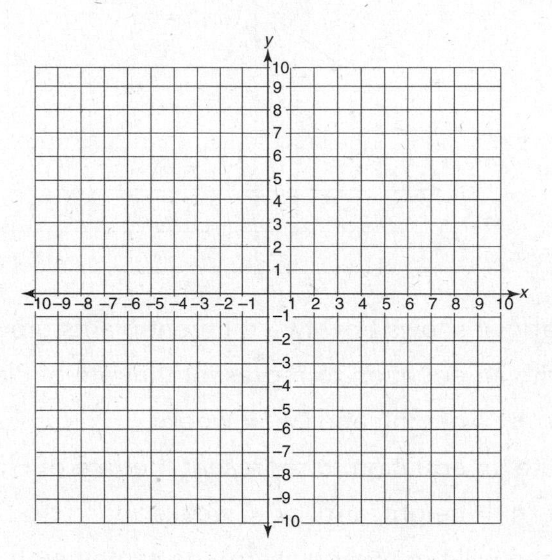

## COMBINING ALGEBRA AND GEOMETRY

### Examples

1. The perimeter of the rectangle drawn is 42 centimeters.

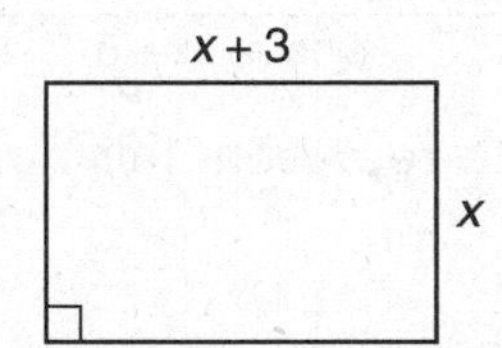


**Part A:** Write an equation to represent this perimeter.

**Part B:** Find the value of $x$.

**Part C:** Find the length and width of the rectangle.

**Part D:** Find the area of the rectangle.

| | Show All Work | Reminders |
|---|---|---|
| Equation | Perimeter = $x + 3 + x + 3 + x + x = 42$ or Perimeter = $2(x + 3) + 2(x) = 42$ | For perimeter, add all sides. |
| Solve for $x$ | $4x + 6 = 42$<br>$-6 \quad -6$<br>$4x = 36$<br>$x = 9$ | Combine like terms.<br>Undo addition.<br>Undo multiplication. |
| Length of each side | Side a = $x + 3 = 9 + 3 = 12$ cm<br>Side b = $x = 9$ cm | Substitute the value for $x$ and solve each equation. |
| Area of rectangle | Area rectangle = (12)(9) = 108 sq. cm | Area = (Side)(Side) |

**2.** The perimeter of the triangle drawn is 80 inches.

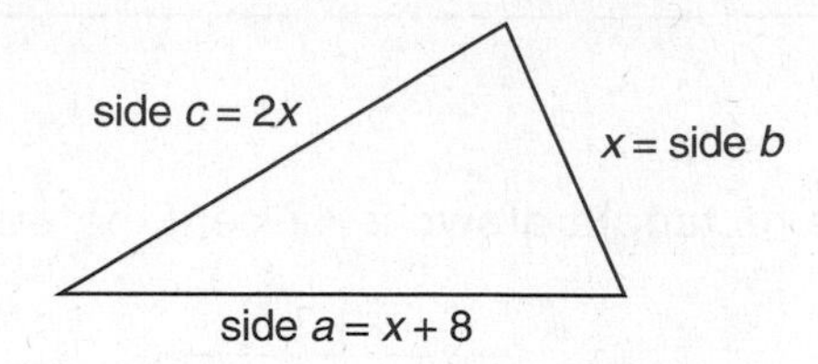


**Part A:** Write an equation to represent this perimeter.

**Part B:** Find the value of $x$.

**Part C:** Find the value of each side.

**Part D:** Check your work.

| | | |
|---|---|---|
| Equation | $2x + x + 8 + x =$ 80 inches | |
| Solve for x | $4x + 8 = 80$<br>$4x = 72$<br>$x = 18$ | |
| Length of each side | Side $a$ = 26 inches<br><br>Side $b$ = 18 inches<br>Side $c$ = 36 inches | side $a = x + 8 = 18 + 8 = 26$ inches<br>side $b = x = 18$ inches<br>side $c = 2x = 2(18) = 36$ inches |
| Check by adding sides | $26 + 18 + 36 = 80$ | Correct, the perimeter is 80! |

**3.** The area of the rectangle drawn is 36 sq. units

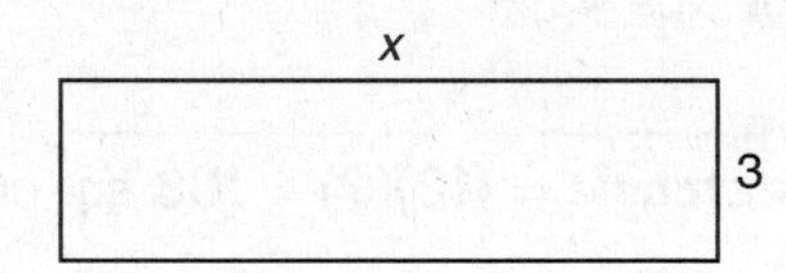


**Part A:** Write an equation to represent the perimeter of this rectangle.

**Part B:** Find the value of $x$.

**Part C:** Find the value of each side.

**Part D:** Check your work.

| | | |
|---|---|---|
| Equation | $3x = 36$ | Area = Length × Width; $(3)(x)$ = Area |
| Solve for $x$ | $x = 12$ | $3x = 36$ (undo multiplication), $x = 12$ |
| Length of each side | Length = $x$ = 12 units<br>Width = 3 units | Top = 12 units long, Bottom = 12 units long<br>Each side = 3 units long |
| Perimeter of rectangle | 30 units | Side + Side + Side + Side = 12 + 12 + 3 + 3 |

## PRACTICE: Combining Algebra and Geometry

(For answers, see pages 263–265.)

The following are multi-step problems. Show all steps and label and circle each of your answers. Each bullet requires an answer. Work on a separate sheet of paper.

**1.** The perimeter of the rectangle drawn is 22 inches.

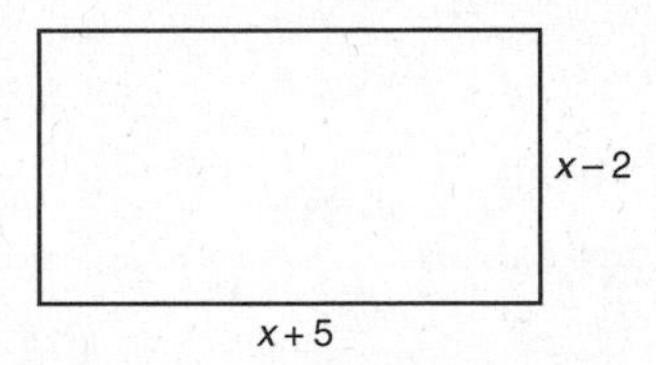


**Part A:** Write an equation to represent this perimeter.

**Part B:** Find the value of $x$.

**Part C:** Find the value of each side.

**Part D:** Check your work.

**2.** The perimeter of the isosceles triangle drawn is 100 feet.

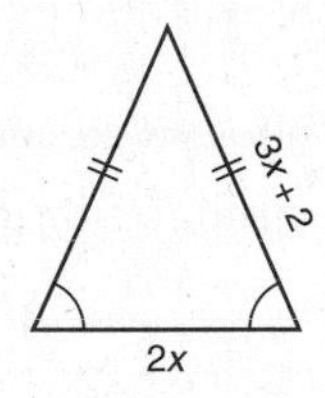


**Part A:** Write an equation to represent this perimeter.

**Part B:** Find the value of $x$.

**Part C:** Find the value of each side.

**Part D:** Check your work.

**3.** The area of the rectangle drawn is 18 square yards.

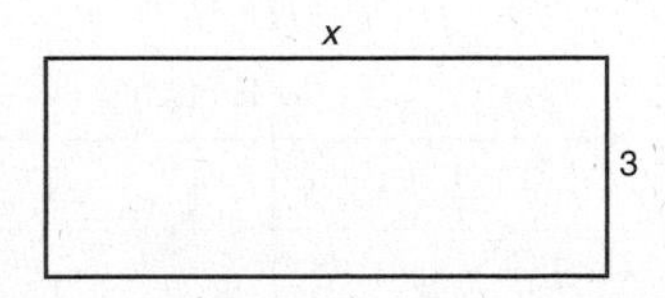


**Part A:** Find the value of $x$.

**Part B:** Write an equation to represent the perimeter of this rectangle.

**Part C:** Find the perimeter of this rectangle.

**Part D:** Check your work.

4. The perimeter of this equilateral triangle is 108 meters.

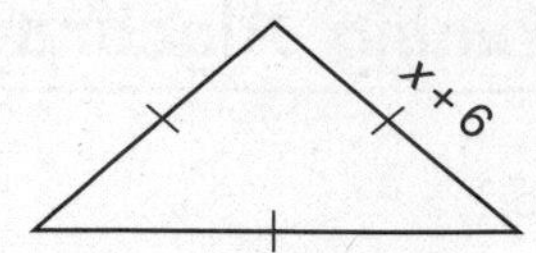


**Part A:** Write an equation to represent the perimeter of this equilateral triangle.

**Part B:** Find the value of $x$.

**Part C:** Find the length of each side.

**Part D:** Check your work.

5. The triangle below shows that
$\angle A = x°$, $\angle B = 2x°$, $\angle C = 3x°$.

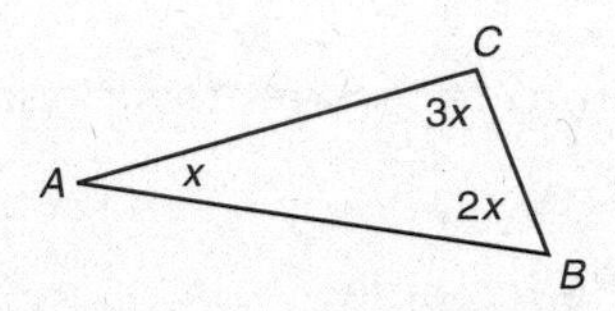


**Part A:** Write an equation to represent the total of these three angles.

**Part B:** Find the value of $x$.

**Part C:** Find the value of $\angle B$.

**Part D:** Find the value of $\angle C$.

6. In the figure below you see two similar triangles, $\Delta EAB$ and $\Delta CED$.
What is the value of $x$?

Answer: ________________

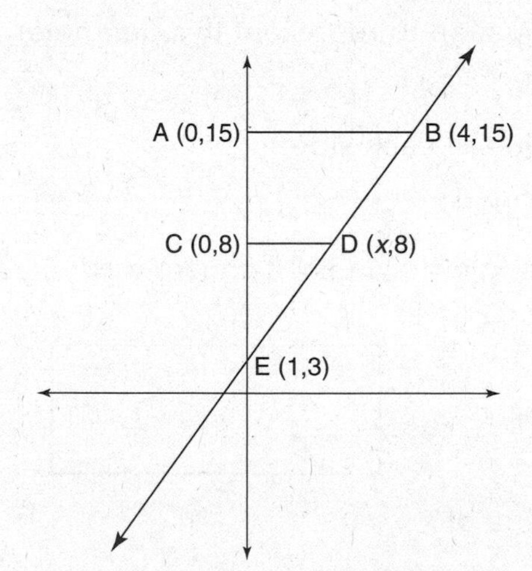

**7.** Stephanie's mom has a square garden. She wants to change the shape of her garden and make it a rectangle with a length that is 2 feet longer than three times its width. Her mom has 100 feet of fencing, so she decides that the perimeter of the new garden should be 100 feet.

What are the dimensions in feet, of her new garden?

**8.** Refer to the right triangle drawn on the coordinate grid below.

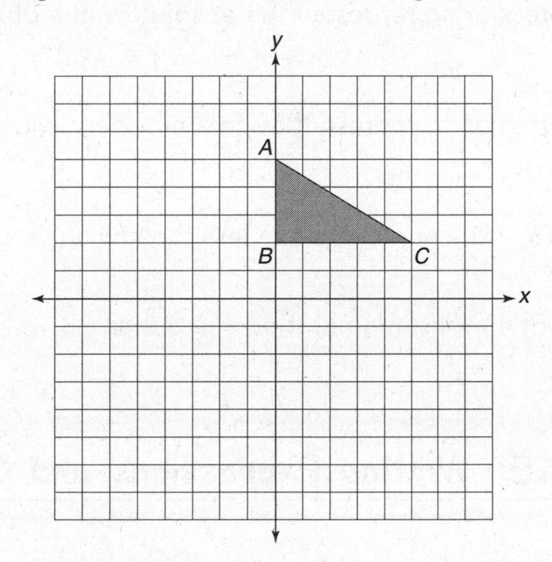


**A.** The length of line segment $AC = 2$. Yes ------ No ------

**B.** The height of the triangle $ABC = 3$ units long. Yes ------ No ------

**C.** You can use the Pythagorean Theorem to find the length of $AC$. Yes ------ No ------

**D.** The area of the $\Delta ABC$ = base × height. Yes ------ No ------

**E.** If $\Delta ABC$ is reflected over the $x$ axis, the coordinates of the new point B′ would be (0, –2). Yes ------ No ------

**F.** The perimeter of this triangle is 12 units. Yes ------ No ------

## WRITING EXPRESSIONS AND EQUATIONS

### Examples

**A.** Write an expression to represent the product of a number and 6. — $(6)(n)$ or $6n$

**B.** Write an expression to represent the absolute value of a number. — $|n|$

**C.** Write an expression to represent twice a number divided by five. — $\frac{2n}{5}$

**D.** The sum of 16 and a number is five less than twice that number. — $16 + n = 2n - 5$

**E.** The product of 4 and some number is less than 25. — 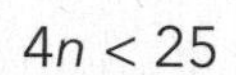 $4n < 25$

## PRACTICE: Writing Expressions and Equations

(For answers, see page 265.)

**1.** Which expression represents 6 less than the product of 4 and some number?

○ **A.** $6 - 4 + n$

○ **B.** $4n - 6$

○ **C.** $6 - 4n$

○ **D.** $4 - 6n$

**2.** Which of the following says that the square root of a number is greater than 4 less than the number?

○ **A.** $\sqrt{n} > n - 4$

○ **B.** $\sqrt{n} < n - 4$

○ **C.** $4 - n \geq \sqrt{n}$

○ **D.** $\sqrt{4 - n} \leq \sqrt{n}$

**3.** Which of the following could NOT represent the perimeter of a square?

☐ **A.** $x + x + x + x$

☐ **B.** $4x$

☐ **C.** $2.5x + 2.5x + x + x$

☐ **D.** $4(3x)$

☐ **E.** $x^2$

☐ **F.** $3.4 + 3.4 + 2(3.4)$

**4.** Which of the following represents that a number squared is equal to 20 more than 9 times that number.

○ **A.** $9n + 20$

○ **B.** $x^2 = (20)(x) + 9$

○ **C.** $x^2 > 9 + 20x$

○ **D.** $x^2 = 9 + x + 20$

## REAL-LIFE APPLICATIONS

Very often it is easiest to solve some basic, everyday situations using algebraic equations and special formulas.

### Example

If Christine put $2,000 into a savings account that received 3% interest *compounded annually*, how much money would she have in that account after 5 years?

The formula to determine compound interest is

$$A = P(1 + r)^t$$

If you do not remember this formula, use the *PARCC Grade 8 Mathematics Assessment Reference Sheet* on page 359. $A$ is the amount she would have after five years, $P$ is the principal amount in the account now, $r$ is the rate (or the percent) of interest she would receive, and $t$ is the time (in this case, $t = 5$ years).

$$A = 2{,}000\ (1 + 0.03)^5 = 2{,}000\ (1.03)^5$$

$$A = 2{,}000\ (1.159274) = 2318.548 \text{ or } \$2{,}318.55$$

At the end of five years, Christine would have $2,318.55 in her savings account.

## PRACTICE: Real-Life Applications

(For answers, see pages 265–266.)

**1.** The freshman class just collected money from their major fund-raiser for the year. They have $5,300 and plan to put this money into a savings account for 3 years; the bank gives 2.9% interest annually.

How much money can they expect to have at the end of the 4 years? (Use the formula $A = P\ (1 + r)^t$.)

○ **A.** $5,329

○ **B.** $5,453.70

○ **C.** $5,774.60

○ **D.** $5,940.07

**2.** This figure shows how a playground is fenced off. Use the figure of the rectangle drawn to answer each question below. The fencing comes in 1-foot lengths.

40'

32'

**Part A:** Could you use the same fencing to fence off a square playground with no fencing left over?

Yes ------ No ------

**Part B:** If you fenced off the new square playground and used all the fencing, what would be the length of *each side* of the new square playground?

------ feet

**Part C:** Is the *perimeter* of the rectangular playground the same as the *perimeter* of the new square playground?

Yes ------ No ------

**Part D:** Is the *area* of the newer square playground larger or smaller than the original rectangular playground?

Larger ------ Smaller ------

**Part E:** What is the difference between the *area* of the rectangular playground and the *area* of the square playground?

The difference is ------ sq. ft.

## PRACTICE: SCR Non-calculator Questions

(For answers, see page 266.)

Write your answers in simplest form.

**1.** Simplify the expression $(-1.5) + (-3.7)$ Answer: ----------------

**2.** Simplify the expression $12x - 2a - 4a - 3x$ Answer: ----------------

**3.** Simplify the expression $10(x + 3) - 3(-x + 2)$

**4.** Distribute the –2.5 $-2.5(x - 10)$ Answer: ----------------

**5.** Multiply. $-3\left(-\frac{1}{3}\right)\left(\frac{2}{3}\right)$ Answer: ----------------

**6.** Divide and simplify. $\dfrac{-6}{\frac{1}{2}}$ Answer: ________________

**7.** Simplify. $\dfrac{9}{3(x+6)}$ Answer: ________________

**8.** Distribute the $\dfrac{-2}{5}$ $\dfrac{-2}{5}\left(25-\dfrac{3}{5}\right)$ Answer: ________________

**9.** Evaluate the expression if $x = -2$; $3x + 8x + 4 - 1 - 11x$ Answer: ________________

**10.** Evaluate the expression when $x=\dfrac{2}{3}$; $6\left(x-\dfrac{1}{2}\right)$ Answer: ________________

**11.** What is the slope of a line parallel to $\dfrac{2}{3}x+y=15$? Answer: ________________

**12.** What is the slope of the line represented by this equation?

$y - 3x = 16 + 2x$ Answer: ________________

**13.** Use the equation $y=\dfrac{3}{4}x-10$ to answer **Parts A, B, and C** below.

**Part A:** The slope of a *perpendicular* line is $\dfrac{-3}{4}$. Yes ______ No ______

**Part B:** This line is rising from left to right. Yes ______ No ______

**Part C:** It is parallel to the line $y=\dfrac{3}{4}x+2$. Yes ______ No ______

**14.** What values of *a* will make this equation true?

$a^2 = 49$ Answer: $a$ = ________________

**15.** The two sides of the rectangle are given as $2x+\dfrac{1}{4}$ and $x+\dfrac{1}{8}$.

What is the perimeter of this rectangle? Answer: ________________

## PRACTICE: Performance-Based Assessment Questions

(For answers, see pages 267–268.)

1. Use the information provided on the table below.

**Part A:** Make a double line graph on the grid provided. One line should represent the electric bills; the other should show the gas bills. Label each axis.

**Part B:** What month were the electric and gas bills approximately the same?

**Part C:** What is the difference between the month with the highest total bills and the month with the lowest total bills? Show and label your work.

**Part D:** Mr. and Mrs. B. use gas for heating their home and electric for air conditioning. Use this information to explain the major changes in their gas and electric bills throughout the year shown.

| Average Monthly Energy Bills of Mr. and Mrs. B's house in central New Jersey | | |
|---|---|---|
| Month | Electric | Gas |
| December | 180.00 | 415.00 |
| January | 140.00 | 330.00 |
| February | 150.00 | 400.00 |
| March | 130.00 | 270.00 |
| April | 140.00 | 110.00 |
| May | 330.00 | 75.00 |
| June | 450.00 | 30.00 |
| July | 300.00 | 20.00 |
| August | 300.00 | 20.00 |
| September | 180.00 | 30.00 |
| October | 130.00 | 80.00 |
| November | 110.00 | 280.00 |

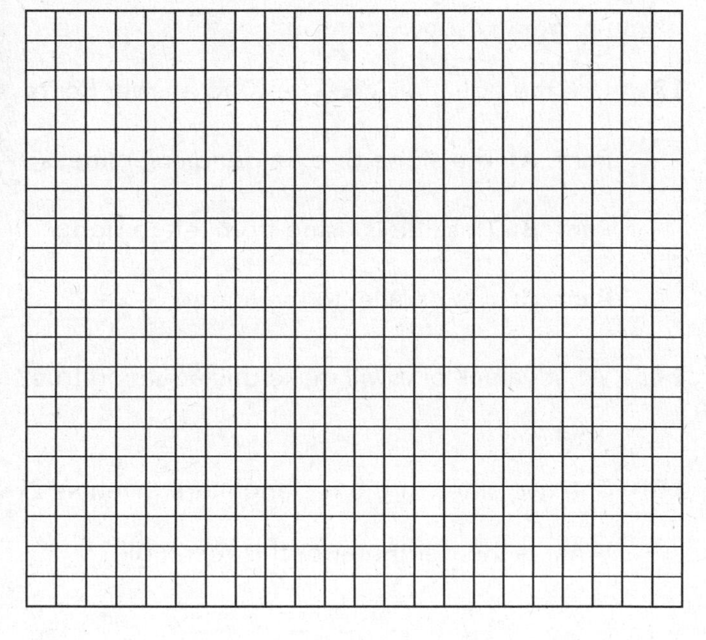

**2.** Nicholas and Miguel solved the same equation. One is correct, and one is incorrect.

| Nicholas | Miguel |
|---|---|
| $2x - 3(4 + 5x) = 2x + 18$ | $2x - 3(4 + 5x) = 2x + 18$ |
| $2x - 12 - 15x = 2x + 18$ | $2x - 12 + 15x = 2x + 18$ |
| $-12 - 13x = 2x + 18$ | $+ 12 \qquad + 12$ |
| $+ 13x \qquad +13x$ | $2x \quad + 15x = 2x + 30$ |
| $-12 = 15x + 18$ | $17x = 2x + 30$ |
| $-18 \qquad -18$ | $-2x \qquad -2x$ |
| $-30 = 15x$ | $15x = 30$ |
| $-2 = x$ | $x = 2$ |

**Part A:** Which student has the correct answer?

**Part B:** Replace this correct value for $x$ in the following equation to show that it is the correct solution: $2x - 3(4 + 5x) = 2x + 18$

**Part C:** How would you explain to the student who was incorrect what he did wrong?

**3.** If $x$ = positive, and $y$ = negative will the sum of $x$ and $y$ be positive, negative or zero? Explain your answer.

**4.** We use the equation $A = P(1 + r)^t$ when calculating compound interest.

$A$ represents the accumulated value of the investment.

$P$ represents the principal (the amount originally invested).

$r$ represents the rate.

$t$ represents the time (in years).

If Tim and his wife, MaryJo, saved $8,000 at 4% compounded annually, what would be the value of their savings after 1 year? After 5 years? Set up the equations and show your work.

**5.** **Part A:** Write an expression to represent three consecutive integers if $x$ is the first integer.

**Part B:** If the sum of the three consecutive integers is 1,356, set up an equation and find the value of each integer.

Name: ............................................ Date: ............................................

## Chapter 3 Test: Expressions, Equations, and Functions

35 minutes

(Use the *PARCC Grade 8 Mathematics Assessment Reference Sheet* on page 359.)

(For answers, see pages 269–270.)

**1.** In the following equation what is the first step in isolating the variable?

$$-8x - 34 = 14$$

- ○ **A.** Subtract 34 from both sides
- ○ **B.** Divide both sides by –8
- ○ **C.** Add 34 to both sides
- ○ **D.** Multiply both sides by 8

**2.** Which equation does *not* have a solution of 12?

- ○ **A.** $4a + 3 = 51$
- ○ **B.** $14 - 2b = -10$
- ○ **C.** $3(x - 5) = 21$
- ○ **D.** $3(6 + y) = 30$

**3.** Judy and Janet took their 3 young children to the movies in nearby Franklin Township. An adult ticket costs 3 times more than a child's ticket. If they paid a total of $27.00, what is the cost of each adult ticket?

- ○ **A.** $3.00
- ○ **B.** $4.00
- ○ **C.** $6.00
- ○ **D.** $9.00

**4.** Which equation matches the following: 6 less than some number is 5 squared.

- ○ **A.** $5^2 = n - 6$
- ○ **B.** $n + 6 = 5^2$
- ○ **C.** $6 - n = 5^2$
- ○ **D.** $5^2 - 6 = n$

**5.** Refer to this equation:

$$y = \frac{-2}{5}x + 10$$

○ **A.** When you graph this line, it will have a very steep slope (like a tall mountain).

○ **B.** When graphed, this line will slant to the right.

○ **C.** When graphed, this line will slant to the left.

○ **D.** When graphed, this line will be parallel to the line $y = \frac{5}{2}x + 10$.

**6.** Write the following equation in slope-intercept form:

$$3x - 9y = 12$$

○ **A.** $3x = 9y + 12$

○ **B.** $y = \frac{1}{3}x - \frac{4}{3}$

○ **C.** $x = 9y + 12$

○ **D.** $x - 3y = 4$

**7.** Solve for $x$.

$$\frac{1}{3}(3x - 11) = 12x$$

Answer: $x =$ ________________

**8.** What is the relationship between the $x$ and $y$ in the graph below?

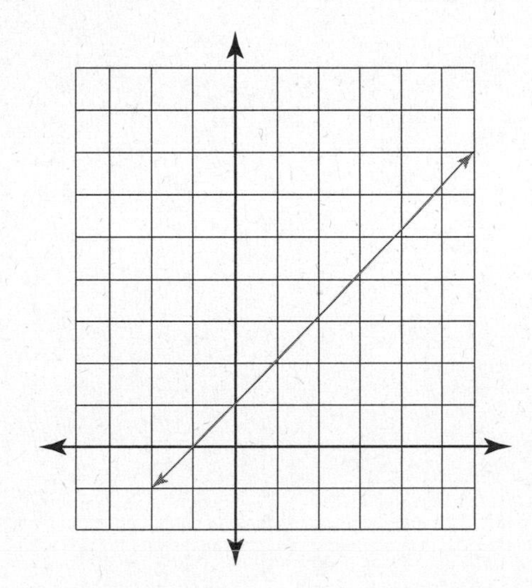

○ **A.** $y = \frac{7}{5}x + 1$

○ **B.** $y = x + 1$

○ **C.** $y = 4x + 2$

○ **D.** $y = -x + 1$

**9.** Solve for $c$ in the following equation:

$$ac + 4 = b$$

○ **A.** $c = \frac{b+4}{a}$

○ **B.** $c = \frac{b}{a} + 4$

○ **C.** $c = \frac{b-4}{a}$

○ **D.** $c = 4b - a$

**10.** Solve the equation for $x$. Show all steps.

$$2x + \frac{1}{4}(x - 24) = 3$$

Write your final answer here: $x =$ ________________

**11.** How many of the following lines have a slope of 4?

$$y = 4x + 3$$

$$y = 4x$$

$$4y = x$$

$$16x - 4y = 24$$

$$y + 4x = 6$$

○ **A.** 1

○ **B.** 2

○ **C.** 3

○ **D.** 4

**12.**

| Time (min.) | Temperature (°F) |
|---|---|
| $x$ | $y$ |
| 0 | 60 |
| 1 | 58 |
| 2 | 56 |
| 3 | 54 |

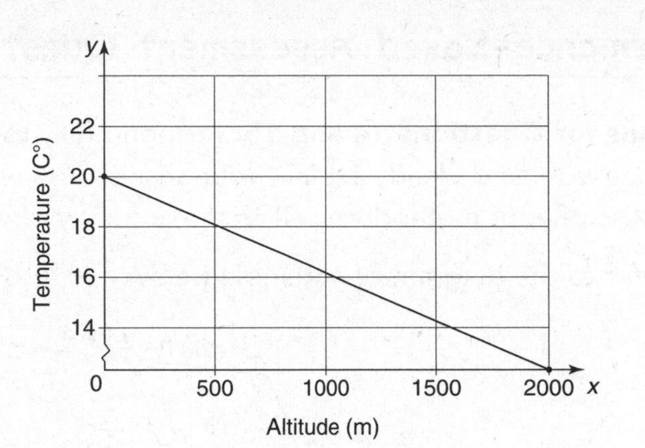


**Part A:** What is the same about the two sets of data?

**A.** They are both increasing.

**B.** They are both decreasing.

**C.** They both have a slope of 1.

**D.** They both have a slope of −1.

**Part B:** What do you know about the *rate of change* in the table data?

**A.** It shows a positive rate of change.

**B.** It shows a negative rate of change.

**C.** There is no rate of change.

**D.** There is not enough information to tell.

**13.** Jaxon's older sister has a large square garden. She wants to change the shape of her garden and make it a rectangle with a length that is 10 feet longer than two times its width. His sister has 200 feet of fencing, so she decides that the perimeter of the new garden should be 200 feet.

What are the dimensions in feet, of her new garden?

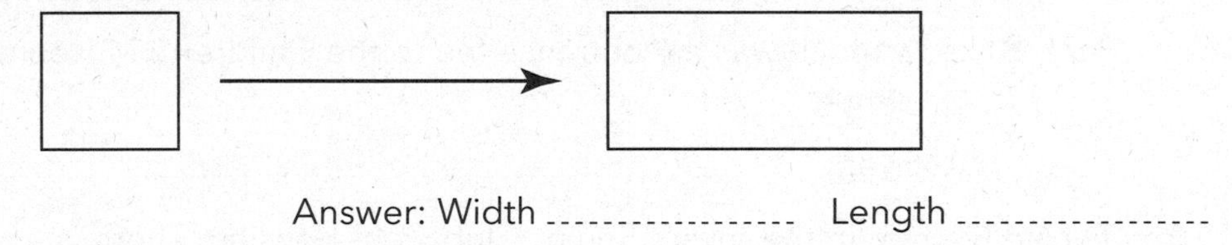

Answer: Width ________________ Length ________________

## Performance-Based Assessment Questions

**Directions for Questions 14 and 15:** Respond fully to the PBA questions that follow. Show your work and clearly explain your answer. You will be graded on the correctness of your method as well as the accuracy of your answer.

**14.** Refer to the diagram of a triangle below.

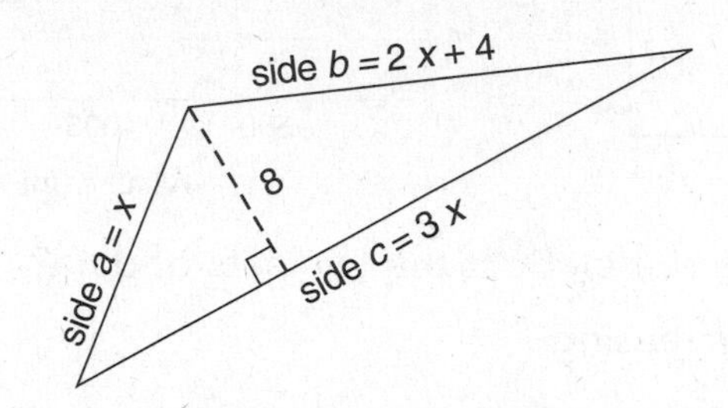


The perimeter of the triangle shown is 100 cm.

**Part A:** Write an equation to represent the perimeter.

**Part B:** Solve for $x$.

**Part C:** Find the length of side $a$, side $b$, *and* side $c$.

**Part D:** If the height of this triangle is 8 cm long, what is the area of the large triangle?

**15.** A class of 24 students went on a field trip to a Children's Museum. There is an admission fee for each student and a $4 fee per student to watch the special 3-D movie. The total cost for the 24 students to enter the museum and to see the movie was $216. What was the fee for each student to enter the museum?

**Part A:** Write an algebraic equation to match this scenario. Let $x$ represent the museum entrance fee.

Answer: ________________

**Part B:** How much was the entrance fee to the Children's Museum for each student?

Answer: ________________

**16.** Twelve women in the town's Senior Club went bowling. There is an entrance fee to the bowling alley and the shoe rental fee was $3 each. The total cost for all the women was $132.

**Part A:** Write an algebraic equation to match this scenario. Let $e$ represent the entrance fee to the bowling alley.

Answer: ________________

**Part B:** How much was the entrance fee to the bowling alley for each woman?

Answer: ________________

# Probability and Statistics

CHAPTER 4

## WHAT DO PARCC 8 PROBABILITY AND STATISTICS QUESTIONS LOOK LIKE?

### Multiple-Choice Questions (MC)

#### Example 1: Probability

Using a six-sided cube numbered from 1 to 6, what is the probability of getting a 4?

○ **A.** 4%

○ **B.** 17%

○ **C.** 25%

○ **D.** 50%

**Strategies and Solutions**

Correct choice is **B**.

$$\frac{\text{Favorable outcomes}}{\text{Total possible outcomes}} = \frac{1}{6}$$

$$\frac{\text{Only one \#4}}{\text{6 different numbers}} = \frac{1}{6} = 0.1666$$

or approximately 0.17 or 17%

Note: Each side of this number cube contains one number: 1, 2, 3, 4, 5, or 6.

## Example 2: Data Analysis and Statistics

The following are scores Jill received on her math quizzes this marking period: 80, 82, 80, 83, 75. If she gets an 80 on her next quiz, which of the following is true?

○ **A.** The mean will change.

○ **B.** The median will change.

○ **C.** The mode will change.

○ **D.** All will remain the same.

**Strategies and Solutions**

Correct choice is **D**.

First, organize original data in order from lowest to highest: 75, 80, 80, 82, 83

- Mean $= 80 \frac{75+80+80+82+83}{5} = \frac{400}{5} = 80$
- Median = 80 (the number in the middle when they are all arranged in numerical order)
- Mode = 80 (the number that appears most often)

If her next test grade is an 80, all will remain the same.

## Example 3: Analyzing Scatter Plots

For a school project, Wade recorded the number of grams of fat and the number of calories in lunches available at the local diner. He created the *scatter plot* shown below.

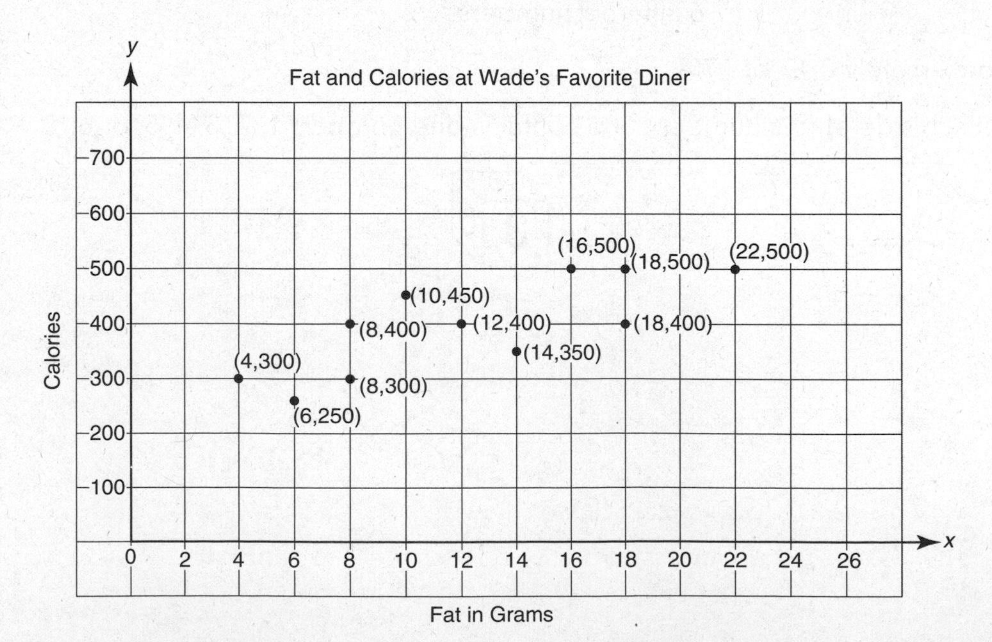

**Part A:** Which diagram below shows an appropriate *line of best fit*?

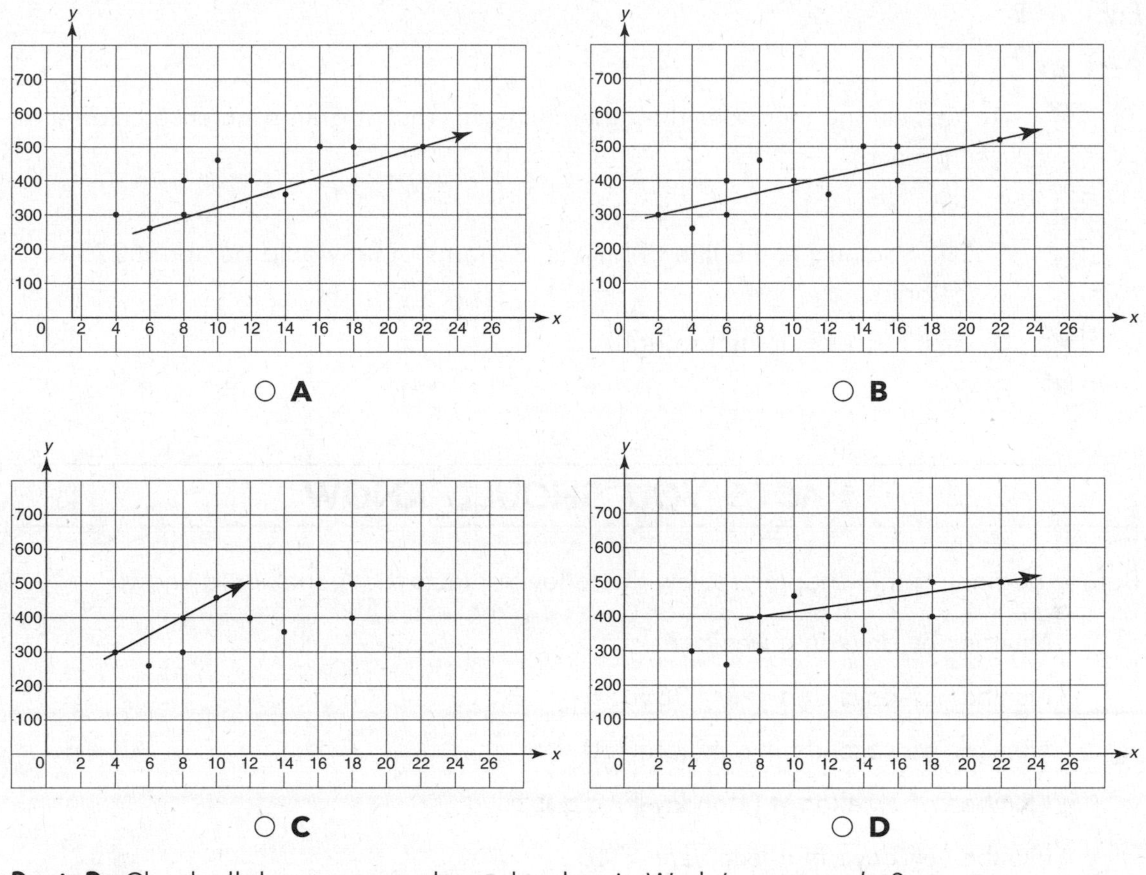


**Part B:** Check all that are true about the data in Wade's *scatter plot*?

☐ **A.** There is a positive correlation between fat and calories in the diner food.

☐ **B.** There is a definite outlier at 16 grams of fat.

☐ **C.** You could estimate that 2 grams of fat would be about 400 calories.

☐ **D.** The slope of the *line of best fit* is positive.

☐ **E.** 10 grams of fat would have about 450 calories.

**Strategies and Solutions** (Notice there is more than one correct answer.)

**Part A: B**

**Part B:**

- [x] **A.** Yes; as one increases the other increases (as fat grams increase calories increase)
- [ ] **B.** No.
- [ ] **C.** No. Looking at the line of best fit, 2 grams of fat would be about 275 calories.
- [x] **D.** Yes; it rises from left to right.
- [x] **E.** Yes

---

## FACTS YOU SHOULD KNOW

---

Before beginning this chapter, review the following facts that you should know:

- [ ] Number of days in a week: 7
- [ ] Number of days in a year: 365
- [ ] Number of seconds in a minute: 60
- [ ] Number of weekdays (Monday–Friday): 5
- [ ] Number of days in a leap year: 366
- [ ] Number of minutes in an hour: 60
- [ ] Number of cards in a regular deck of playing cards: 52
- [ ] A deck of cards is numbered 1–10 plus face cards (Jacks, Queens, and Kings)
- [ ] Number of suits in a deck of cards: 4 (spades, clubs, hearts, and diamonds)
- [ ] Number of colors in a deck of cards: 2 (spades and clubs are black; hearts and diamonds are red)
- [ ] Each suit contains 13 different cards (1, 2, 3, 4, 5, 6, 7, 8, 9, 10, Jack, Queen, King)

# PROBABILITY AND STATISTICS

Probability is used to make predictions. Probabilities can be written in many forms such as fractions, decimals, or percents.

- If an event will never happen (an elephant flies on its own), the probability of that happening is 0.
- If the event will definitely happen (it will rain in the rainforest this year), then the probability of that happening is 1.
- Sometimes percent is used, for example, if the weather forecaster says there is a 70% chance of snow tomorrow, then there is a 70% probability that it will snow tomorrow.

## PRACTICE: Probability (8.SP)

(For answers, see pages 270–271.)

**1.** What is the probability that you will pick one card that is a number 12 out of a regular deck of playing cards?

- **A.** 0
- **B.** 0.50
- **C.** 1
- **D.** not enough information given

**2.** What is the probability that you will pick an even number out of the numbers 2, 4, 6, 8, 10?

- **A.** 0
- **B.** 0.50
- **C.** 1
- **D.** not enough information given

**3.** What is the probability that it will snow in Alaska this year?

- **A.** 0
- **B.** 0.50
- **C.** 1
- **D.** not enough information given

**4.** What are your chances of spinning a number less than 2 using the following spinner?

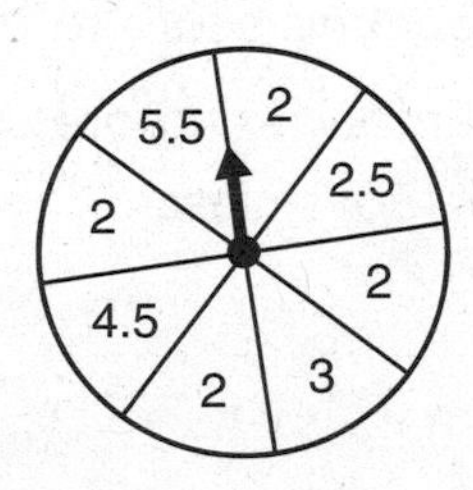


- **A.** 0
- **B.** 0.50
- **C.** 1
- **D.** not enough information given

5. Arnot and his brother Erik created a similar survey to see the relationship between the amount of time students studied and their score on an upcoming test.

They surveyed students in different classes.

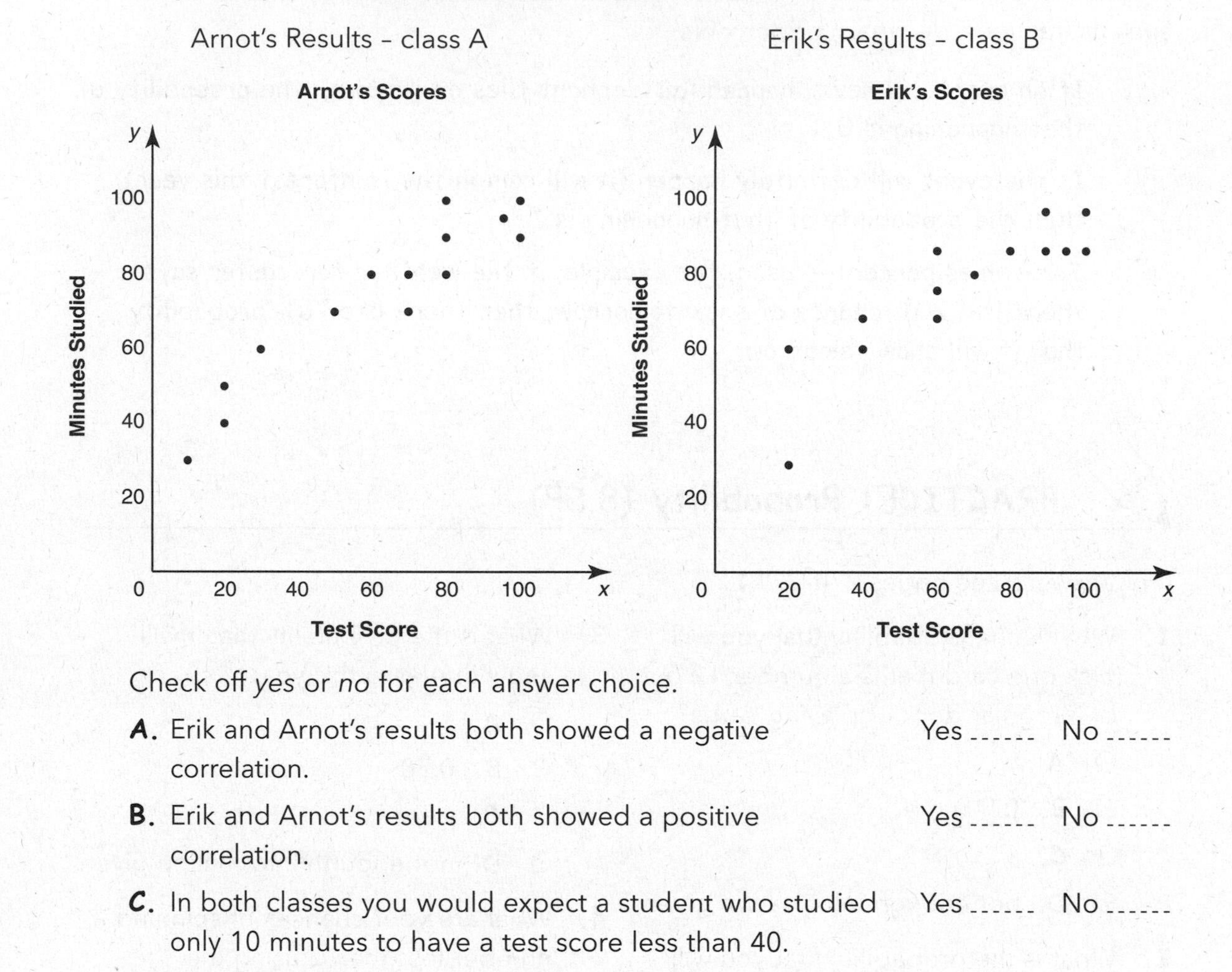


Check off *yes* or *no* for each answer choice.

**A.** Erik and Arnot's results both showed a negative correlation. Yes ------ No ------

**B.** Erik and Arnot's results both showed a positive correlation. Yes ------ No ------

**C.** In both classes you would expect a student who studied only 10 minutes to have a test score less than 40. Yes ------ No ------

**D.** If Sam had studied for about 80 minutes you would expect his score to be between 80 and 100. Yes ------ No ------

**E.** In Erik's class more students scored $\geq 60$ points than in Arnot's class. Yes ------ No ------

6. The Woodside movie theater manager collected data about their popcorn sales and beverages sold. He put the data into a scatter plot to see it more clearly. Which conclusions can you draw from the scatter plot below? Check all that are true.

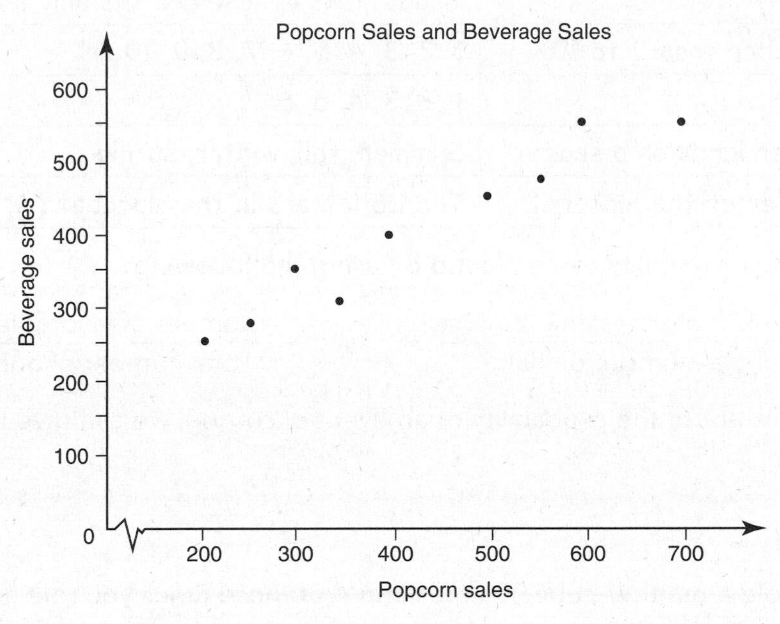


- [ ] **A.** There is a positive correlation between popcorn sales and beverage sales.
- [ ] **B.** There is a negative correlation between popcorn sales and beverage sales.
- [ ] **C.** There is no correlation between popcorn sales and beverage sales.
- [ ] **D.** The coordinate point (500, 450) is an outlier in this data.
- [ ] **E.** If 100 boxes of popcorn were sold you would expect about 200 beverages to be sold.

## EXPERIMENTAL PROBABILITY

An *experiment* is an activity where the results are observed. Each time you repeat the activity, it is called a *trial*, and the result of each trial is called the *outcome*.

> **Flip a coin 5 times (5 trials); the outcome is tails, heads, heads, tails, heads (2T, 3H)**

> **Spin a spinner 4 times (4 trials); the outcome is red, blue, red, yellow (2R, 1B, 1Y)**

The *sample space* is the set of all possible outcomes of an experiment.

| Experiment | Sample Space (all the different possible outcomes) |
|---|---|
| Flip a penny | Heads, tails (These are the only possibilities.) |
| Pick a number from 1 to 10 | 1, 2, 3, 4, 5, 6, 7, 8, 9, 10 |
| Roll a die | 1, 2, 3, 4, 5, 6 |
| The spinner lands on a season | Summer, fall, winter, spring |
| Pick a letter of the alphabet | The 26 letters in the alphabet (A, B, C, ...) |

Experimental probability can be found by using the following:

$$\frac{\text{The number of times an event occurs}}{\text{Total number of trials}} \text{ or } \frac{\text{The number of favorable outcomes}}{\text{Total number of outcomes}}$$

When we talk about the probability of an event occurring, we can give the answer as a percent.

## Examples

**A.** If you role a number cube 30 times and 4 of those times you role a 2, then based on that experiment, the probability of rolling a 2 is $\frac{4}{30} = 0.1\overline{333} \sim 13\%$.

**B.** If a spinner lands on green 5 times out of 20 spins, the *experimental probability* of landing on green is $\frac{5}{20} = \frac{1}{4} = 0.25 = 25\%$.

## PRACTICE: Experimental Probability

(For answers, see page 271.)

**1.** If Janet picks 3 red jellybeans out of the 10 that she picked from the jar, what is the experimental probability of picking a red jellybean from that jar?

- ○ **A.** 3%
- ○ **B.** 30%
- ○ **C.** 300%
- ○ **D.** 70%

**2.** If John takes a handful of mixed nuts from a can and picks 6 salted peanuts out of the 12 he picked, what is the experimental probability of picking a salted peanut?

- ○ **A.** 10%
- ○ **B.** 20%
- ○ **C.** 40%
- ○ **D.** 50%

**3.** Jack has a Halloween bag filled with small pieces of candy. If he takes a handful and gets 4 chocolate kisses out of the 8 pieces he picked, what is the experimental probability of picking a chocolate kiss from that bag?

- ○ **A.** 5%
- ○ **B.** 50%
- ○ **C.** 24%
- ○ **D.** 40%

**4.** Approximately what is the probability that Mrs. Davis' birthday will be on a Monday?

- ○ **A.** 8.33%
- ○ **B.** 14.29%
- ○ **C.** 20%
- ○ **D.** 3.33%

**5.** Ronnie is correct when she says that there is about an 8% chance that anyone's birthday is

- ○ **A.** in the spring
- ○ **B.** on a Wednesday
- ○ **C.** in May
- ○ **D.** an even number

**6.** If you spin the spinner below, what is the probability that it will **not** land on an odd number?

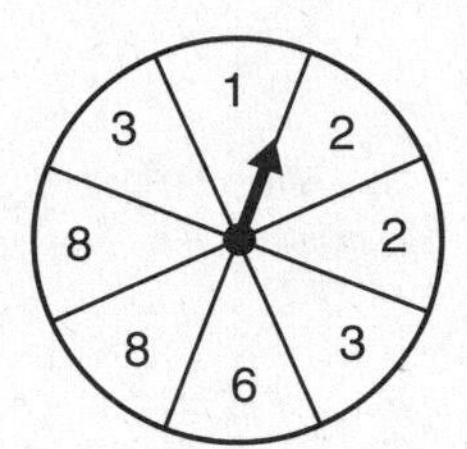


- ○ **A.** $\frac{5}{8}$ or 62.5%
- ○ **B.** $\frac{4}{8}$ or 50%
- ○ **C.** $\frac{3}{8}$ or $37\frac{1}{2}$%
- ○ **D.** $\frac{2}{8}$ or 25%

## FINDING PROBABILITIES OF OUTCOMES IN A SAMPLE SPACE

### Examples

**A.** What is the probability of landing on yellow using the spinner and table below?

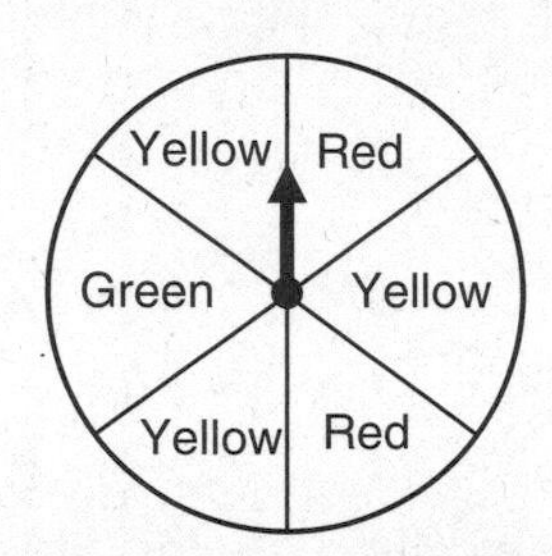


| Outcome | Red | Yellow | Green |
|---|---|---|---|
| Probability | 2 | 3 | 1 |
| Sample Space | Red, yellow, green (6 spaces) | Red, yellow, green (6 spaces) | Red, yellow, green (6 spaces) |

The probability of the spinner landing on yellow is $\frac{3}{6}$ or $\frac{1}{2}$ or $P(\text{yellow}) = \frac{1}{2}$.

The probability of the spinner landing on red is $\frac{2}{6}$ or $\frac{1}{3}$ or $P(\text{red}) = \frac{1}{3}$.

The probability of landing on green is $\frac{1}{6}$. Check to be sure the sum of all the possibilities equals 1. (Make common denominators and add.)

$\frac{1}{2}+\frac{2}{6}+\frac{1}{6}=\frac{3}{6}+\frac{2}{6}+\frac{1}{6}=\frac{6}{6}=1$ ✔ Correct!

**B.** If you have a regular deck of cards, what is the probability of choosing a red 5? Since there are only 2 red 5s (a 5 of hearts and a 5 of diamonds) and since there are 52 cards in a regular deck of cards, we can create the fraction of

$$\frac{\text{Favorable outcomes}}{\text{Total possible outcomes}}.$$

This is $\frac{2}{52} = 0.03846$ or ~3.8% chance of picking a red 5.

## FINDING PROBABILITIES OF EVENTS

### Examples

**A.** When you are finding the probability of *one or more* events, you add the probabilities. What is the probability that Rodney or John will win the race?

| Runner | Rodney | Kevin | Joe | John |
|---|---|---|---|---|
| **Probability of winning** | 30% | 15% | 20% | 35% |

$$P\text{ (Rodney or John)} = 30\% + 35\% = 65\%$$

There is a 65% chance that Rodney or John will win the race.

**B.** When you find the probability that *one and another* event will happen, then you find the probability of each one and *multiply* them.

What is the probability that it will rain on Monday and Wednesday? Remember, there is less of a chance that it will rain on both days than on one day.

| Day of week | Monday | Tuesday | Wednesday | Thursday | Friday |
|---|---|---|---|---|---|
| **Probability of rain** | 10% | 15% | 20% | 25% | 70% |

$$P\text{ (Mon and Wed)} = (10\%)(20\%) = (0.10)(0.20) = 0.02 = 2\%$$

There is only a 2% probability that it will rain on Monday and Wednesday. *Remember*: You are multiplying decimal numbers less than 1, so your product will be a smaller number. Be careful multiplying small decimal numbers!

## PRACTICE: Probability of Events

(For answers, see pages 271–272.)

**1.** The sports announcer said the New Jersey Devils have a 60% chance of winning their next game at the Meadowlands. Complete the chart (chances of winning, chances of losing).

| Outcome | The Devils win | The Devils lose |
|---|---|---|
| **Probability** | ? | ? |

○ **A.** 0.06, 0.04 ○ **B.** 0.60, 0.40 ○ **C.** $\frac{4}{10}, \frac{6}{10}$ ○ **D.** 40%, 60%

**2.** The table below charts the probability of the Trenton Thunder minor league baseball team getting a certain number of runs. What is the probability they will get 1 or 2 runs?

| Runs | 0 runs | 1 run | 2 runs | 3 runs | 4 runs |
|---|---|---|---|---|---|
| **Probability of getting a run** | 0.10 | 0.30 | 0.25 | 0.15 | 0.05 |

○ **A.** 15% ○ **B.** 55% ○ **C.** 65% ○ **D.** 0.075%

**3.** Kyle and Zak are having similar success in basketball this year. According to the latest school basketball statistics, there is a 50% probability that Zak or Kyle will make a basket during the last quarter of the game.

**Part A:** From the choices shown, Kyle's chances of making a basket are most likely

○ **A.** 25% ○ **B.** 50% ○ **C.** 75% ○ **D.** 100%

**Part B:** Estimate the probability that Zak and Kyle will both make a basket?

○ **A.** 6%-7% ○ **B.** 25%-30% ○ **C.** 50% ○ **D.** 75%-80%

**4.** The latest weather forecast gave a 50% chance of rain in New York City tomorrow, and a 20% chance of rain the next day. What is the probability it will rain on both days?

○ **A.** 70% ○ **B.** 50% ○ **C.** 30% ○ **D.** 10%

## DATA COLLECTION AND ANALYSIS (A Review From Grade 6.SP.5C)

### Mean, Median, Mode, and Range

The *mean* is the average of a set of data. 74 is the average of the following set of data: 60, 80, 70, 90, 70. You find the *mean* by adding the data and dividing by the number of numbers.

$$60+80+70+90+70=370; \quad \frac{370}{5}=74$$

The *median* is the number in the middle after a set of data is organized from lowest to highest.

60, 70, 70, 80, 90 The median is 70.

The *mode* is the number that appears most frequently. (*Remember*: mode = most often.) In this case, it is 70.

In a set of data, the spread of the numbers given from the lowest to the highest is the *range*.

$$90 - 60 = 30$$

There are 30 whole numbers between 90 and 60.

## PRACTICE: Mean, Median, Mode, and Range

(For answers, see page 272.)

**1.** Find the mean, median, and mode of the data set: 15, 7, 9, 12, 21, 11, 13, 12, 8

**(a)** The mean ○ **A.** 11.2 ○ **B.** 12 ○ **C.** 13 ○ **D.** 14

**(b)** The median ○ **A.** 11 ○ **B.** 12 ○ **C.** 12.5 ○ **D.** 13

**(c)** The mode ○ **A.** 12 ○ **B.** 21 ○ **C.** 13 ○ **D.** none

**2.** Find the range of the data set: 40, 33, 46, 50, 36, 51, 47, 35, 51, 47, 52, 53

○ **A.** 12 ○ **B.** 13 ○ **C.** 19 ○ **D.** 20

**3.** The following are the amounts Jane saved each week in her new savings account: $50, $75, $60, $100, $60, $70, $90.

**Part A:** What is her approximate average savings each week (the mean average)?

○ **A.** $57.85 ○ **B.** $69.39 ○ **C.** $70.71 ○ **D.** $72.10

**Part B:** What is the mode of this data set?

○ **A.** $50 ○ **B.** $60 ○ **C.** $70 ○ **D.** $90

**Part C:** How much should she save the following week to raise her mean average by $10?

○ **A.** $152.10 ○ **B.** $82.10 ○ **C.** $92.10 ○ **D.** $89.10

**4.** The following are the batting averages of the local school's best hitters: 310, 285, 300, 270, 300, 275. If a new hitter joins the team with a batting average of 290, what would change?

○ **A.** the mean, the mode, and the median

○ **B.** the mean and the median

○ **C.** the median only (mean remains at 290, mode stays at 300)

○ **D.** nothing would change

**5.** The teacher asks you to create a graph of the class' grades on a recent test.

The data set is 79, 66, 82, 98, 86, 75, 85, 80, 92, 78, 82, 85, 77, 63, 88, 100, 88, 92, 82, 80, 77, 98. What is the range you would use to set up the axis to represent these grades?

○ **A.** 47 ○ **B.** 37 ○ **C.** 50 ○ **D.** 100

## SCATTER PLOTS (SCATTERGRAMS)

A scatter plot looks like a graph on a grid with many points that are not connected.

### Examples

**A.** First, we'll study the points to understand the basic information.

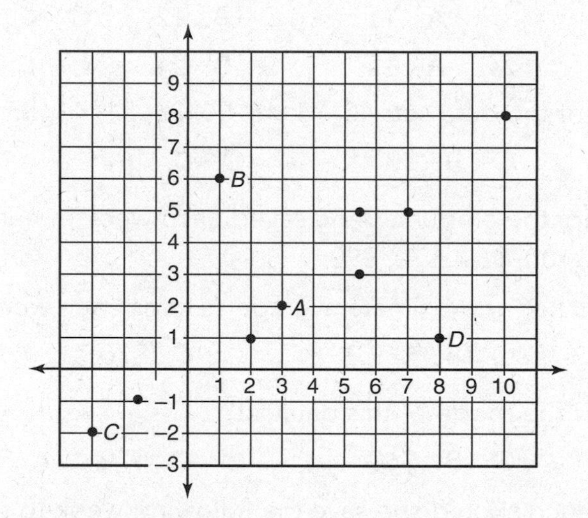


**Point A:** (3, 2) **Point B:** (1, 6) **Point C:** (-3, -2) **Point D:** (8, 1)

The points in a *scatter* plot (sometimes called a *scattergram*) do not lie on a particular line. But, many points seem to lie near a line. Sometimes you can take a thin piece of spaghetti and place it on top of a scattergram to see where the line seems to cover the most number of points. This line is called a *trend line* or *line of best fit*. Do *not* connect the points to make a trend line. There is no data between the points; the points in a *scatter plot* do not match a particular linear equation. The trend line is used to estimate and to make predictions about the data. Look at the data that seems to be far away from this line, like points B(1, 6) and D (8, 1). We call these points *outliers*. Some scatter plots show a *positive correlation* among the data graphed, and some show a *negative correlation*. Study the next two graphs to see examples of each.

**B.** Negative correlation A local gym took a survey to compare the number of hours a group of women worked out at the gym compared to their weight. We would expect to see that the more hours worked out each week would show the person weighed less. If you were to draw a line of best fit, you would see a line with a *negative slope* (more hours, less weight). We can also say: this line of best fit is *falling*.

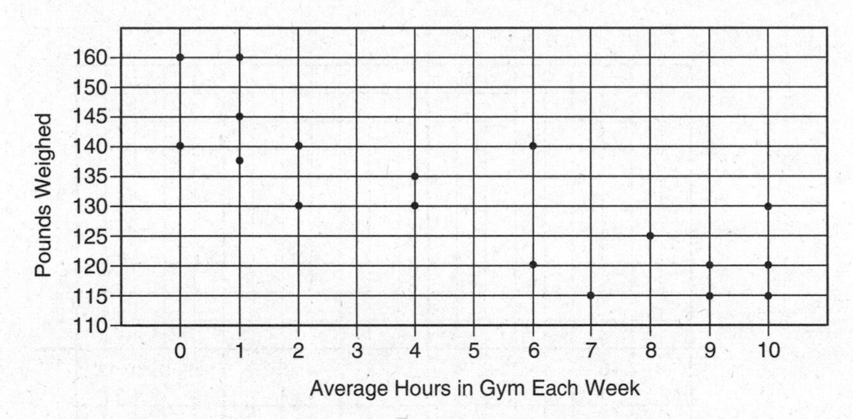


**C.** Positive correlation We can take similar data and create a scatter plot with a *positive slope*. Look at the graph below. Here we take the number of hours a group of women worked out and compare this to the number of pounds lost. If you were to draw a line of best fit, you would see a line with a *positive slope* (more hours, more pounds lost). We also can say this line of best fit is *rising*.

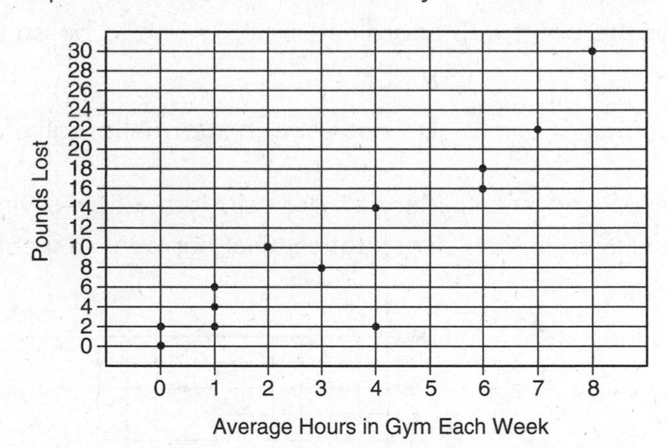

## PRACTICE: Scatter Plots

(For answers, see page 272.)

**1.** Use the scatter plot below to answer **Parts A and B**.

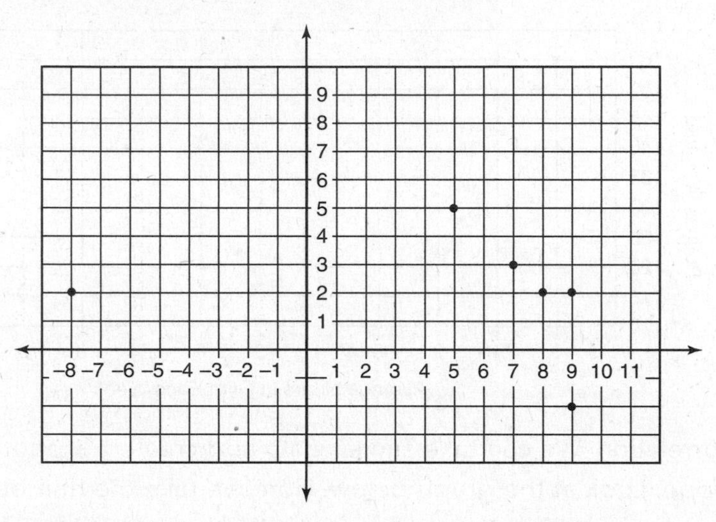


**Part A:** Which of the following points seems to be an outlier?

○ **A.** (8, –2) ○ **B.** (–8, 2) ○ **C.** (8, 2) ○ **D.** (9, –2)

**Part B:** Does the point with coordinates (5, 5) seem to be on the trend line?

○ **A.** yes ○ **B.** no

○ **C.** sometimes ○ **D.** not enough information given

**2.** If this scatter plot represents Margie's deposits into a new savings account, how much has she saved so far? (*Hint*: Notice what happens at week #8.)

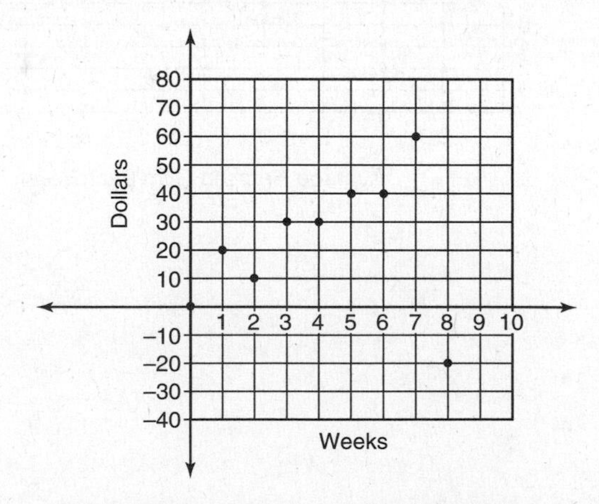


○ **A.** \$260 ○ **B.** \$240 ○ **C.** \$210 ○ **D.** \$200

**3.** Mrs. Teller, an 8th grade math teacher, was concerned that her students had not been doing well on their quizzes and tests lately. On yesterday's test, she asked the students to write at the top of their tests the amount of time they studied. She then created a scatter plot to see if there was a positive correlation between the time students studied and their testing scores.

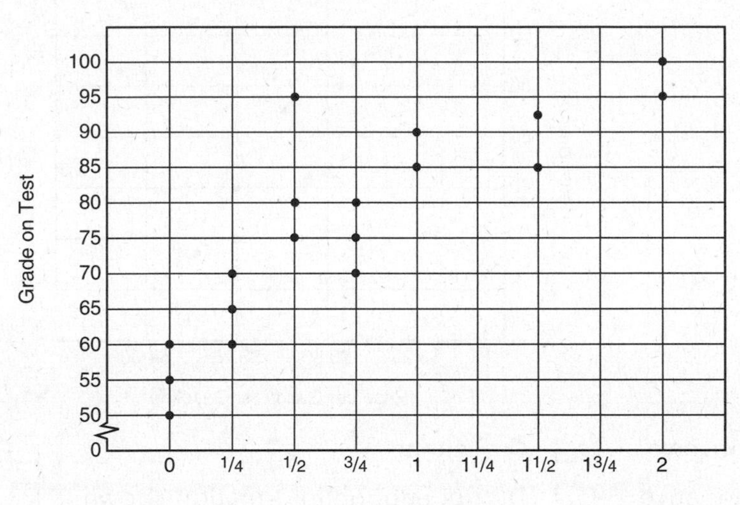


**Part A:** Does this graph show a positive or negative correlation?

○ **A.** positive ○ **B.** negative

○ **C.** neither ○ **D.** not enough information shown

**Part B:** If you draw a line of best fit are there any outliers?

○ **A.** yes (0 hours and 55%) ○ **B.** yes ($\frac{1}{2}$ hr and 95%)

○ **C.** yes (2 hours and 95%) ○ **D.** no

**Part C:** Which point might represent a student who didn't study very long the night before but still received a high grade because he always does his homework, takes good notes in class, and is rarely absent?

○ **A.** (2, 95) ○ **B.** (2, 90) ○ **C.** ($\frac{1}{2}$, 95) ○ **D.** ($1\frac{1}{2}$, 90)

**Part D:** Would this line of best fit be described as rising or falling?

Answer: ________________

**4.** This scatter plot below shows the relationship between the number of hours a family takes for its Sunday car rides compared to the price of gasoline for their car. Does this show a positive or negative correlation? (There is more than one correct answer.)

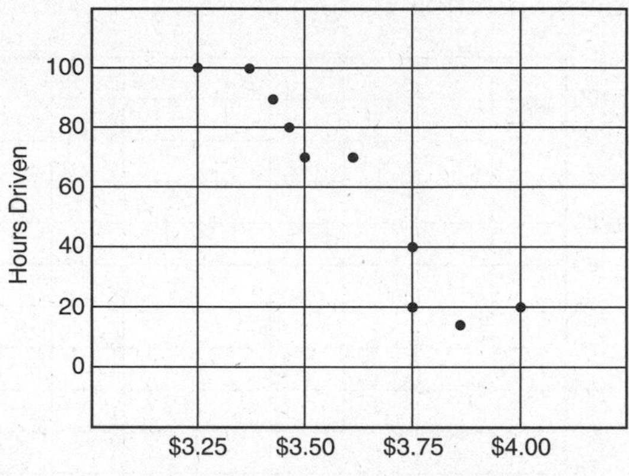


- [ ] **A.** positive
- [ ] **B.** negative
- [ ] **C.** neither
- [ ] **D.** not enough information shown
- [ ] **E.** rising
- [ ] **F.** falling

## CIRCLE GRAPHS

*A circle graph* compares parts of the circle to the whole circle. Sometimes a circle graph is divided into degrees. There are 360° in a whole circle.

**Example 1:**

**A.** How many degrees are in $\frac{1}{2}$ of a circle? $\frac{360}{2} = 180°$

**B.** How many degrees are in $\frac{1}{5}$ of a circle? $\frac{360}{5} = 72°$

**C.** Sometimes a circle is divided into uneven sections. How many degrees are in the empty section?

$180° + 30° + 110° = 320°$

$360° - 320° = 40°$

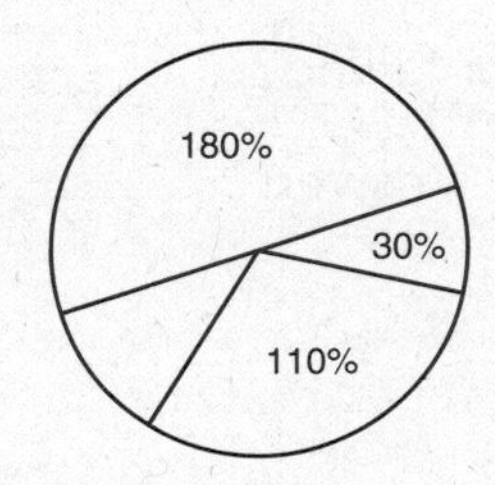

Other times a circle graph is divided into percents. The whole circle equals 100%.

**Example 2:**

**A.** What percent is $\frac{1}{2}$ of a circle? $\frac{100\%}{2} = 50\%$

**B.** What percent is $\frac{1}{5}$ of a circle? $\frac{100\%}{5} = 20\%$

**C.** What percent of the whole circle is the empty section?

60%

10%

$60\% + 10\% = 70\%$

$100\% - 70\% = 30\%$

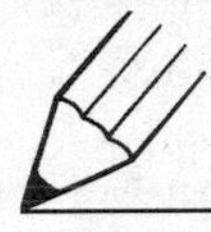

## PRACTICE: Circle Graphs

(For answers, see page 273.)

Use the circle graph below for questions 1-4.

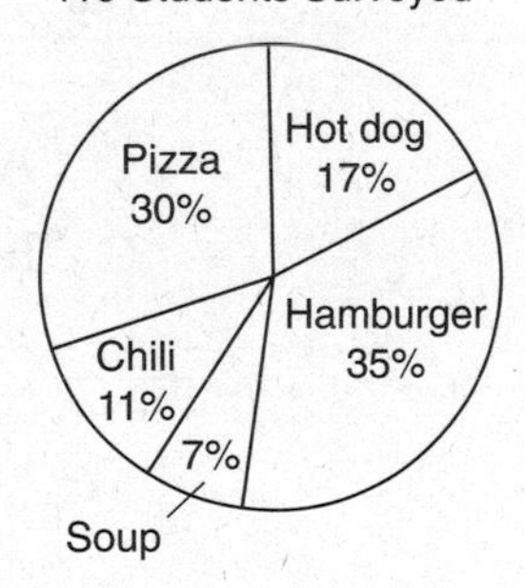


**1.** What percent of the students surveyed selected hamburger or pizza as their favorite?

- ○ **A.** 30%
- ○ **B.** 35%
- ○ **C.** 65%
- ○ **D.** 5%

**2.** Approximately how many students chose chili as their favorite?

- ○ **A.** 11
- ○ **B.** 41
- ○ **C.** 4.5
- ○ **D.** 45

3. Which equation shows how many more students chose hamburger than soup?

- ○ **A.** (410)(0.35) – (410)(0.07)
- ○ **B.** (410)(0.35 – 0.07)
- ○ **C.** (410 – 0.35) + (410 – 0.07)
- ○ **D.** $\frac{410}{0.35} - \frac{410}{0.07}$

4. More than half the students chose

- ○ **A.** chili, pizza, or hot dog
- ○ **B.** hamburger or soup
- ○ **C.** hot dog, chili, or soup
- ○ **D.** pizza, chili, or soup

5. Use the circle graph below for **Parts A, B, and C.**

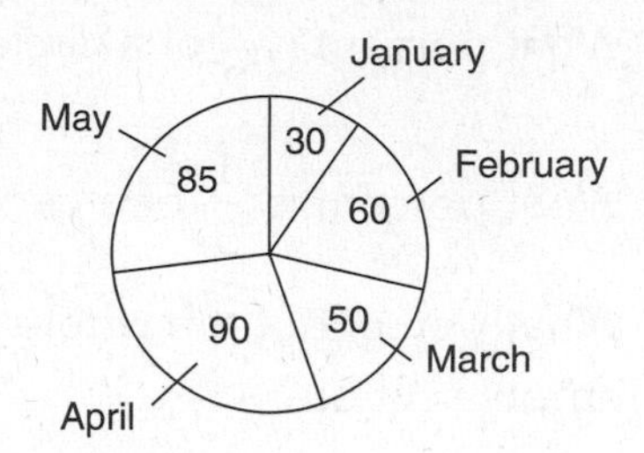


**Part A:** What percent of students surveyed had birthdays in February?

- ○ **A.** 19%
- ○ **C.** 40%
- ○ **B.** 20%
- ○ **D.** 60%

**Part B:** Approximately what percent of students surveyed have birthdays either in April or May?

- ○ **A.** 175%
- ○ **C.** 56%
- ○ **B.** 180%
- ○ **D.** 58%

**Part C:** What percent of student birthdays are in the fall?

- ○ **A.** 0%
- ○ **C.** 25%
- ○ **B.** 50%
- ○ **D.** 100%

## PRACTICE: SCR Non-calculator Questions

Each question is worth 1 point. No partial credit is given.

(For answers, see page 273.)

**1.** Jessica's four tests scores are: 80 75 80 85.

If she scores 80 on her fifth test, will the *mean*, the *median*, or the *mode* change?

Answer: ------------------

**2.** In northern New Jersey, Mr. and Mrs. B. use Orange & Rockland Light & Power Company for their home electricity. This scatter plot shows their monthly use. Is there a positive or negative correlation between the number of kilowatt hours (KWH) used and the cost of their monthly bill?

Cost in Dollars ($)
300
250
200
150
100
50
0
600 650 700 750 800 850 900 950 1,000 1,050 1,100 1,150 1,200
Kilowatt Hours (KWH) Used

Answer: ------------------

**3.** The school cafeteria sells frozen yogurt. You can get chocolate, vanilla, or swirl. You can choose chocolate sprinkles, multicolored sprinkles, or no sprinkles. How many different combinations of frozen yogurt are available?

Answer: ------------------

**4.** When you flip a fair coin (heads on one side and tails on the other), what is the probability of getting *heads* on your 10th flip?

Answer: ------------------

**5.** Bonnie and her sister each have a bag of colored marbles as described below.

| | RED | BLUE | YELLOW | GREEN | TOTALS |
|---|---|---|---|---|---|
| BONNIE (bag-B) | 5 | 4 | 3 | 4 | **16** |
| HER SISTER (bag-S) | 2 | 5 | 4 | 3 | **14** |
| **TOTALS** | **7** | **9** | **7** | **7** | **30** |

Select all that are true below.

- [ ] **A.** Bonnie has a better chance of selecting a red marble from her bag-B than her sister has using her bag-S?
- [ ] **B.** Bonnie has a 50% chance of selecting a *blue* or a *green* marble from her bag-B.
- [ ] **C.** Her sister has a 50% chance of selecting a *red* or a *blue* marble from her bag-S.
- [ ] **D.** If they mixed all their marbles and put them in one bag, they would have a better chance of selecting a red marble than a yellow one.
- [ ] **E.** Bonnie has more marbles than her sister.

**6.** At the R. Brown Junior High School, Ms. Crowley's class is planning to take a survey. They want to find out what 8th grade students prefer. They will distribute their survey to 30 people. Which would be the best group of people to use if they want a fair and useful sample?

○ **A.** 30 people coming out of the local supermarket.

○ **B.** 30 eighth-grade students as they leave the auditorium.

○ **C.** 30 teachers in the school.

○ **D.** 30 students as they leave school at the end of the day.

## PRACTICE: Performance-Based Assessment Questions

(For answers, see pages 274-275.)

**1.** Suppose two different varieties of tomatoes are growing in pots. The regular tomato plant is already 12 inches tall and is growing at a rate of $\frac{3}{2}$ inches per week. The cherry tomato plant is 6 inches tall and is growing at a rate of 2 inches per week.

**Part A:** Complete the chart below to show how their heights change each week. Show all work.

| | Regular Tomato Plant | | Cherry Tomato Plant | |
|---|---|---|---|---|
| | Show work! | Height of Plant (inches) | Show work! | Height of Plant (inches) |
| Now | 12 | | 6 | |
| Week 1 | $12 + \frac{3}{2} =$ | | $6 + 2 =$ | |
| Week 2 | | | | |
| Week 3 | | | | |
| Week 4 | | | | |
| Week 5 | | | | |
| Week 6 | | | | |
| Week 7 | | | | |
| Week 8 | | | | |
| Week 9 | | | | |
| Week 10 | | | | |
| Week 11 | | | | |
| Week 12 | | | | |
| Week 13 | | | | |
| Week 14 | | | | |

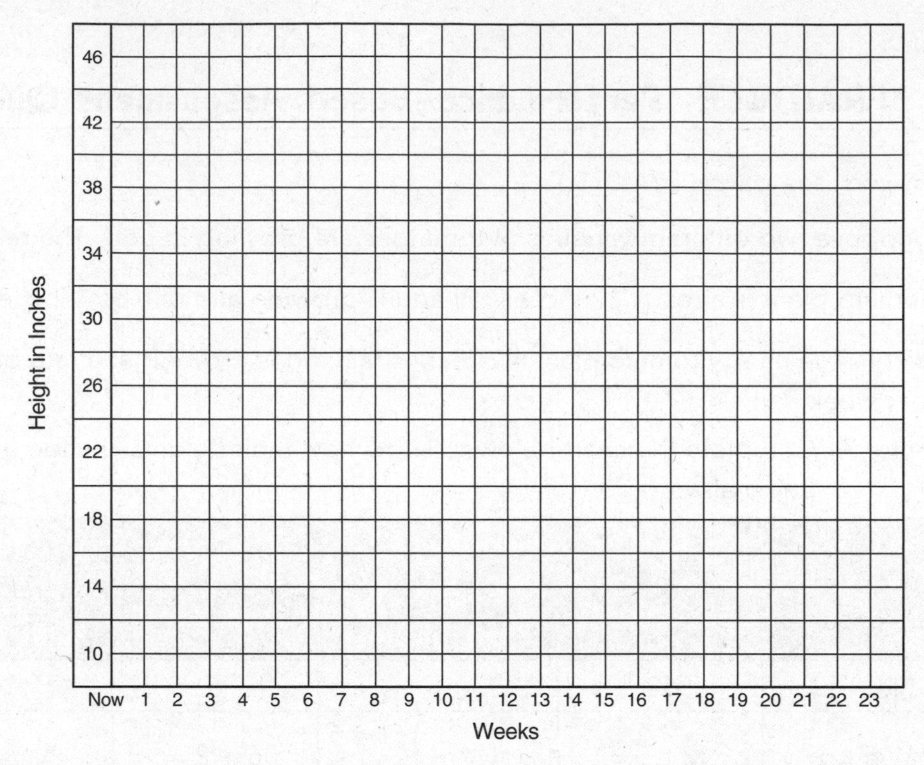


**Part B:** Create a double line graph on the coordinate grid above. Label each line regular tomato or cherry tomato. Graph each plant's height for at least the next 14 weeks.

**Part C:** In how many weeks will both plants be the same height?

**Part D:** What happens where the two lines intersect?

**Part E:** What happens after this?

2. Refer to the circle graph of *JB's Bakery: Annual Sales Breakdown* shown below.

JB's Bakery: Annual Sales Breakdown

Bread 16%
Cookies 8%
Miscellaneous 3%
Cakes 26%
Donuts 22%
Bagels 25%

**Part A:** If the bakery sold $23,769 in May 2014, how much money did it make from bread and bagels?

**Part B:** After taking a survey of his customers, the owner expects to sell 50% more cookies than he sold during the year shown. If his annual total remains the same, how much money would he expect to make from cookies next year?

3. Consider working with a deck of cards.

**Part A:** How do you find the probability that a card drawn at random from a full deck of 52 cards will be a 10?

**Part B:** If one card is chosen at random, how do you find the probability that you will choose a black King? What is that probability?

**Part C:** What is the probability of selecting a card that is a prime number? Show how you found your answer.

Name: ........................................ Date: ........................................

# Chapter 4 Test: Probability and Statistics

35 minutes

(Use the *PARCC Grade 8 Mathematics Assessment Reference Sheet* on page 359.)

(For answers, see pages 275–278.)

**1.** After 500 spins of the spinner, the following information was recorded. Estimate the probability of the spinner landing on blue.

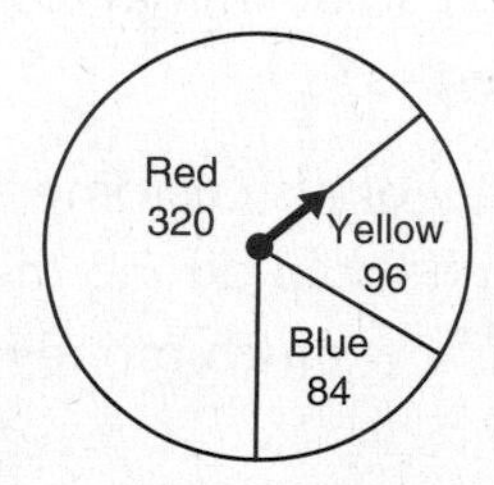


| Outcome | Red | Yellow | Blue |
|---|---|---|---|
| Spins | 320 | 96 | 84 |

- ○ **A.** approximately 84%
- ○ **B.** approximately 17%
- ○ **C.** approximately 8.4%
- ○ **D.** approximately 1.7%

**2.** What is the probability of landing on a perfect square number using a six-sided die with the following numbers: 4, 5, 6, 7, 8, 9 ?

- ○ **A.** $\frac{1}{6}$ or approximately 17%
- ○ **B.** $\frac{2}{6}$ or approximately 33%
- ○ **C.** $\frac{3}{6}$ or approximately 50%
- ○ **D.** $\frac{4}{6}$ or approximately 67%

**3.** If 2 coins are tossed, what is the probability of getting exactly 2 heads $P(2H)$?

Answer: ________________

**4.** Brianna collected data about the number of students registered in her school in 2000, in 2005, in 2010, and again this year. What type of graph would be best to show this data?

○ **A.** a line graph
○ **B.** a bar graph
○ **C.** a scatter plot
○ **D.** a box-and-whisker plot

**5.** Use the data below to answer the following questions.

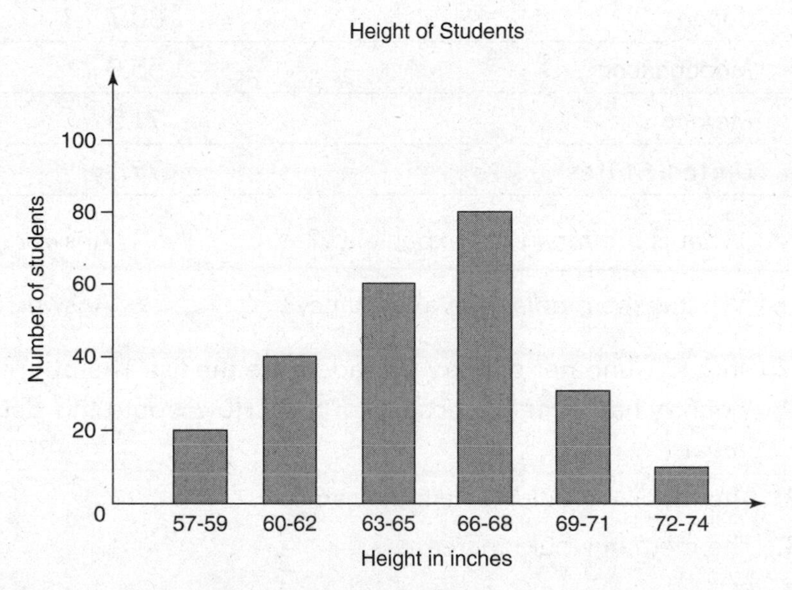


**Part A:** How many students are taller than 65 inches?

Answer: ________________

**Part B:** Can you tell how many students are exactly 59 inches tall?

Yes ______ No ______

Explain your answer to Part B above. Write in the box below.

**6.** Mario and his group are collecting data for their Health class project. They went online and collected the following data about the life expectancy of people living in different countries.

| Country | Life Expectancy (years) |
|---|---|
| Afghanistan | 45.9 |
| Australia | 79.8 |
| Brazil | 62.9 |
| Canada | 79.4 |
| France | 78.8 |
| Haiti | 49.2 |
| Japan | 80.7 |
| Madagascar | 55.0 |
| Mexico | 71.5 |
| United States | 77.1 |

**Part A:** What is the *mean* life expectancy? Answer: ________________

**Part B:** What is the *median* life expectancy? Answer: ________________

**Part C:** In 2000, another country was added to the list. The people in this country had a life expectancy of 57.5. How would this data affect the *mean?*

○ **A.** The *median* would remain the same.

○ **B.** The *median* would increase.

○ **C.** The *median* would decrease.

○ **D.** The *mean* would increase.

**7.** Your brother is the student manager of the high school's basketball team. He is working on the team's program guide and has recorded the *height* and *weight* of the eleven varsity players listed in the table below. Use the data to make a *scatter plot* and draw the *line of best fit.*

| Height | 72 | 70 | 71 | 70 | 69 | 70 | 69 | 73 | 66 | 70 | 76 |
|---|---|---|---|---|---|---|---|---|---|---|---|
| Weight | 190 | 170 | 180 | 175 | 160 | 160 | 150 | 180 | 150 | 150 | 200 |

**Part A:** Plot the coordinate points from the above table on the coordinate grid below.

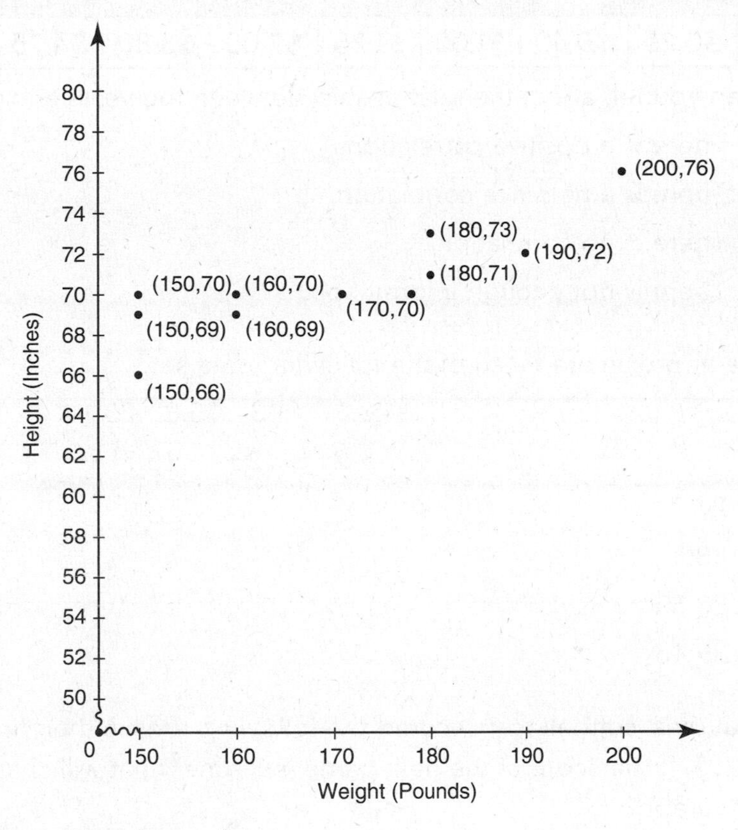


**Part B:** Draw a line of best fit.

Is there a positive, negative, or no correlation? Answer: ------------------

**Part C:** Select two points on the line of best fit and find the slope.

Answer: The slope is ------------------

**Part D:** Use your line of best fit and make a prediction about future events.

What would you expect a player who is 65 inches tall to weigh?

Answer: about ------------------ pounds

8. In a scatter plot, a trend line is also called

- ○ **A.** a parallel line
- ○ **B.** a diagonal line
- ○ **C.** an average line
- ○ **D.** the line of best fit

9. The table below shows the U.S. federal minimum wage in past years. (from the U.S. Dept. of Labor: http://www.dol.gov/whd/minwage/chart.htm)

| Year | 1938 | 1945 | 1956 | 1963 | 1974 | 1990 | 1996 | 2007 | 2009 |
|---|---|---|---|---|---|---|---|---|---|
| Min. Wage | $0.25 | $0.40 | $1.00 | $1.25 | $2.00 | $3.80 | $4.75 | $5.85 | $7.25 |

What can you tell about the relationship between the years and minimum wage?

- ○ **A.** There is a positive correlation.
- ○ **B.** There is a negative correlation.
- ○ **C.** There is no correlation.
- ○ **D.** There is not enough information.

10. Find the approximate mean of the following data set.

42.6 41.3 35.8 23.1

21.5 20.4 15.7

- ○ **A.** 27.2
- ○ **B.** 58.3
- ○ **C.** 28.62
- ○ **D.** 29.15

11. The local girls' softball team scored the following runs in their first 5 games: 2, 1, 4, 2, 3. If the score of the next game is 3 runs, what will change in this team's average?

- ○ **A.** Only the mean will change.
- ○ **B.** Only the median will change.
- ○ **C.** Only the mode will change.
- ○ **D.** All of them–the mean, median, and mode–will change.

**12.** What is the probability that you will select a heart from a regular deck of cards?

○ **A.** $\frac{13}{52}$ or 25%

○ **B.** $\frac{4}{20}$ or 20%

○ **C.** $\frac{4}{52}$ or about 7.6%

○ **D.** $\frac{4}{50}$ or 8%

## Performance-Based Assessment Questions

**Directions for Questions 13, 14, and 15:** Respond fully to the PBA questions that follow. Show your work and clearly explain your answer. You will be graded on the correctness of your method as well as the accuracy of your answer.

**13.** The Guidance Department in your middle school has surveyed the 8th graders to see what world language they would like to study when they enter high school next year. They put the data collected into a two-way table. See below.

| | Spanish | French | Italian | German | Mandarin | Total |
|---|---|---|---|---|---|---|
| Boys | 70 | 20 | 55 | 15 | 40 | **200** |
| Girls | 50 | 40 | 50 | 10 | 40 | **190** |
| **Totals** | **120** | **60** | **105** | **25** | **80** | **390** |

Round all answers to the nearest tenth.

**Part A:** What percent of *boys* selected Italian? Answer: ________________ %

**Part B:** What percent of *students* selected French? Answer: ________________ %

**Part C:** What percent of *girls* selected either Spanish or Mandarin?

Answer: ________________ %

**Part D:** If they keep class size to between 20–25 students, how many classes should they plan for each language? The total number of classes should be 18.

Spanish: ________________ classes

French: ________________ classes

Italian: ________________ classes

German: ________________ classes

Mandarin: ________________ classes

**14.** Use the circle graph below to answer the following questions:

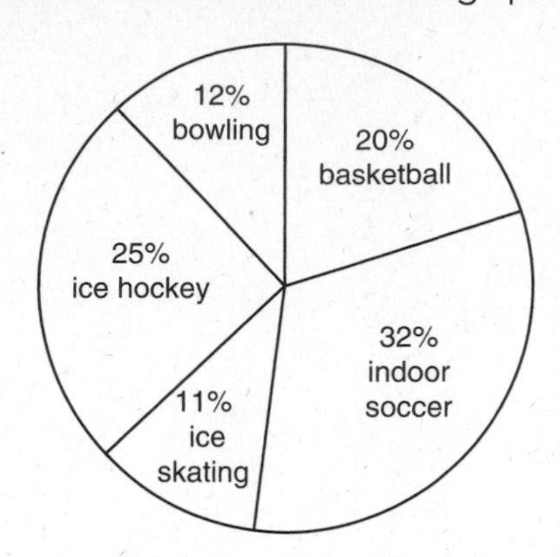


**Part A:** Of the 520 students surveyed, how many said their favorite winter sport was ice hockey?

**Part B:** How many students chose bowling or basketball as their favorite winter sport?

**Part C:** How many more students selected indoor soccer than ice skating?

**Part D:** One of the students who collected the survey data said that all was correct except the percentage of students selecting ice skating. He said that should have been 20%. Explain why this could not be possible if all the other data are true.

**15.** The local supermarket is planning to do a survey. They want to find out what fruit teenagers prefer. They will distribute their survey to 100 people. Which would be the best group of people to use if they want a fair and useful sample?

○ **A.** 100 people entering their local supermarket.

○ **B.** 100 students as they walk out of the local high school.

○ **C.** 100 teachers in the local elementary school.

○ **D.** 100 people getting gas for their car at the corner gas station.

# Chapters 1–4 Solutions

## CHAPTER 1: THE NUMBER SYSTEM (8.NS)

### Compare and Order Integers (page 17)

1. **A.** –2, 2, 6 and **C.** –10, –4, 0, 3 Both show least to greatest.
2. **A.** –4, –6, –8, **B.** 4, –3, –5, and **E.** 5, 1, –6, –8 All show greatest to least.
3. **A.** 6, 4, –8, –10 and **D.** 5, –1, –6, –8

### Absolute Values (page 18)

1. **A.** $|-16|$ $|-16| = 16$; $|5| = 5$; $16 > 5$; 16 is the greater distance.
2. **B.** $|-20|$ $|-20| = 20$; $|-6| = 6$; $20 > 6$; 20 is greater.
3. **B.** $|-20|$ The –20 would be measuring beneath the water, and the absolute-value sign tells you it is distance with a positive value. This could represent 20 feet.

### Rational and Irrational Numbers (page 19)

1. **B.** $\sqrt{6}$ The calculator shows [2.44948743]. There seems to be no repeating pattern. This is an irrational number.
2. **B.** $\sqrt{8}$ The calculator shows [2.828427125]. There seems to be no repeating pattern. This is an irrational number.
3. **D.** 12 $12 \times 12 = 144$ therefore $\sqrt{144}$ is 12 exactly.
4. **B.** $9.27 \sim \sqrt{86}$ Your calculator will show 9.27**3**618; rounding to the nearest hundredth gives you 9.27. (Since the next number is only a 3, you do not change the 7 in the hundredths place.)
5. **C.** $7.35 \sim \sqrt{54}$ Your calculator will show 7.34**8**46922; rounding to the nearest hundredth gives you 7.35. (Since the next number is > 5, you round up the "4" to 5.)

6. **A.** $10.49 \sim \sqrt{110}$ Your calculator will show 10.48**8**08848; rounding to the nearest hundredth gives you 10.49. (Since the next number is > 5, you round up the 8 to the next higher number which is 9.)
7. **K.** $\sqrt{49} = 7$
8. **B.** $\sqrt{4} = 2$
9. **A.** $\sqrt{2}$ is an irrational number; it cannot be written as a fraction.

   Since we know the $\sqrt{1} = 1$, and we know the $\sqrt{3}$ is less than 2,

   we know $\sqrt{2}$ is somewhere between 1 and 2.

   If you check with a scientific calculator, you will see that

   $\sqrt{2} = 1.414213562$ or approximately 1.4 (almost 1½).
10. **G.** $\sqrt{25} = 5$ exactly, because 5 × 5 = 25 exactly.
11. **D.** $\sqrt{9} = 3$ exactly, because 3 × 3 = 9 exactly.
12. **E.** $\pi$ is approximately 3.14. (You should memorize this.)
13. **J.** $\pi + 3$ is about 3.14 + 3 = approximately 6.14.
14. **C.** $\sqrt{7}$ is between $\sqrt{4}$ and $\sqrt{9}$.

    We pick these because we know the square roots of these numbers.

    $\sqrt{4} = 2$ and $\sqrt{9} = 3$. So $\sqrt{7}$ is between 2 and 3.

## Adding, Subtracting, Multiplying, and Dividing Integers and Fractions (page 22)

1. **C.** 5
2. **A.** 9
3. **B.** −15.6
4. **D.** 10
5. 5
6. 5
7. $\frac{11}{12}$ $\quad -\frac{1}{2}+\frac{2}{3}+\frac{3}{4} = -\frac{6}{12}+\frac{8}{12}+\frac{9}{12} = \frac{-6}{12}+\frac{17}{12} = \frac{11}{12}$
8. $5\frac{3}{5}$ $\quad 8-2\frac{2}{5} = 6-\frac{2}{5} = 5+\frac{5}{5}-\frac{2}{5} = 5\frac{3}{5}$
9. $-2\frac{11}{20}$ $\quad -4\frac{3}{4}+2\frac{1}{5} = -4\frac{15}{20}+2\frac{4}{20} = -2\frac{11}{20}$

10. 0 $16-\frac{1}{2}(8)-\frac{3}{4}(16)=16-4-12=16-16=0$

## (8.EE.1-3) Exponents (page 24)

1. **B.** $6^2$
2. **C.** five cubed
3. **D.** $3 \times 3 \times 3 \times 3$
4. **D.** $8^2 - 4^2 = 8 \times 8 - 4 \times 4 = 64 - 16 = 48$ is the largest number.

   $4^2 = 16$; $(4)(2) = 8$; $240 = 1$
5. **C.** $(9)(3)(3) = (9)(9) = 81$ $9^2$ also $= (9)(9) = 81$
6. $2^3$
7. $(-8)^2$
8. $8^3$
9. **A.** $6^{-2} \times 6^4$ It does not equal 1/36. It equals 36.
10. **A.** No **B.** Yes **C.** Yes **D.** Yes **E.** No **F.** Yes

## (8.EE.2) Perfect Square and Perfect Cube Numbers (page 29)

1. **C.** 10
2. **B.** 11
3. **B.** $\frac{1}{4}$
4. 9
5. 162 $(-\sqrt{81})(-\sqrt{81})(2)=(-9)(-9)(2)=(81)(2)=162$
6. 4 *You cannot use –4 since (–4)(–4)(–4) = (16)(–4) which = –64, not 64*
7. **C.** 9
8. **C.** 12 $\sqrt{81}+3=9+3=12$
9. **B.** $4^3$ and **C.** $(\sqrt{16})(4^2)=(4)(16)=64$

   and **D.** $(-\sqrt{16})(-\sqrt{16})(4)=(-4)(-4)(4)=(16)(4)=64$

## Square Roots and Cube Roots (page 30)

1. 5 and –5 $5 \times 5 = 25$ and $(-5) \times (-5) = 25$
2. 9 and –9 $9 \times 9 = 81$ and $(-9) \times (-9) = 81$
3. 6
4. 12
5. 7
6. 10
7. **B.** $5 + 10$
8. **D.** $\sqrt{36} + \sqrt{25} = 6 + 5 = 11$
9. **A.** $16 + 5$ $4^2 + \sqrt{25} = 16 + 5$
10. **C.** $8 + 7$ $2^3 + \sqrt{49} = 2 \times 2 \times 2 + 7 = 4 \times 2 + 7 = 8 + 7$
11. **D.** $\sqrt{100} - 1^2 = 10 - 1 = 9$
12. **B.** $\sqrt{4} = 2$

## (8.EE.2) More Practice with Radicals (Square Roots and Cube Roots) (pages 30–31)

1. ±9 2. ±12 3. –11

4. **A.** $\frac{1^0}{2^1} = \frac{1}{2}$, **C.** $0.5 = \frac{5}{10} = \frac{1}{2}$, and **D.** $\frac{\sqrt{25}}{\sqrt{100}} = \frac{5}{10} = \frac{1}{2}$

5. **B.** and **C.** are correct $\frac{\sqrt{9}}{\sqrt{144}} = \frac{3}{12} = \frac{1}{4}$

## (8.EE.2) More Practice with Multiple-Choice Questions (Square Roots and Cube Roots) (page 31)

1. **B.** 2. **C.** 3. **D.** 4. **A.** 5. **C.**

6. **A.** Yes **B.** No **C.** Yes **D.** No **E.** No

## (8.EE.2) Mixed Practice with Square Roots and Cube Roots (pages 31–32)

1. $x = 5$ 2. $y = 2$ 3. $a = 10$ 4. $m = 3$
5. **B.** 4 because $4^3 = \mathbf{(4)(4)}(4) = \mathbf{(16)}(4) = 64$
6. **B.** 2 because $2^3 = \mathbf{(2)(2)}(2) = \mathbf{(4)}(2) = 8$

## (8.EE.2) Estimating Square Roots (pages 32–33)

1. **D.** $\sqrt{50}$ is between $\sqrt{49}$ and $\sqrt{64}$, $\sqrt{49} = 7$, and $\sqrt{64} = 8$.
2. **B.** $\sqrt{15}$ is a little less than 4. $\sqrt{16}$ is 4. $\sqrt{15}$ is about 3.87.
3. **C.** $\sqrt{30}$ because $\sqrt{25} = 5$ and $\sqrt{36} = 6$; 30 is between 25 and 36.
4. **C.** $\sqrt{115}$ is between $\sqrt{100}$ and $\sqrt{144}$. (It is between 10 and 12.)
5. **B.** $\sqrt{100} = 10$.
6. **A.** Yes **B.** No **C.** Yes **D.** Yes **E.** No
7. **B.** It is less than 7.0. $\sqrt{38}$ is between $\sqrt{36}$ *and* $\sqrt{49}$, between 6 and 7
   **D.** It is greater than π. π is approximately 3.14 and $\sqrt{38}$ is between 6 and 7.
8. **A.** It is equivalent to $(-6)^2$. This means $(-6)(-6)$ which = 36.
   **C.** It is equivalent to $(2^2)\,[(-18) \div (-2)]$ which $= 4(-18 \div -2) = 4(9) = 36$.
   **D.** It is equivalent to $(-72)\,(-1/2)$ which $= \frac{-72}{-2}$ which = 36.

## (8.EE.4) Scientific Notation (pages 35–36)

1. See the bold answers in the chart.

| Standard Form | How Many Places Will You Move the Decimal Point? | Scientific Notation Form |
|---|---|---|
| 650,000,000,000,000 | 14 | $\mathbf{6.5 \times 10^{14}}$ |
| 2,000,000,000 | **9** | $2.0 \times 10^{9}$ |
| **450,000,000,000** | 11 | $4.5 \times 10^{11}$ |
| **3,250,000,000** | **9** | $3.25 \times 10^{9}$ |
| 12,000,000,000 | **10** | $\mathbf{1.2 \times 10^{10}}$ |

2. **A.** $9.45 \times 10^{12}$ because you move the decimal point 12 places 9,450,000,000,000.
3. **C.** $6.5 \times 10^{11}$

4. **B.** 110,000,000

5. **B.** $2.8 \times 10^8$ Barbara is correct. A number in scientific notation format has only one digit to the left of the decimal point.

6. **Part A:** **C.** They are both correct.

   **Part B:** 250 $2.5 \times 10^2 = (2.5)(100) = 250$

   0.062 $(6.2)(10^{-2}) = (6.2)\frac{1}{10^2} = (6.2)\frac{1}{100} = \frac{6.2}{100} = 0.062$

   **Part C:** $\underline{1.2 \times 10^2}$ definitely represents a larger quantity than $\underline{3.6 \times 10^{-2}}$.

   $1.2 \times 10^2 = 1.2 \times 10 \times 10 = 1.2(100) = \mathbf{120}$

   $3.6 \times 10^{-2} = 3.6 \times \frac{1}{10^2} = \frac{3.6}{10^2} = \frac{3.6}{(10)(10)} = \frac{3.6}{100} = \mathbf{0.036}$

   **Part D:** It is equivalent to a fraction or decimal number less than one.

   $8 \times 10^{-2}$ is equivalent to $8 \times \frac{1}{10^2} = 8 \times \frac{1}{100} = \frac{8}{100} = \frac{4}{50} = \frac{2}{25} = 0.08$

## Comparing Data in Real-Life Situations (pages 39–41)

1. The land area of Saudi Arabia is about **5** times as large as the land area of Sweden.

   $\frac{8.64869 \times 10^5}{1.73732 \times 10^5} = 4.97817903\ldots \times 10^{5-5}$ which is equivalent to $4.97817903 \times 10^0$

   Since any number to the *zero power* = 1, $4.97817903 \times \mathbf{10^0} = 4.97817903 \times \mathbf{1}$; which is **5** when rounded to the nearest whole number.

2. **Part A:** **Russia** (6.6 million sq. miles) **India** (1.24 million sq. miles)

   **Part B:** Russia's land area is about **5.3** times as large as India's land area.

   $\frac{6.6}{1.24} = 5.3225$ *or about* 5.3 *times greater*

   **Part C:** U.S. land area is about **3 times as large as** India's land area.

   $\frac{3.8}{1.24} = 3.064516129$ *or about* 3 *times as large*

3. **Part A:** China: $\mathbf{1.3042 \times 10^3}$ South Korea: $\mathbf{4.77 \times 10^1}$

   **Part B:** Answer: China's population was $\mathbf{.27341719 \times 10^2}$ times greater than South Korea's population.

   $\frac{1304.2}{47.7} = \frac{1.3042 \times 10^3}{4.77 \times 10^1} = \frac{1.3042}{4.77} \cdot 10^{3-1} = 0.273 \times 10^2$

**Part C:** In 2006 China's population was about **27** times greater than South Korea's population.

$.27341719 \times 10^2 = .27341719 \times 100 = 27.341719$

which is **about 27 times larger**

4. **Part A:** China: **$1.3042 \times 10^3$ million** United States: **$2.94 \times 10^2$ million**

**Part B:** In 2006, China's population was **.44x 605442 x $10^1$** times as large as the U.S. population.

$$\frac{1304.2}{294} = \frac{1.3042 \times 10^3}{2.94 \times 10^2} = \frac{1.3042}{2.94} \cdot 10^{3-2} = .443605442 \times 10^1$$

**Part C:** In 2006 China's population was about **4.4** times as large as the U.S. population.

$44.3605442 \times 10^1 = 4.43605442$ or about 4.4 (rounded to nearest tenth)

## The New Look of Multiple-Choice Questions (pages 42-43)

**A.** No Remember *PEMDAS, Please Excuse My Dear Aunt Sallie* reminds you to work inside parentheses first.

**B.** No $3^2 = 3 \times 3 = 9$

**C.** No $(5 - 2)^2 = 3^2 = 9$ $5^2 - 2^2 = 25 - 4 = 21$

**D.** Yes $4\,(3^2) = 4 \times 3 \times 3$ $4\,(3^2) = (4)(9) = 36$

$4 \times 3 \times 3 = (4)(9) = 36$

**E.** Yes $3\,(5 - 2)^2 + 4\,(3^2) - 9 \div 3\,(2)$

$3\,(3)^2 \quad + 4\,(9) \quad - 3\,(2)$

$3\,(9) \quad + 36 \quad - 6$

$27 \quad + 36 \quad - 6 = 27 + 30 = 57$

## (6.EE.3, 7.NS.1-2) Algebraic Order of Operations (pages 43-44)

1. **B.** Divide and **B.** 24 because $27 - 3 = 24$
2. **C.** Divide and **D.** 55 $36 + 9 \times 3 - 8 = 36 + 27 - 8 = 63 - 8 = 55$
3. **A.** 12 $(72 \div 9 - 2) + 2 \times 3 = (8 - 2) + 6 = 6 + 6 = 12$
4. **C.** 58 $(20 + 8 \div 4) + (3^2 \times 4) = (20 + 2) + (9 \times 4) = 22 + 36 = 58$
5. **B.** 16 $[(2(4) + 3(4)] - 4 = 8 + 12 - 4 = 20 - 4 = 16$

6. **C.** 4 — $(6-4)^3 - 4 = 2^3 - 4 = 8 - 4 = 4$
7. **D.** $2 \times 4$ — Work inside **P**arentheses first; **M**ultiply before **S**ubtracting.
8. **A.** — $(2+3) \times 2 \times 5 - 1 = 5 \times 2 \times 5 - 1 = 10 \times 5 - 1 = 50 - 1 = 49$
9. **D.** — $4 + 3 \times 2 \times (5-1) = 4 + 3 \times 2 \times (4) = 4 + 6 \times (4) = 4 + 24 = 28$
10. Billy — In the expression $2(2 + 5 \times 2)$ you should work inside the **P**arentheses first, then **M**ultiply before **A**dding: $2(2 + 10) = 2(12) = 24$

## (6.RP.1–3, 7.RP.3) Ratio and Proportion (pages 46–48)

1. **B.** girls/total $= \frac{14}{25}$
2. **B.** 3,012
3. **B.** $\frac{60}{100}$
4. **C.** 40 cm
5. **D.** 9 feet — $\frac{3}{4} = \frac{x}{12}$; therefore, $4x = (3)(12)$, $4x = 36$, $x = 9$
6. **B.** $\frac{28}{15} = \frac{7}{x}$
7. **A.** $m = 10$ — $2m = (2.5)(8)$, $2m = 20$, $m = 10$
8. **C.** 180 — $0.3c = (16)(9)$, $0.3c = 54$, $c = \frac{54}{0.3}$, $c = 180$
9. **B.** 1.35″ — $\frac{24}{5.4} = \frac{6}{x}$, $(6)(5.4) = 24x$, $32.4 = 24x$, $1.35 = x$
10. **B.** Leoh Ming Pei 1.25 or $1\frac{1}{4}$ — $\frac{5}{j} = \frac{12}{3}$, $12j = 15$ $\quad j = \frac{15}{12}$ or $\frac{5}{4}$ or $1\frac{1}{4}$ or 1.25

## SCR Non-calculator Questions (pages 48–50)

1. (5 – 2) Remember PEMDAS (*Please Excuse My Dear Aunt Sally*); work inside Parentheses first
2. Sarah
3. $2.389 \times 10^5$ miles away
4. $\sqrt{3}$ $\quad\sqrt{9}$ $\quad\pi$ $\quad 2^3$ $\quad 3^2$ $\quad\quad \sqrt{3} \sim 1.7$ $\quad \sqrt{9} = 3$ $\quad \pi \sim 3.14$ $\quad 2^3 = 8$ $\quad 3^2 = 9$

5. $\frac{85}{100}$

6. 50 miles $\frac{0.5}{10} = \frac{2.5}{50}$ or think that $0.5 \times 5 = 2.5$ miles
then $10 \times 5 = 50$ miles

7. 800 e-mails (5 days)(4 weeks) = 20 days; (40 e-mails)(20 days) = (40)(20) = 800

8. Yes burger = \$2.50, fries = \$1.50, small soda = \$1.00, for a total of \$5.00

9. \$3.25 chicken nuggets = \$3.25, fries = \$1.50, medium juice = \$2.00, for a total of \$6.75; \$10.00 – \$6.75 = \$3.25

10. **A.** 24 inches wide

## PBA Questions (pages 50–52)

1. **Part A:** They should *not* use the whole-day rate. The whole-day rate would cost \$40.00; the other option would cost only \$22.00 [6 + 2(8)].

   **Part B:** The total with the second option would be \$6.00 + \$48.00 = \$54.00.
   The first hour would cost \$6.00.
   The next 6 hours would cost \$8.00 each, so (8)(6) = \$48.00.

2. **Part A:** His new hourly rate would be \$11.04 instead of \$12.
   (12)(0.08) = \$.96, then \$12.00 – \$.96 = \$11.04, or (12)(.92) = \$11.04.

   **Part B:** His weekly salary would be \$38.40 less with the 8% reduction.
   Original weekly salary: (12)(40 hrs.) = \$480.00
   Reduced weekly salary: (11.04)(40 hrs.) = \$441.60
   The difference is \$480 – 441.60 = \$38.40

3. **Part A:** He would save \$1.45 buying the *Morning Special* instead of buying each item separately.
   1.20 + 1.20 + 1.25 + 2.25 + 1.50 + 1.00 = \$8.40 if items are purchased separately.
   \$8.40 – \$6.95 (*Morning Special*) = \$1.45 difference.

   **Part B:** His total bill, including 6% sales tax and a \$1.50 tip, would be \$8.87.
   (\$6.95)(1.06) = 7.367 (round up to nearest penny); \$7.37 + \$1.50 = \$8.87

   **Part C:** His change from \$10.00 would be \$1.13. 10.00 – 8.87 = 1.13

4. **Part A:** 37.5% (or $37\frac{1}{2}$%) also attended the park the week before. $\frac{3}{8} = 0.375$ or 37.5%

   **Part B:** Approxmiately 1,500 people who attended the park this week also attended the park the previous week. (0.375)(4,000 people) = 1,500 people

5. **Part A:** He would make 10 trips in all.

   Oil changes at miles

   500, **1,000**, 1,500, **2,000**, 2,500, **3,000**, 3,500, **4,000**, 4,500, **5,000**

   Tires rotated at miles

   **1,000** **2,000** **3,000** **4,000** **5,000**

   Notice that there are 5 times when he gets only his oil changed and there are 5 more times when he gets both an oil change and the tires rotated at the same time.

   **Part B:** He would spend $19.60 on gasoline for the 10 round trips to the service station.

   (10 trips)(16 miles each round trip) = 160 miles total

   (160 miles)(20 miles per gallon) = 8 gallons of gasoline used

   (8 gallons)($2.45 per gallon) = $19.60 on gasoline

6. **Part A:** Sweater Barn at the upstate New York mall where there is an 8% sales tax on clothing:

   Cost for sweater:

   $24 less 15% is same as $24 × 0.85 = $20.40 per sweater

   Cost for cardigan:

   $20.40 × 50% is (1/2 of $20.40) = $10.20 per cardigan

   Total cost for one sweater and one cardigan before taxes = $30.60 per set

   Three sets = (3)($30.60) = $91.80

   Total for 3 sets with sales tax = (1.08)($91.80) = **$99.15** ($99.14 also accepted as correct) total with sales tax

**Part B:** Teen Outlet in New Jersey (where there is no sales tax on clothing):

Cost for sweater: $18 less 10% = $16.20 for each sweater

Cost for cardigan: Each cardigan is the same price = $16.20 per cardigan

Total cost for one sweater and one cardigan = (2)($16.20) = $32.40 per set

Total for 3 sets = (3)($32.40) = **$97.20** total. This would would be the final price because there is no sales tax on clothing in New Jersey.

**Part C:** The Teen Outlet store in New Jersey would be a slightly better buy. The cost at the New Jersey store would be **$1.95** (or $1.94) less than the New York store.

## Chapter 1 Test: The Number System (pages 53-57)

1. −12 $-3\left(\frac{1}{2}\right)^{-2}=-3\left(\frac{2}{1}\right)^{2}=-3(4)=-12$

2. 23½ $\sqrt[3]{27}+\sqrt{\frac{36}{144}}+|-20|=3+\frac{\sqrt{36}}{\sqrt{144}}+20=3+\frac{6}{12}+20=23\frac{1}{2}$

3. **D.** 0.45

4. **C.** $\sqrt{115}$ is between numbers that I know are perfect square numbers.

   $\sqrt{115}$ is between $\sqrt{100}$ and $\sqrt{121}$.

   Since $\sqrt{100}$ is 10, and $\sqrt{121}$ is 11, then $\sqrt{115}$ is between 10 and 11.

5. **Part A:** 5 $\frac{1450}{300}=4.8\overline{333}$ which is about 5.

   **Part B:** 72.5 or 72½ $\frac{1450}{300}=4.8\overline{333}$ $\frac{1450}{300}(15)=72.4\overline{9999}$

6. **D.** Subtract 2 from 8 (8 – 2)

   Use the correct *Algebraic Order of Operations*. Work inside parentheses first. Remember, PEMDAS or *Please Excuse My Dear Aunt Sally*.

7. **A.** Yes **B.** No **C.** No **D.** No **E.** Yes

8. **A.** Yes $\sqrt{144}=12$

   **B.** Yes Because $\sqrt{144}=12$ and $\sqrt{100}+\sqrt[3]{8}=10+2=12$

   **C.** No $\left(\frac{1}{12}\right)^{2}=\frac{1^2}{12^2}=\frac{1}{144}$

   **D.** No 12

   **E.** Yes 12 is an even number

9. **D.** Floor with carpets. Look at the given choices in **A, B, C,** and **D** and write them down or circle their values.

**A.** double brick wall filled with foam 0.5

**B.** roof with no insulation 2.2

**C.** double-glazed window with air gap 2.7

**D.** floor with carpets 0.3

Then, reread the question to see that you need the *smallest U-value* (the smallest number), which is 0.3.

10. **A.** 5 In this case it is easiest to use the choices given and plug in the numbers. To find 5% of $7,000, multiply ($7,000)(.05) = $350

11. **A.** 61.5 inches First, notice that each side of the large shape is 3 times larger than the smaller shape. You can find the perimeter of the large shape two ways.

1. Find the perimeter of the smaller shape and multiply by 3:

$$3 + 5 + 5.5 + 7 = 20.5;\ (20.5)(3) = 61.5$$

2. Multiply each side of the small shape by 3 to get the dimensions of the large shape, and then add them together:

$$9 + 15 + 16.5 + 21 = 61.5$$

12. **A.** 429 students

**B.** 416 students

**C.** 64% of students Pizza + chili = 56% + 8% = 64%

**E.** > $(20)^2$ students 416 > 400 $(20)^2 = (20)(20) = 400$

13. **B** and **D** are equivalent to 64.

**B.** $8^{-2} \times 8^4 = 8^{-2+4} = 8^2 = \mathbf{64}$ or $8 \cdot 8 = \frac{8 \cdot 8}{1} = \frac{64}{1} = 64$

**D.** $8^6 \times 8^{-4} = \frac{8^6}{8^4} = \frac{8^2}{1} = \frac{64}{1} = 64$ or $8^6 \times 8^{-4} = 8^{6-4} = 8^2 = \mathbf{64}$

14. **B.** Yes **C.** Yes **D.** Yes

**Performance-Based Assessment Questions**

15. **Part A:** Alco is the least expensive. Gas at Alco costs $1.92 per gallon.

Total price ÷ Number of gallons = Price per gallon

| | | |
|---|---|---|
| 18.24 ÷ 9.5 = | $1.92 per gallon | Alco |
| 18.31 ÷ 9.2 = | $1.99 per gallon | Bright |
| 17.02 ÷ 8.3 = | $2.05 per gallon | Custom |
| 16.80 ÷ 8 = | $2.10 per gallon | Dixon |
| 18.45 ÷ 9 = | $2.05 per gallon | Extra |

**Part B:** Marco would save $3.60.

$$\begin{array}{ll} \underline{20\text{ gal} \times 2.10 =} & \underline{\$42.00}\text{ most expensive} \\ 20\text{ gal} \times 1.92 = & -38.40\text{ least expensive} \\ & \\ \text{savings:} & \$3.60 \end{array}$$

16. Again, it is easy to solve this problem if you organize your information in a table or chart. Show your work.

| If on sale for 5% off, then she would pay 95% or 0.95 times previous price | | Price Each Hour |
|---|---|---|
| Before 1:00 P.M. Original price | | $89.50 |
| Price at 1:00 P.M. less 5% of 89.50 | 89.50 × 0.95 | $85.03 |
| Price at 2:00 P.M. less 5% of 85.03 | 85.03 × 0.95 | $80.78 |
| Price at 3:00 P.M. less 5% of 80.78 | 80.78 × 0.95 | $76.74 |
| Price at 4:00 P.M. less 5% of 76.74 | 76.74 × 0.95 | $72.90 |
| Price at 5:00 P.M. less 5% of 72.90 | 72.90 × 0.95 | $69.26 |

**Part A:** Yes, Susan will be able to buy the jacket today.

**Part B:** It will cost her $69.26.

**Part C:** She will be able to buy the jacket at 5:00 P.M.

# CHAPTER 2: GEOMETRY (8.G)

## (7.G.2–6) Points, Lines, and Planes (pages 70–75)

1. **A.** a line
2. **C.** a circle
3. **D.** a rectangle, **F.** a square, and **G.** a parallelogram
4. **D.** a right triangle and **F.** a polygon
5. **B.** a circle
6. **D.** $m\angle a = m\angle b$
7. **D.** Vertical angles are equal.
8. **B.** $m\angle 1 = 46°$
9. **A.** line

The statements that are true are choices are

10. **A.**, **B.**, **C.**, **E.**, and **F.**
11. **B.** It is a triangular prism with a rectangular base.

    **C.** The base has four right angles. (Yes, it is a rectangle.)

    **E.** It has four triangular faces (front, back, and two sides).
12. **A.** a point
13. **D.** a circle
14. 1. **C** 4. **A** 7. **F**

    2. **K** 5. **D** 8. **H**

    3. **E** 6. **J** 9. **G**

    10. **B**

## (8.G.A.5) Parallel Lines Cut by a Transversal Line (pages 77–79)

1. **A.** No **B.** Yes **C.** Yes **D.** Yes **E.** Yes **F.** Yes
2. **Part A:** No

**Part B:** If they were parallel the measure of angle-2 would equal the measure of angle-6 since they are *corresponding* angles. And the measure of angle-6 would equal the measure of angle-8 because they are *vertical* angles. In the diagram angle-2 measures 80° and angle 8 measures 85°. They are not equal.

3. **B.** $A'(2, 4)$, $B'(6, 8)$, $C'(12, 8)$, $D'(8, 4)$
4. **C.** $P'(4, 6)$, $Q'(12, 6)$, $R'(8, 0)$
5. There is more than one way to explain this. One explanation is:

   If the lines are parallel, then *corresponding angles would be equal.*

   Angle 2 and angle 6 are corresponding angles. If lines *a* and *b* are parallel, then angle 2 should = angle 6.

   Angle 5 + angle 6 are supplementary and = 180°, therefore angle 6 = 80°.

   Angle 2 measures 70°, and angle 6 measures 80°. They are not equal.

## Area of Flat Shapes (pages 84–85)

1. 59.45 or 59 square feet
2. 30 square feet
3. a. **B.** 7.5 in.
   b. **D.** 8.6 yd.
   c. **D.** 113 square feet
   d. **D.** 255 square feet
   e. **B.** 113 square feet
4. Area of circle is approximately 50.24 sq. cm.

   First, find height of the rectangle. The height = 8 cm.

   Note: there is a right triangle here that has a base of 6 (15 – 9 = 6) and a hypotenuse of 10. Using the Pythagorean theorem you have:

   $a^2 + b^2 = c^2 \qquad 6^2 + x^2 = 10^2 \qquad 36 + x^2 = 100 \qquad x^2 = 64 \qquad x = 8$

   You now know the *diameter* of the circle is also 8 cm. You need to use the *radius* of the circle to find the area of the circle. The radius is 4 cm.

   The formula for *area of a circle* = $\pi r^2$

   The area of this circle is ~$(3.14)(4^2) = (3.14)(16)$ = about 50.24 sq. cm.
5. **D.** $r^2 \sim 81$; therefore, $r \sim 9$
6. $1 \times 72 \quad 2 \times 36 \quad 3 \times 24 \quad 4 \times 18 \quad 6 \times 12$

## Area of the Shaded Region (pages 88–91)

1. 30.96
2. **D.** 20-21 square feet
3. 20 square inches
4. 22 square units
5. **C.** $25 \times 25 = 625$ (area not 100)
6. **C.** 102 square units
7. **B.** $24 + \sim 6.28 = 30.28$; use 30 sq. cm
8. **A.** (using 3.14 for $\pi$) 497.44 sq. ft
9. **B.** $54 \times 44 = 2{,}376$ sq. ft
10. **D.** $(9 \times 4)/2 = 36/2 = 18$
11. 300 square inches; $A_{rectangle} = 30 \times 20 = 600$; $A_{triangle} = \frac{1}{2}(30)(20) = 300$

## Surface Area and Volume (pages 96–98)

1. **C.** $3 \times 3 = 9$; $9 \times 6$ sides $= 54$ square units
2. **B.** 3 in.
3. **B.** 4 cm
4. **C.** 9 sq. ft.
5. **B.** 8 feet
6. **A.**, **B.**, and **C.** are correct expressions of the sphere.
7. **A.** Yes — Volume of a cylinder $= \pi r^2 h = (\pi)(2)(2)(8) = \pi(4)(8) = 32\pi$

   **B.** Yes — Volume of a cube = side cubed $= s^3 = (5)(5)(5) = (25)(5)$

   **C.** No — Volume of a rectangular solid = (length)(width)(height)

   Volume $= (2.5)(5)(8) = (2.5)(40) = 100$ cubic units

   **D.** No — Volume of the cube = side cubed $= (2)(2)(2) = (4)(2) = 8$

   Volume of the cylinder = (area base)(height)

   $= \pi r^2 h = \pi(2)(2)(4) = (4)(4) = 16$

   **E.** Yes — Volume shorter cylinder $= \pi r^2 h = \pi 3^2(2) = 18\pi$ cubic units

   Volume taller cylinder $= \pi r^2 h = \pi 2^2(4) = 16\pi$ cubic units

## (6.G.2, 8.G.9) Perimeter (pages 101–104)

1. **B.** $63\frac{2}{3}$  $2\left(12\frac{1}{3}\right)+2\left(19\frac{1}{2}\right)=24\frac{2}{3}+39=63\frac{2}{3}$

2. 42 inches  (8.4) (5) = 42

3. **Part A:** 42 feet  12 + 12 + 9 + 9 = 24 + 18 = 42

   **Part B:** **36** sq. feet and **24** sq. feet

   **Part C:** First, divide this irregular shape into rectangles.

Then find the area of each section and add them together. Either way the areas measure 36 sq. feet and 24 sq. feet. The total area is 36 + 24 = **60 square feet.**

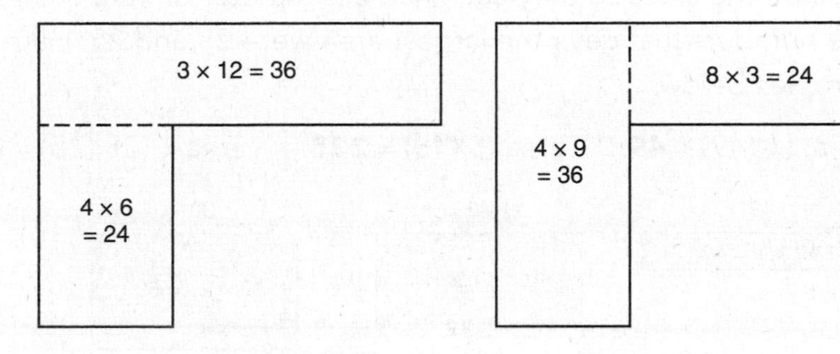


4. The correct answers are **B.** 9.5 × 3 and **F.** 9.5 + 9.5 + 9.5

5. **B.** 27.6  10.6 + **8.5** + **8.5** = 27.6

   Remember, in an *isosceles* triangle, the **two *sides*** are the same length.

6. 42 units long

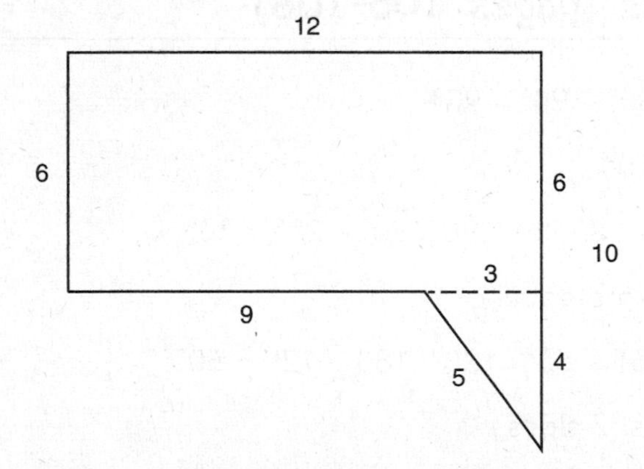


Notice that you can create a 3,4,5 special right-triangle to calculate the length of the missing segment. It is 5 units long.

Now just add all the five sides to calculate the *perimeter*: 6 + 12 + 10 + 5 + 9 = 42

7. 39.564 feet is the approximate *circumference* of the circle.
   The formula for *circumference* is $\pi d$ or (3.14)(12.6) = 39.564

8. **Part A:** Willy is correct.
   **Part B:** The formula for circumference of a circle is $\pi$***d***.
   Here, the diameter is given as 16.
   Therefore, the circumference is **$\pi$16** or **16$\pi$.**

9. **Part A:** It should be shaped more like a **square.**
   **Part B:** The dimensions should be **25 × 25.**
   **Part C:** I tried different combinations of numbers. I drew some rectangles and labeled the sides so they each had a perimeter of 100. Then I saw that the numbers that gave the largest area were 25 and 25. Here are some samples below.
   area: (1)(49) = **49**  area (5)(45) = **225**  area = (25)(25) = **625**

49
49 × 1 = 49
1

40
40 × 10 = 400
10

30
30 × 20 = 600
20

10. ***C.*** 88  Diameter = 12, height = 32
    12 × 2 = 24 + 32 × 2 = 64
    24 + 64 = 88.

## (8.G.4) Triangles (pages 106–108)

1. 40° + 50° = 90°, 180° – 90° = 90°
2. 120°  180° – 60°
3. 70° 180° – 90° (right angle) = 90°; 90° – 20° = 70°
4. 55° Vertical angles are equal.
5. Angle B = 60°  65 + 55 = 120°; 180 – 120 = 60°
6. **Part A:** isosceles (2 sides ≅)
   **Part B:** equilateral (all sides ≅)
   **Part C:** scalene (no sides ≅)
7. **Part A:** acute (all angles less than 90°)
   **Part B:** obtuse (one angle = 102°)
   **Part C:** right (50° + 40° = 90°)

8. $\angle y$ $\angle y$ is opposite the longest side.

9. Triangle *A* is larger. Triangle *A* $(5 \times 5)/2 = 25/2 = 12.5$ sq. units
   Triangle *B* $(4 \times 5)/2 = 20/2 = 10$ sq. units

10. **Part A:** 80°, vertical
    **Part B:** 100°, 180°
    **Part C:** 50°, 180° $\angle 2 = 100°$, $\angle 7 = \angle 2 = 100°$, $\angle 4 = 30°$
    **Part D:** 130° $\angle 6 + \angle 5 = 180°$; $\angle 6 = 50°$, $\angle 5 = 130°$
    **Part E:** Answers may vary. $\angle 1$ and $\angle 2$, $\angle 2$ and $\angle 3$, $\angle 3$ and $\angle 7$, $\angle 7$ and $\angle 1$, or $\angle 5$ and $\angle 6$.

## Other Polygons (pages 109–111)

1-4, answers will vary:

| | |
|---|---|
| 8-sided polygon | 6 triangles |
| 6-sided polygon | 4 triangles |
| 5-sided polygon | 3 triangles |
| 4-sided polygon | 2 triangles |

5. 180°

6. Answers will vary: (number of triangles × 180°)
   8 sided = 6 × 180 = 1080°
   6 sided = 4 × 180 = 720°
   5 sided = 3 × 180 = 540°
   4 sided = 2 × 180 = 360°

7. **C.** 84 sq. units Explanations may vary:
   One example: $(6 \times 8) = 48$; $(12 \times 3) = 36$; $48 + 36 = 84$
   Other example: $2(3 \times 2) = 12$; $(9)(8) = 72$; $12 + 72 = 84$

8. **A.**

## Right Triangles (pages 113–115)

1. **C.** 15 cm long Use $a^2 + b^2 = c^2$: $12^2 + 9^2 = c^2$; $144 + 81 = c^2$, $225 = c^2$, $15 = c$

2. **D.** 7.8 inches long Use $a^2 + b^2 = c^2$: $6^2 + 5^2 = c^2$; $36 + 25 = c^2$, $61 = c^2$, $7.8 = c$

3. **A.** 15 cm Use $a^2 + b^2 = c^2$: $8^2 + b^2 = 17^2$; $64 + b^2 = 289$, $225 = b^2$, $15 = b$

4. **B.** 26 feet tall — Use $a^2 + b^2 = c^2$: $24^2 + 10^2 = c^2$; $576 + 100 = c^2$, $676 = c^2$, $26 = c$

5. ~10.24 yards — Use $40^2 + 12^2 = \text{Hypotenuse}^2$: $1{,}600^2 + 144^2 = \text{Hypotenuse}^2$; $1{,}744 = \text{Hypotenuse}^2$, $41.76 \sim \text{Hypotenuse}$

   Walking around the 2 sides = 40 + 12 = 52 yards;

   Walking along the diagonal line (hypotenuse) = 41.76 yards

   52 – 41.76 is approximately a 10.24-yard shorter walk.

6. ~7 meters — $5^2 + 5^2 = c^2$: $25 + 25 = c^2$; $50 = c^2$, approximately $7 = c$

7. ~13.7 ft — $x^2 + 6^2 = 15^2$: $x^2 + 36 = 225$; $x^2 = 225 - 36$, $x^2 = 189$, $x \sim 13.7$ ft.

8. ***C.*** ~11

9. 15 inches exactly

   Redraw the figure.

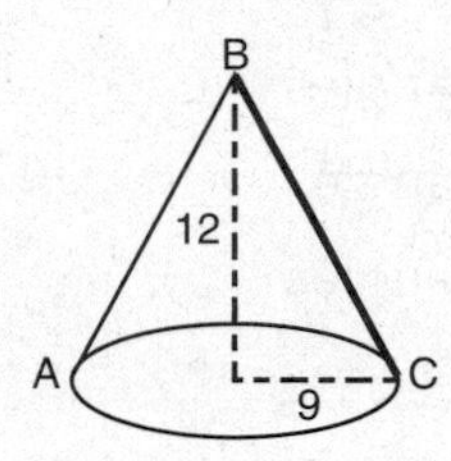


   Label with the information you know.

   You have a right triangle. Use the Pythagorean Theorem to solve for the length of BC.

$$9^2 + 12^2 = (\textit{segment } BC)^2$$

$$81 + 144 = 225 = (\textit{segment } BC)^2$$

   Or, you might notice that the sides are multiples of 3 and 4. This is a version of the special 3-4-5 right triangle; this one measures **3** × 3 **4** × 3 then **5** × 3

   9 12 15

10. about 18 feet

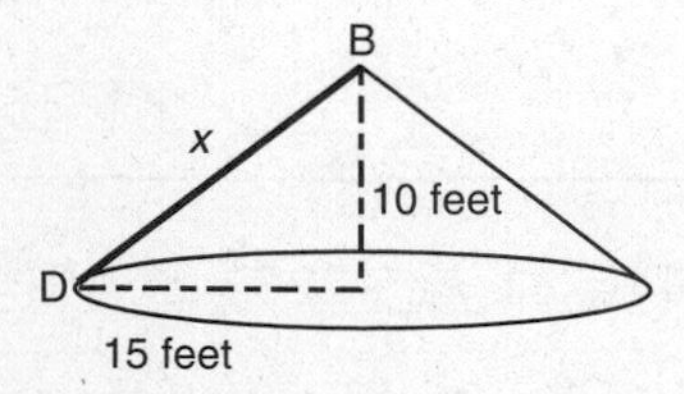


   Use the Pythagorean Theorem to solve for $x$.

$$10^2 + 15^2 = x^2$$

$$100 + 225 = x^2$$

$$325 = x^2$$

$\sqrt{325} = x$ *x is approximately 18 feet*

## (6.G.3-6, 8.G.1-4) Coordinate Geometry (pages 118–121)

1. 6
2. 5
3. 10 From –4 to 6 is 10 units; or think $|-4| + |6| = 4 + 6 = 10$.
4. 9 from –3 to 6 is 9 units; or think $|-3| + |6| = 3 + 6 = 9$.
5. **D.** 32
6. **A.** Area of rectangle = (10)(9) = 90 sq. units
7. **Part C:** 20 units $AB = 6$, $BC = 4$, $CD = 3$, $DE = 2$, $EF = 3$, $FA = 2$

   Add all sides: 6 + 4 + 3 + 2 + 3 + 2 = 20 units is perimeter.

   **Part D:** 18 sq. units Divide the shape into rectangles and find the area of each rectangle and add them together. One solution is: (3)(4) = 12 and (2)(3) = 6; 12 + 6 = 18.
8. **B.** Area = 8 × 9 = 72 sq. units
9. **A.** $BC = 6$, $CA = 8$, $AB = 10$ (hypotenuse)

   Use $a^2 + b^2 = c^2$: $6^2 + 8^2 = 100^2$; 36 + 64 = 100;

   Perimeter (add all sides) = 6 + 8 + 10 = 24 units

## (6.G.3) Congruency (pages 122–124)

1. **C.** The corresponding angles are =, but we don't have information about their sides. To be congruent their corresponding sides must be the same length.
2. **B.** No, because their corresponding sides are not equal.
3. No Answers will vary. The following two shapes have the same perimeter, but they are not congruent: A rectangle could have dimensions of 12, 10, 12, 10 ($P = 44$); a trapezoid could have dimensions 11, 14, 11, 8 ($P = 44$).
4. Yes
5. **C.** The length of the short sides.
6. Yes
7. No
8. No Example: Rectangle $R$ has length = 40 and width = 10; Rectangle $B$ has length = 30 and width = 20
9. No They could be different sizes.
10. Answers will vary.

11. Answers will vary.

12. (1, –5)

## Translating Points (page 125)

1.

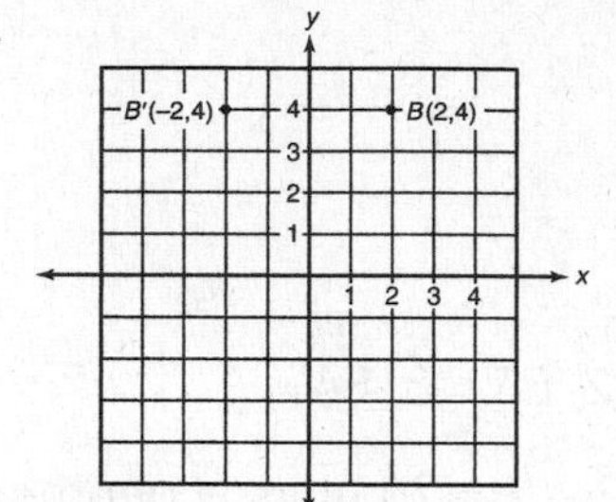


2.

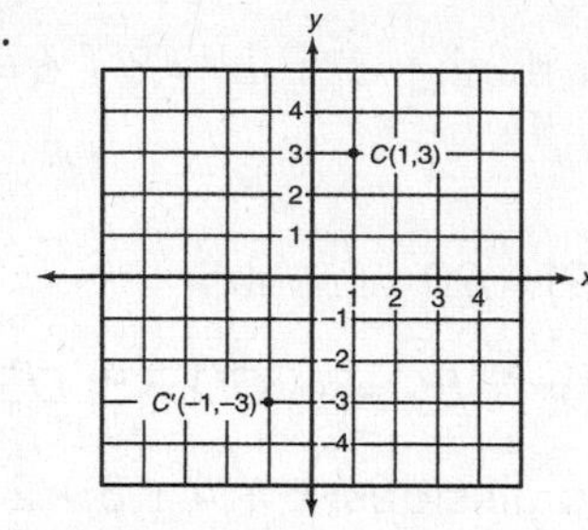


3.

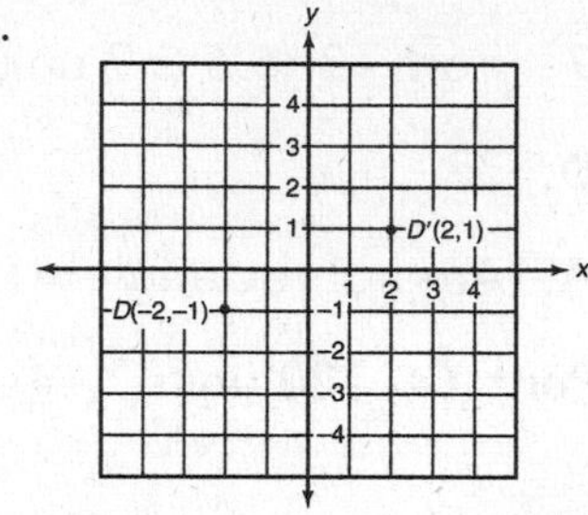


4. Show $D'$ moved to the right at coordinate 10.

5. Show $A'$ moved to the left at coordinate –2.

## Translating Polygons (pages 133–134)

1. **D.** 

2. **D.** Dilation (The same shape is enlarged.)

3. **Part A:**

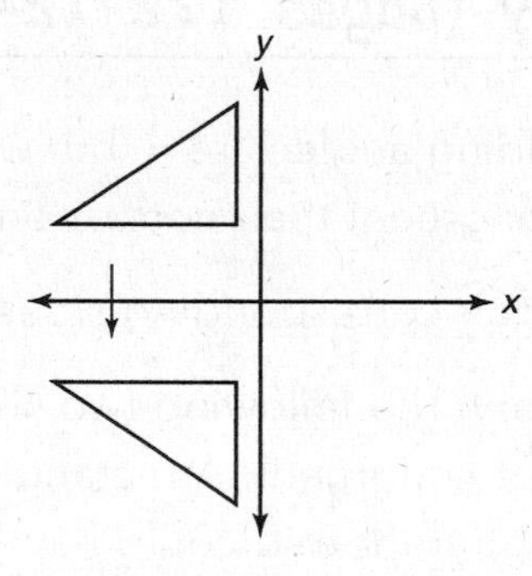


**Part B:**

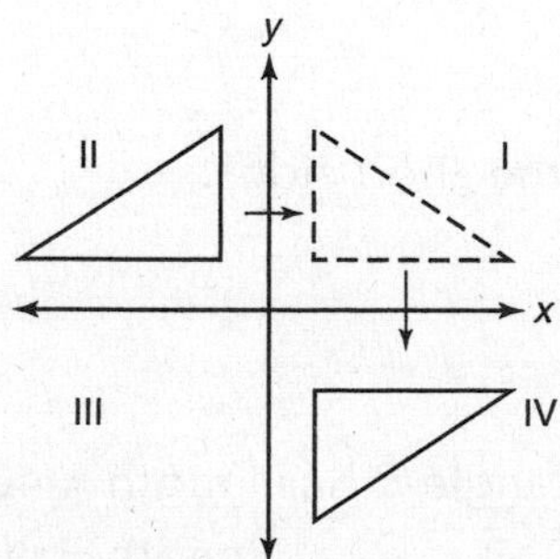


IV Bottom right, quadrant IV

**Part C:** **A.** translation (just a slide)

4. 4 $12/16 = 3x$, $(3)(16) = (12)x$, $48 = 12x$, $4 = x$
5. $320 \times 75$ $20/50 = 30/x$, $(30)(50) = (20)x$, $1{,}500 = 20x$, $75 = x$
6. **Part A:** No

   **Part B:** In similar triangles corresponding angles are equal.

   That is not true here.
7. **A.** (1, 4) becomes $A'(2, 8)$

   Multiply each original coordinate by 2.

## Non-calculator Questions (pages 135–138)

1. Quadrant IV

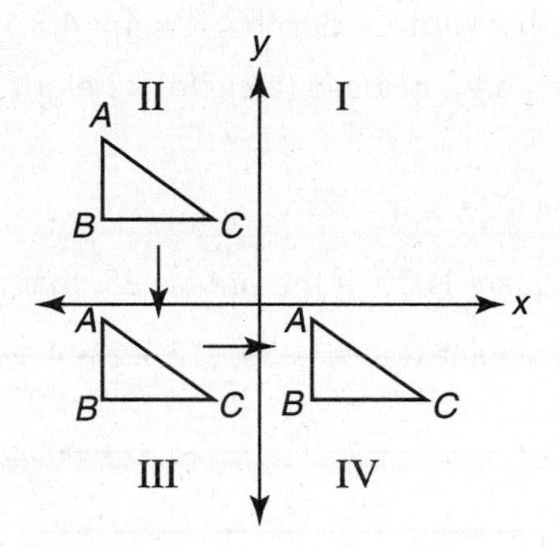


2. There are two possible correct answers: (–1, 4) and (3, 4) or (–1, –2) and (3, –2).
3. The area of the obtuse traingle is different.

   Area of square: $6^2 = 6 \times 6 = 36$

   Area of obtuse triangle: $\frac{1}{2}(\text{base})(\text{height}) = \frac{1}{2}(9)(4) = \frac{1}{2}(36) = 18$

   Area of trapezoid: $[\frac{1}{2}(3 + 5)] \times 9 = \frac{1}{2}(8) \times 9 = 4 \times 9 = 36$

   Area of tall rectangle: $4 \times 9 = 36$
4. The polygons all have the same perimeter, 12.
5. $x = 100°$
6. Triangles #1 and #2 are similar.

7. The shortest distance from Devan's house to Carl's house is 5 miles. You might recognize this as a special right triangle, a 3-4-5 right triangle. You could also use the Pythagorean theorem:

$$a^2 + b^2 = c^2$$
$$3^2 + 4^2 = c^2$$
$$9 + 16 = c^2$$
$$25 = c^2$$
$$5 = c$$

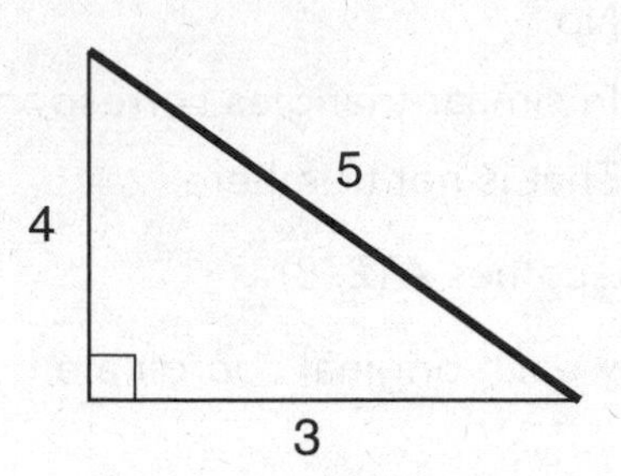


8. The volume of the cube is smaller.

   Volume of cube = length × width × depth = $4 \times 4 \times 4 = 16 \times 4$

   Volume of the cylinder = area of base (a circle) × height

   $\pi r^2 \times 4$

   $(3.14) \times 4 \times 4 \times 4 = 3.14 \times 16 \times 4$

9. The perimeter of the square is 20. If the area is 25, that means each side = 5.

10. The perimeter of the irregular figure is 34. $(12 + 2 + 4 + 3 + 8 + 5)$

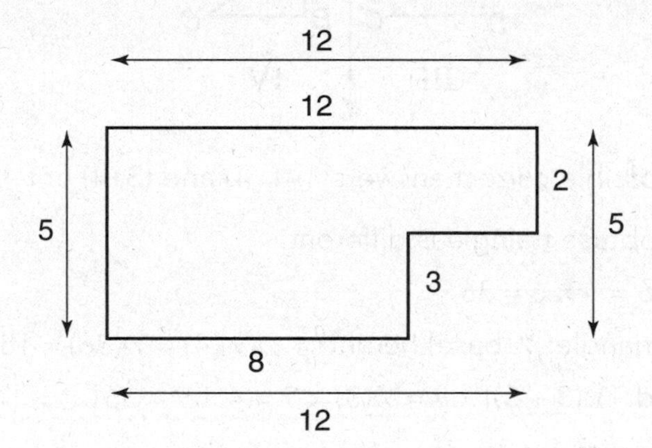


11. $36\pi$ $\quad \frac{4}{3}\pi r^3 = \frac{4}{3}(3)(3)(3)\pi = 4(3)(3)\pi = 4(9)\pi = 36\pi$

    Or, 113.04 if you used 3.14 instead of $\pi$. $3.14 \times 36 = 113.04$

12. The correct answers are **A.**, **B.**, **C.** and **D.**

13. **C.** $M'(-12, 6)$, $N'(-12, 16)$, $O'(-4, 16)$, and $P'(-4, 6)$

14. 15,700 square inches of wrapping paper for the 25 cylinders.

    The surface area of one cylinder consists of 2 circles (the top and bottom) and the side that wraps around the circle (a rectangle).

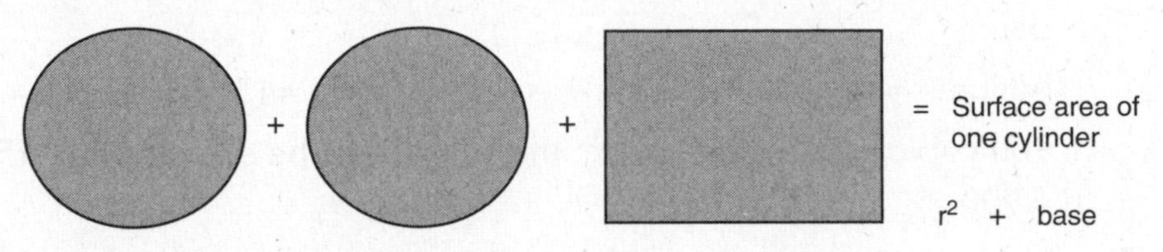


(circumference) (height) = Surface area

$\pi r^2$ + base (circumference) (height) = Surface area

$\pi 5^2 \quad + \quad \pi 5^2 \quad + \quad (\pi d)\ (h)$

$25\pi \quad + \quad 25\pi \quad + \quad (10\pi)\ (15) \quad = \quad 50\pi + 150\pi = 200\pi$

$(200)\pi = (200)(3.14) = 628$ sq. inches $\quad 628\ (25) = 15{,}700$ sq. inches

## PBA Questions (pages 139–141)

1. **94 sq. feet**

   Area of front and back = area of 2 triangles $= (2)\frac{bh}{2} = (2)\frac{4(4)}{2} = 16$ sq. ft.

   Area of 2 sides (rectangles) = 2 (length)(width) = 2(6)(4.5) = 54 sq. ft.

   Area of floor (also a rectangle) = (length)(width) = (4)(6) = 24 sq. ft.

   Total surface area = 16 + 54 + 24 = 94 sq. ft.

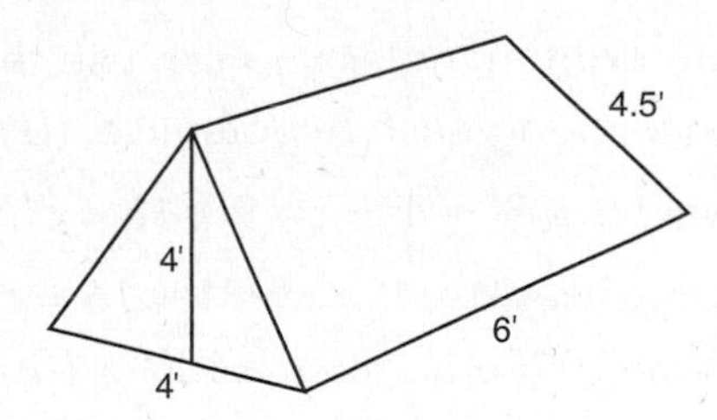


2. **Board B**

   All the boards are the same size and shape. The total area of each dartboard is the area of a square = (side)(side) = (8) (8) = 64 square inches. The dartboard with the largest shaded area would be the best.

   Board A: Area of unshaded region (circle) $= (\pi r^2) = (\pi)(4) = 12.566$ sq. in.

   Board B: Area of shaded region (square – circle) = 64 – 12.566 = 51.434 sq. in.

   Board C: Area of shaded region is $\frac{1}{2}$ total area of entire board $= \frac{64}{2} = 32$ sq. in.

   The correct answer is Board B, since its shaded region has the largest area.

3. **Part A:** 960 cubic inches = the maximum volume of the rectangular pan: (12)(16)(5) = 960.

   1004.8 cubic inches = the approximate maximum volume of the round pan:

   [Volume = (area of the base)(height) = ($\pi r^2$)($h$) = (3.14)(8)(8)(5) = 1004.8.]

   900 cubic inches = the maximum volume of the square pan: (15)(15)(4) = 900.

   **Part B:** The round pan would hold the most batter (1004.8 cubic inches).

   **Part C:** The square pan filled only 3" high would hold 675 cubic inches of batter.

   Volume = (side)(side)(depth) = (15)(15)(3) = 675 cubic inches.

4. **Part A:** 120 tiles that are 6 square inches would be needed to cover this floor.

   Change the dimensions of the floor from feet to inches: 5 ft × 6 ft = (5)(12 in./ft.) × (6)(12 in./ft.) = 60" × 72" = 4, 320 sq. in. = area of floor.

   Find the area of one tile: (6")(6") = 36 sq. in. = area of one tile.

   4,320 (area of floor) ÷ 36 (area of one tile) = 120 tiles.

   **Part B:** The total cost to purchase the tiles would be $228.96 including sales tax.

   120 sq. ft. (area of floor) ÷ 12 tiles per box = 10 boxes would be needed.

   (10 boxes)($21.60 per box) = $216.00 + 6% sales tax.

   ($216,00)(1.06) = $228.96, the total cost including the sales tax.

5. **Part A:** The answer is **A,** Twice as long (**less than** twice as long)

   **Part B:** Since these are right triangles you can use the Pythagorean theorem to find the distance along each hypotenuse. (Use $a^2 + b^2 = c^2$).

   For Katlyn's walk use $3^2 + 4^2 = c^2$; $9 + 16 = c^2$; $25 = c^2$ so $\mathbf{c = 5}$

   For Ashley's walk use $9^2 + 4^2 = c^2$; $81 + 16 = c^2$; $97 = c^2$; so $\mathbf{c = \sqrt{97}}$ which is less than10 (*Remember,* $\sqrt{100} = 10$; $\sqrt{97}$ *is less than 10).*

   Ashley's walk is less than twice as long as Katlyn's walk to school.

   **Part C:** Ashley's walk would be longer if she walked to Katlyn's house first.

   Ashley to Katlyn's = 6; Katlyn's to school = 5; 6 + 5 = **11**

   Ashley's usual way to school directly = $\sqrt{97}$ which is a little less than **10.**

6. **Part A:** **6 miles shorter**

Since you have a right triangle you can use the Pythagorean formula:

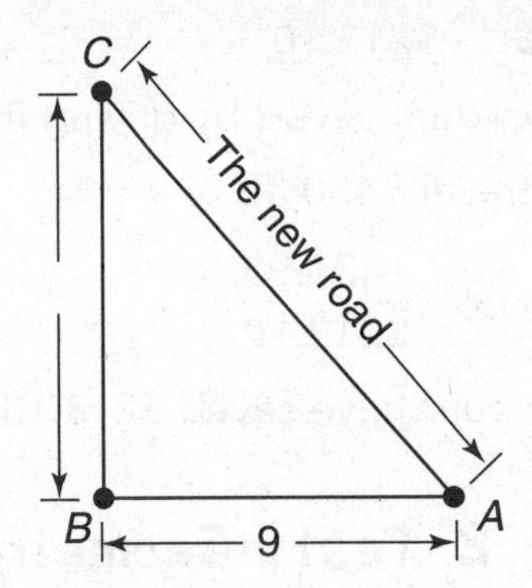


$a^2 + b^2 = c^2$

$12^2 + 9^2 = c^2$

$144 + 81 = c^2$

$225 = c^2$; $\sqrt{225} = \sqrt{c^2}$; 15 miles = $c$, the distance from "C" to "A"

The longer distance (from "C" to "B" to "A") is 12 + 9 or 21 miles.

The shorter distance would be 21 – 15 (from "C" to "A") = 6 miles shorter.

**Part B:** 436.90

| Travel Information | Your work space to determine total cost |
|---|---|
| • 1,072 miles from Clarke, New Jersey to Orlando, Florida.<br>• 1,072 miles from Orlando back home to New Jersey<br>Mom's car would have an average of 24 miles per gallon, and we'll estimate gasoline costs $2.40 per gallon. | Gasoline Cost:<br>1,072 + 1,072 = 2,144 total miles<br>$\frac{2{,}144}{24}$ miles per gallon = 89.33 gallons<br>(89.33 gallons) ($2.40/gallon) = **$214.40** |
| • Tolls round-trip would be approximately $16.00. | Tolls: **$16.00** |
| • $68.50 hotel for one night while driving down to Florida<br>• $59.00 hotel for one night on the drive home | Hotel Cost:<br>68.50 + 59.00 = **$127.50** |
| • Food purchased on trip from NJ to Florida ($15.00 + 28.00)<br>• Food purchased on trip from Florida back to NJ ($12.00 + $6.00 + $18.00) | Food Cast:<br>15 + 28 = $43.00<br>12 + 6 + 18 = $36.00<br>Total food cost = **$79.00** |
| | **Total Cost by car: $436.90** |

**Part C:** $213.10

Total amount saved by driving instead of flying

Plane travel $650.00

Car travel $\underline{-436.90}$
$213.10

They would have saved $213.10 by driving.

## Chapter 2 Test: Geometry (pages 142–149)

1. **D.** 2. **C.** 3. **B.** 4. **A.** 5. **C.** 6. **D.** 7. **B.**

8. **Part A:** No

**Part B:** Container B

**Part C:** The wider, shorter cylinder "B" will hold more.
The volume of container B:
(area of base)(height) = $\pi r^2 h = \pi(6^2)(3) = \pi(36)(3) = \mathbf{108\pi}$

The volume of container A:
(area of base)(height) = $\pi r^2 h = \pi 3^2(6) = \pi(9)(6) = \mathbf{54\pi}$

9. $288\pi$

$x^2 + 8^2 = 10^2$

$x^2 + 64 = 100$

$x^2 = 36$

$x = 6$

$(\frac{4}{3})(6^3)(\pi) = 288\pi$

10. **A.** 11. **C.** 12. **A.**

13. **A.** No **B.** No **C.** Yes **D.** Yes **E.** Yes **F.** Yes

14. **Part A:** 5 units long (length of base of triangle) Use the Pythagorean theorem:
$12^2 + X^2 = 13^2$ $144 + X^2 = 169$ $X^2 = 25$ $X = 5$ length of base

**Part B:** 30 square units (area of triangle)

$\frac{1}{2}(\text{base})(\text{height}) = \frac{5 \cdot 12}{2} = \frac{60}{2} = 30$ *or* $5 \cdot 6 = 30$

15. **D.** $E'$ (3, 12) $F'$ (3, 18) $G'$ (12, 18) $H'$ (12, 12)

**Performance-Based Assessment Questions**

16. ■ The irregular shape should be divided into geometric shapes. There are two possible solutions drawn below. The **bold** lines show how you might divide this shape.

■ Next, you should demonstrate how you found the area of the various shapes. (You need to show your work)

Area rectangle = 45 × 30 = 1,350

Area triangle = (45)(8)/2 = 180

Rectangle + triangle = Total area

1,350 + 180 = 1,530 sq. ft

Area large rectangle = 45 × 38 = 1,710

Area one triangle = (22.5)(8)/2 = 90

Area large rectangle – 2 triangles = Total area

1,710 – 2(90) = 1,710 = 180 = 1,530 sq. ft

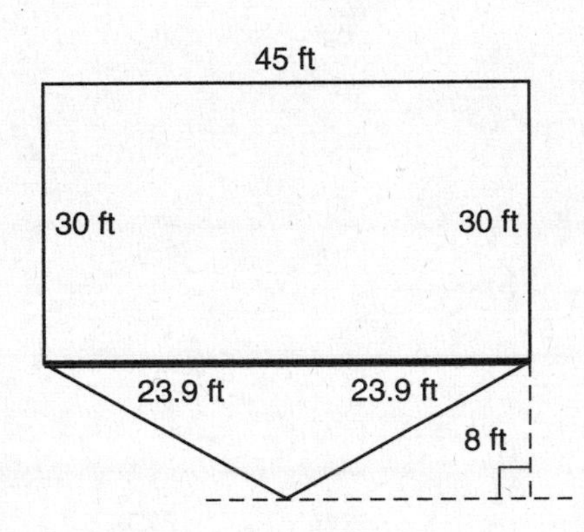


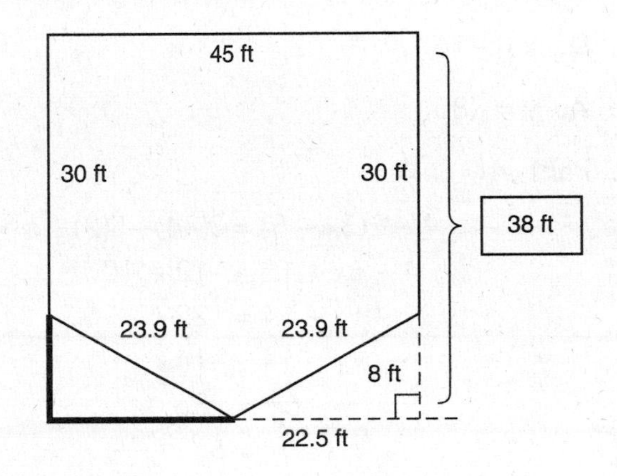


17.

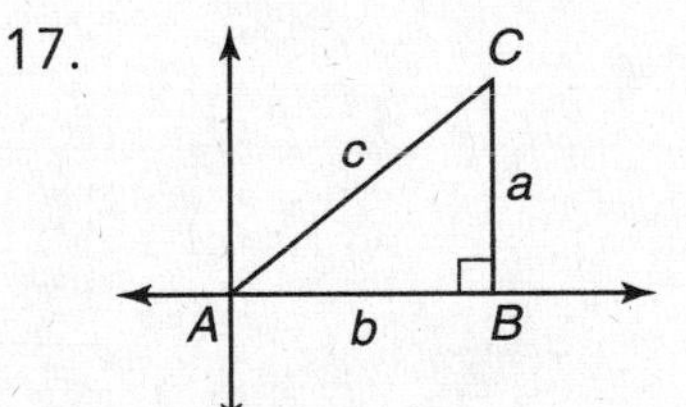


1. See the diagram.
2. See the diagram.
3. This is a right triangle.
4. 7.8 = length of hypotenuse

   $5^2 + 6^2 = x^2$

   $25 + 36 = 61 = x^2$

   if $61 = x^2$ then $\sqrt{61} = x$, $7.8 = x$

5. 6 + 5 + 7.8 = 18.8 units = Perimeter of triangle
6. $\frac{(5)(6)}{2} = \frac{30}{2} = 15$ sq. units = Area of triangle

# CHAPTER 3: EXPRESSIONS, EQUATIONS, AND FUNCTIONS (8.EE, 8.F)

## Solving Equations (pages 159–160)

1. **B.** $w = 15$
2. **C.** add $\frac{1}{2}x$ *to* $2x$
3. **C.** $-9 = a$
4. **C.** $z = 80$
5. **D.** $x = -16$
6. **A.** $y = -80$
7. **Part A:** $x = 6$

$$4 - 2(3x - 5) = 3(-4) - 5(2)$$
$$4 - 6x + 10 = -12 - 10$$
$$-6x + 14 = -22$$
$$-14 \quad -14$$
$$-6x = -36$$
$$x = 6$$

**Part B:**

The correct answers are:

**A.** $-3x = -18$ then $-x = -6$ and $\mathbf{x = 6}$

**C.** $4^2 - 2(\sqrt{25})$ then $16 - 2(5) = 16 - 10 = \mathbf{6}$

**D.** $\sqrt{81} - \sqrt{9}$ then $9 - 3 = \mathbf{6}$

8. Answer: $x = -18$

$$\frac{-3}{2}(2+x) = 24 \text{ multiply both sides by the reciprocal } \frac{-2}{3}$$

$$\left(\frac{-2}{3}\right)\frac{-3}{2}(2+x) - 24\left(\frac{-2}{3}\right)$$

$$2 + x = -\frac{48}{3} \qquad 2 + x = -16 \qquad x = -18$$

9. Answer: $\frac{31}{3} = 10\frac{1}{3}$

$\frac{2}{3}(w-9) = \frac{8}{9}$ first multiply both sides by the reciprocal $\frac{3}{2}$

$\left(\frac{3}{2}\right)\frac{2}{3}(w-9) = \frac{8}{9}\left(\frac{3}{2}\right)$ $w-9 = \frac{4}{3}$ add 9 $\left(\text{or } \frac{27}{3}\right)$ to both sides

$w = \frac{4}{3} + \frac{27}{3} = \frac{31}{3} = 10\frac{1}{3}$

10. **B.** $a = -1\frac{1}{2}$ *or* $-\frac{3}{2}$

$-\frac{1}{3}(a+1) = \frac{1}{6}$ Multiply both sides by the reciprocal, $\frac{-3}{1}$

$\left(\frac{-3}{1}\right) - \frac{1}{3}(a+1) = \frac{1}{6}\left(\frac{-3}{1}\right)$ $a+1 = \frac{-3}{6}$ $a+1 = -\frac{1}{2}$

add –1 to both sides; $a = -1\frac{1}{2}$ *or* $-\frac{3}{2}$

11. Answer: $x = -110$

$-.05(x - 10) = 6$

(It is easiest to work with whole numbers so we multiply each side by 100)

$-5(x - 10) = 600$ distribute the –5

$-5x + 50 = 600$ subtract 50 from both sides

$-5x = 550$ divide both sides by –5

$x = -110$

12. Answer: $w = 96.5$

$0.02(w - 1.5) = 1.9$

(It is easiest to work with whole numbers so we multiply each side by 100)

$2(w - 1.5) = 190$ distribute the 2

$2w - 3 = 190$ add 3 to both sides

$2w = 193$ divide both sides by 2

$w = 96.5$

## Lines and Slope (pages 164–165)

1. **D.** line with slope of $\frac{4}{2}$ or 2

2. **A.** line with slope of $\frac{2}{3}$

3. **A.** parallel line with slope of $\frac{2}{3}$

4. **C.** slope of line $c$ is $\frac{2}{3}$; slope of line $d$ is $\frac{2}{3}$

5. **C.** The two lines are parallel; they have the same slope.

6. $\frac{-2}{5}$

7. **C.** The lines are perpendicular. One has a slope of $\frac{-3}{2}$ and the other has a slope of $\frac{2}{3}$. They are negative reciprocals of each other.

8. **A.** No **B.** Yes **C.** Yes **D.** No **E.** Yes **F.** No

## Extra Practice: Slope (pages 171–176)

1. **B.** slope $=\frac{-2}{3}$

2. **A.** slope $=\frac{1}{2}$

3. **C.** $\frac{2}{3}$

4. **B.** 3 $2y = 6x + 10$ becomes $y = 3x + 10$; slope is 3

5. **D.** 4 $y - 6 = 4x$ becomes $y = 4x + 6$; slope is 4

6. **B.** 2 $y - 2x = 14$ becomes $y = 2x + 14$; slope is 2

7. **A., B.,** and **D.** All are equivalent to $\frac{4-0}{2-3}=\frac{4}{-1}$

8. **B.** and **D.** Choice **B** shows that $\frac{(3-4)}{(5-2)}=\frac{-1}{3}$ or **D** $\frac{1-4}{5+4}=\frac{-3}{9}=\frac{-1}{3}$.

9. **Part A:**

| Table A | |
|---|---|
| $x$ | $2x + 1$ |
| 4 | $2(4) + 1 = 9$ |
| 7 | $2(7) + 1 = 15$ |

| Table B | |
|---|---|
| $x$ | $x^2$ |
| 3 | $(3)(3) = 9$ |
| 5 | $(5)(5) = 25$ |
| 7 | $(7)(7) = 49$ |

**Part B:** Table A: \$50 would give you (2)(50) + 1 = \$101.00

**Part C:** Table B: \$50 would give you (50)(50) = \$2,500.00

**Part D:** Answers will vary: In the Table B bank the money amount is squared, which increases your money much more quickly than if you just doubled it and added \$1.00 (as in Table A). In Table A, the money is just multiplied by the number 2, but in Table B the number is multiplied by a higher and higher number each time.

10. **A.** No **B.** Yes **C.** No **D.** Yes **E.** Yes **F.** Yes

11. **A.**

slope using the table is $\frac{3}{1}=3$ $\quad \frac{-10--4}{-2-0}=\frac{-6}{-2}=\frac{3}{1}=3$

slope of line A is –3 $\quad y=-\mathbf{3}x-4$

slope of line B is 3 $\quad y=\mathbf{3}x-4$

slope of line C is 3 $\quad 2y-6x=-8,\ 2y=6x-8,\ y=\mathbf{3}x-4$

slope of line D is 3 $\quad f(x)=3(x-2)+2=3x-6+2=\mathbf{3}x-4$

12. **C.** $y=\frac{-9}{4}x-14$

13. The linear function **C** has the smallest rate of change.

**Part A:** The rate of change (or slope) is $\frac{-4-+4}{2--2}=\frac{-8}{4}=-2$

**Part B:** The rate of change (or slope) is $\frac{4}{2}=2$ (use points (0, 0) and (4, 2))

**Part C:** The rate of change (or slope) is 2/3.

$$3y+6=2x$$
$$3y=2x-6$$
$$y=\frac{2}{3}x-2$$

None of the functions have the same rate of change.

See how slopes were determined in **A, B, C** above; none are the same.

14. **B.** The slope is $\frac{-5}{2}$ and the $y$-intercept is + (0, 4).

## Systems of Equations (pages 179–180)

1. **Part A:** The equation of the line for plant A: $\frac{3}{2}x + 7$

   The slope of the line for plant A is $\frac{2}{3}$ and the *y*-intercept is 7. You know this just by looking at the graph.

   **Part B:** The equation of the line for plant B: $\boldsymbol{h} = \frac{x}{2} \boldsymbol{= 11}$

   The slope of the line for plant B is 2/4 or ½ and the *y*-intercept is 11. You can see this just by looking at the graph

   **Part C:** The two plants will be the same height at week **4**

   The two lines intersect at the coordinate (4, 13). Therefore the two plants will be the same height at week 4.

2. **Part A:** Yes, line *p* intersects with line *q*.

   **Part B:** Explanation: When I found the slope of each line they were NOT the same, so the lines are NOT parallel and will meet.

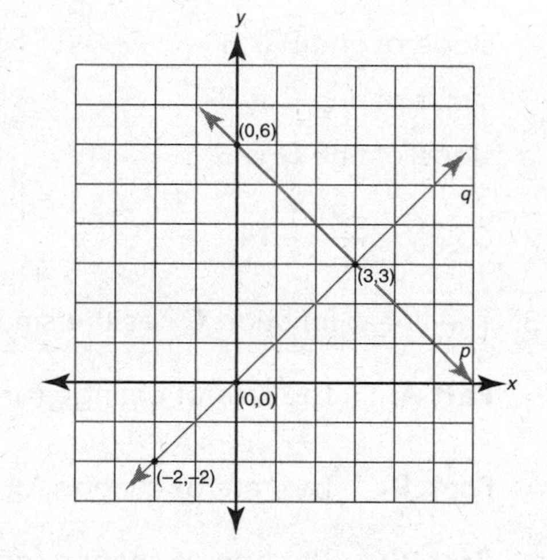


*Line p*: $\text{slope} = \frac{y_1 - y_2}{x_1 - x_2} = \frac{5-1}{4-0} = \frac{4}{4} = \frac{1}{1}$

$\frac{3-0}{2-4} = \frac{3}{-2}$

*Line q*: $\text{slope} = \frac{y_1 - y_2}{x_1 - x_2} = \frac{3-0}{2-4} = \frac{3}{-2}$

Model: See the graph. You now can count to find the $\frac{rise}{run}$ of each line and will see the slope of one line is $\frac{1}{1}$ and the slope of the other is $-\frac{3}{2}$. The slopes are not the same; the lines are not parallel. You can see the lines meet at about (2, 3).

3. **Part A:** No

   **Part B:** Explanation: When I found the slope of each line they were the same; the lines are parallel and will never meet.

Line $p$: slope $= \dfrac{y_1 - y_2}{x_1 - x_2} = \dfrac{8-4}{4-2} = \dfrac{4}{2} = \dfrac{2}{1}$

Line $q$: slope $= \dfrac{y_1 - y_2}{x_1 - x_2} = \dfrac{7-5}{5-4} = \dfrac{2}{1}$

See the diagram. You now can count to find the $\dfrac{rise}{run}$ of each line and will see that the slope of each line is $\dfrac{2}{1}$

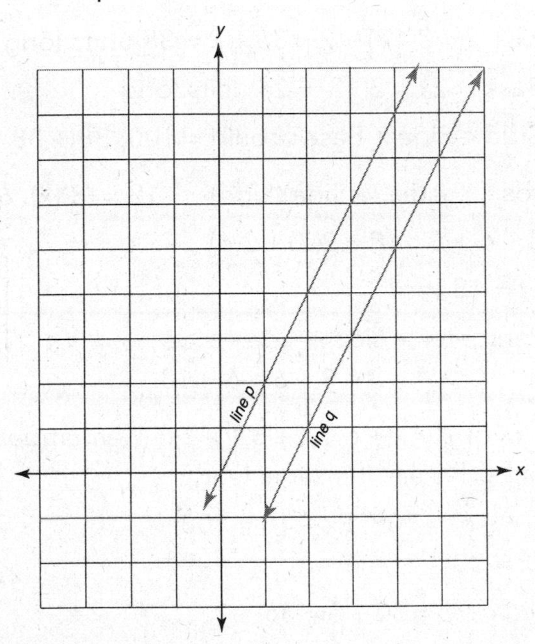


## Combining Algebra and Geometry (pages 183–185)

1. **Part A:**

   Equation: $P = 2(x + 3) + 2x = 22$ or $P = x + 3 + x + 3 + x - 2 + x - 2 = 22$

   **Part B:** $x = 2x + 6 + 2x = 22$ or $4x + 10 - 4 = 22$

   $4x + 6 = 22$ $\quad$ $4x + 6 = 22$ (Combine like terms.)

   $4x = 16$ $\quad$ $4x = 16$ (Undo multiplication)

   $x = 4$ $\quad$ $x = 4$

   **Part C:** One side is 7 units long. $\quad$ $x + 3 = 4 + 3 = 7$

   The other side is 4 units long. $\quad$ $x - 2 = 4 - 2 = 2$ units long

**Part D:** Check: Perimeter $= 2(x + 3) + 2x$
$= 2(4 + 3) + 2(4)$
$= 2(7) + 2(4)$
$= 14 + 8 = 22$, which is correct!

2. *Remember:* In an isosceles triangle, two sides are equal in length.

**Part A:** Equation: $2(3x + 2) + 2x = 100$ or $3x + 2 + 3x + 2 + 2x = 100$

**Part B:** $x = 12$

$6x + 4 + 2x = 100$
$8x + 4 = 100$
$8x = 96$
$x = 12$

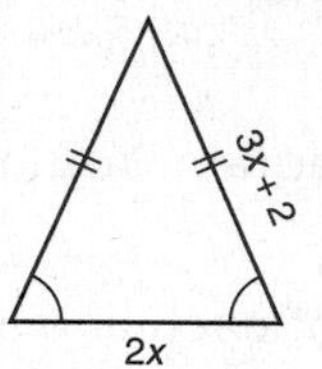


**Part C:** Side: $3x + 2 = 3(12) + 2 = 36 + 2 = 38$ units long

Base: $2x = 2(12) = 24$ units long

**Part D:** Check: Side + Side + Base should = 100; 38 + 38 + 24 = 100, correct!

3. **Part A:** $x = 6$ yards Area = (Side)(Side) $18 = (3)(x)$; $6 = x$

**Part B:** $P = 3 + 3 + 6 + 6$ or $P = 2(3) + 2(6)$

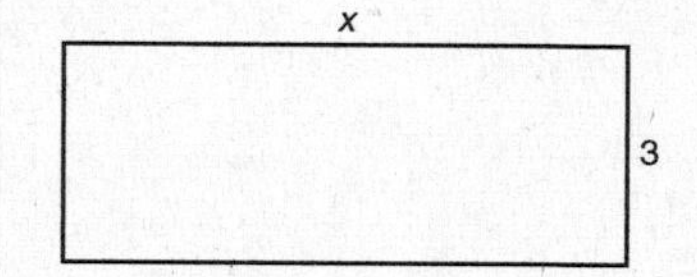


**Part C:** Perimeter = 18 yards

**Part D:** Check: Perimeter = side + side + side
$= 3 + 3 + 6 + 6 = 18$.

4. **Part A:** $P = 3(x + 6)$ or $P = x + 6 + x + 6 + x + 6$ (*Remember:* In an equilateral triangle all sides are the same length.)

**Part B:** $x = 30$ $3(x + 6) = 108$; $3x + 18 = 108$;
$3x = 90$; $x = 30$

x + 6

**Part C:** Each side: $x + 6 = 30 + 6 = 36$

**Part D:** Check: Perimeter = Side + Side + Side = 36 + 36 + 36 = 108, correct!

5. *Remember:* The sum of the three angles in any triangle will be 180°.

**Part A:** Equation: $x° + 2x° + 3x° = 180°$

**Part B:** $6x = 180°$, $x = 30°$

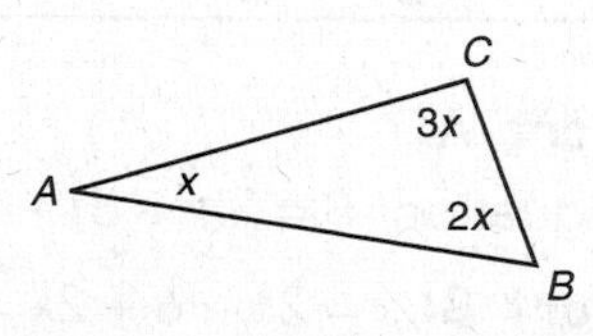


**Part C:** $\angle B = 2x = (2)(30°) = 60°$

**Part D:** $C = 3x = (3)(30°) = 90°$

6. $x = \frac{3}{2}$

Compute slope of $EB = \frac{15-3}{4-1} = \frac{12}{3} = 4$ Compute slope of $ED = \frac{8-3}{x-1} = \frac{5}{x-1}$

Since lines $EB$ and $ED$ are parts of the same line, their slopes are equal.

$\frac{5}{x-1} = \frac{4}{1}$, $4x - 4 = 5$, $4x = 9$, $x = \frac{9}{4}$

7. 12 feet wide and 38 feet long

$x + x + 3x + 2 + 3x + 2 =$ or think $2(x) + 2(3x + 2) =$ $8x + 4 = 100$

$8x = 96$

$\mathbf{x = 12}$ **width**; and

**length** $= 3x + 2 = 3(12) + 2 = 36 + 2 = \mathbf{38}$

8. **A.** No **B.** Yes **C.** Yes **D.** No **E.** Yes **F.** No

## Writing Expressions and Equations (page 186)

1. **B.** $4n - 6$
2. **A.** $\sqrt{n} > n - 4$
3. **C.** The sides of a square all have the same measurement. Here two sides are shown as "2.5$x$" each, and two other sides as "$x$" each...This could not represent the perimeter of a square.

   **E.** $x^2$ This could represent the **area of a square** (side)(side); but not the perimeter. The perimeter means you add the four sides, not square or multiply them.
4. **A.** $9n + 20$ 20 more becomes 9 times a number.

## Real-Life Applications (pages 187-188)

1. **C.** $A = 5{,}329\ (1 + .029)^3 = 5{,}300\ (1.029)\ (1.029)\ (1.029)$

   $= 5{,}300\ (1.0895473) = \$5{,}774.60$

2. **Part A:** Yes $2(40 + 32) = 144$ = rectangle perimeter and 144 is divisible by 4

   **Part B:** 36 ft. $144 \div 4 = 36$ ft.

   **Part C:** Yes The *perimeter* of the rectangle is 144 ft.

   The *perimeter* of the square is $4(36) = 144$ ft.

**Part D:** larger The *area* of the rectangle is (32)(40) = 1280 sq. ft.;

The *area* of the square is (36)(36) = 1296 sq. ft.

**Part E:** 16 sq. feet 1296 – 1280

## SCR Non-calculator Questions (pages 188–189)

1. $-5\frac{1}{5}$ $\left(-1\frac{5}{10}\right)+\left(-3\frac{7}{10}\right)=-4\frac{12}{10}=-5\frac{2}{10}=-5\frac{1}{5}$ or –5.2

2. $9x - 6a$ $(12x-3x)+(-2a-4a)=9x-6a$

3. $13x + 24$ $10(x + 3) - 3(-x + 2) =$

   $10x + 30 + 3x - 6 = 10x + 3x + 30 - 6$ $13x + 24$

4. $-2.5x + 25$ $(-2.5)(x) + (-2.5)(-10) = -2.5x + 25$

5. $\frac{2}{3}$ $\frac{-3}{1}\times\frac{-1}{3}\times\frac{2}{3}=\frac{6}{9}=\frac{2}{3}$

6. $-12$ $\frac{-6}{\frac{1}{2}}=-6\left(\frac{2}{1}\right)=-12$

7. $\frac{3}{x+6}$ $\frac{9}{3(x+6)}=\frac{3}{1(x+6)}=\frac{3}{x+6}$

8. $-8\frac{4}{5}$ $\frac{-2}{5}\left(25-\frac{3}{5}\right)=\frac{-50}{5}+\frac{6}{25}=-10+\frac{6}{25}=-9\frac{19}{25}$

9. 3 $3x + 8x + 4 - 1 - 11x = 11x - 11x + 4 - 1 = 3$

10. 1 $6\left(x-\frac{1}{2}\right)=6\left(\frac{2}{3}-\frac{1}{2}\right)=6\left(\frac{4}{6}-\frac{3}{6}\right)=6\left(\frac{1}{6}\right)=\frac{6}{6}=1$

11. $-\frac{2}{3}$ $\frac{2}{3}x+y=15$; change to slope intercept form: $y=-\frac{2}{3}x+15$

12. 5 $y - 3x = 16 + 2x$ $y = 16 + 2x + 3x$ $y = 5x + 16$

13. **Part A:** No The slope of a perpendicular line would be $\frac{-4}{3}$

    **Part B:** Yes It has a positive slope; it is rising from left to right.

    **Part C:** Yes They have the same slope, $\frac{3}{4}$.

14. $a = 7$ *or* $-7$ Remember 7 × 7 = 49, and (–7) × (–7) also = 49

15. $6x+\frac{3}{4}$ $2(2x + ¼) + 2(x + 1/8) =$

    $4x+\frac{2}{4}+2x+\frac{2}{8}=6x$ *and* $\frac{2}{4}+\frac{2}{8}=6x+\frac{6}{8}=6x+\frac{3}{4}$

## PBA Questions (pages 190–191)

1. **Part A:**

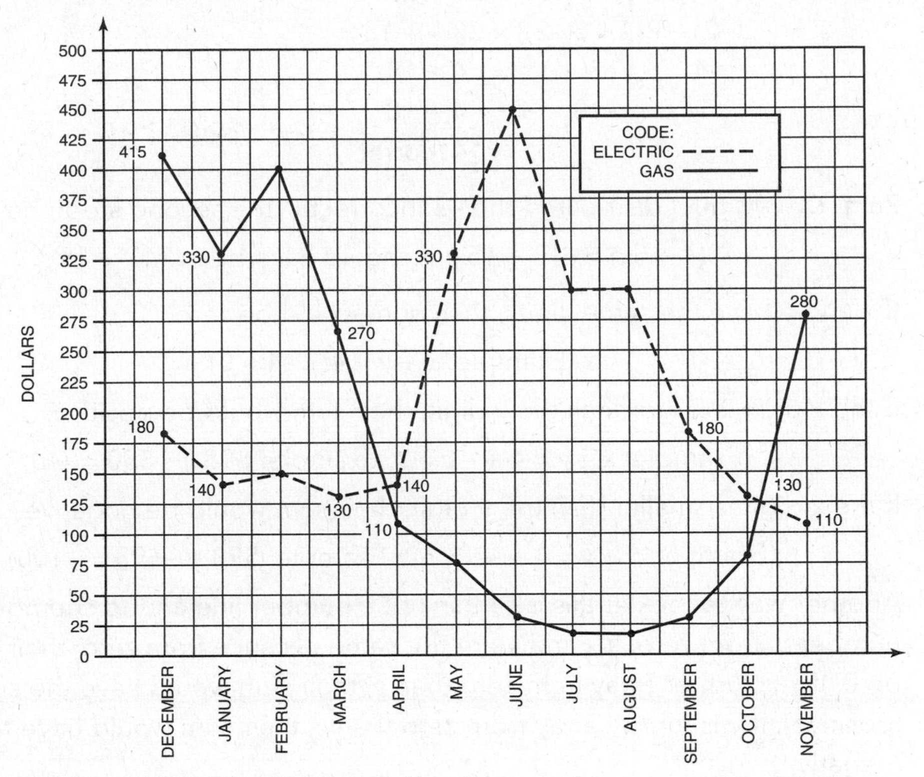


**Part B:** **April** If you look at the numbers on the chart, you see that in April the electric bill was $140 and the gas bill was $110; there was only a $30 difference
(140 – 110 = 30).

However, if you look at the double line graph above you notice that the two lines intersect twice; the gas and electric usage between March and April at one point were each about $40, and the usage between October and November were each about $125.

**Part C:** **June** In June the electric bill was $450 and the gas bill was only $30; this was a $420 difference (450 – 30 = 420).

**Part D:** The electric bills were high (over $300) in May, June, July, and August; these were the hotter months when Mr. and Mrs. B. had their air conditioning working more often. The colder months, when Mr. and Mrs. B. used their heating system often, seems to be from November to March, when the gas bills were at least $130 per month.

2. **Part A:** Nicholas is correct; $x = -2$

   **Part B:**

$$2x = 3(4 + 5x) = 2x + 18$$
$$2(-2) - 3(4 + 5(-2)) = 2(-2) + 18$$
$$-4 \quad -3(-6) \quad = -4 + 18$$
$$-4 \quad +18 \quad = -4 + 18$$
$$14 \quad = 14 \text{ (correct)}$$

   **Part C:** Miguel distributed the $-3$ incorrectly (the second step); he said: $-3(4 + 5x) = -12 + 15x$; it should have been $-12 - 15x$

3. If $x$ and $y$ have the same digits, their sum would be zero.

   (Example: $x + y = 4 + -4 = 0$)

   If the $x$ digit is larger than the $y$ digit, their sum would be positive.

   (Example: $x + y = 4 + -1 = 3$; Example: $598 + -550 = 48$)

   If the $x$ digit is smaller than the $y$ digit, their sum would be negative.

   (Example: $x + y = 4 + -6 = -2$; Example: $598 + -698 = -100$)

   Another way to look at this is to think of a number line and to see how far each number is from zero. If $x$ and $y$ are the same distance from zero, their sum would be 0. If $x$ is further away from zero than $y$, their sum would have the sign of $x$ (positive). If $y$ is further away from zero than $x$, their sum would have the sign of $y$ (negative).

4. After one year their savings would equal $8,320.

   $A = P(1 + r)^t$

   $A = 8{,}000(1 + 0.04)^1 = 8{,}320$

   After five years their savings would equal $9,733.22

   $A = \$8{,}000(1 + 0.04)^5 = 9.733.22$

5. **Part A:** $x \quad + \quad x + 1 \quad + \quad x + 2$

   **Part B:**

$$x + x + 1 + x + 2 = 1{,}356$$
$$3x + 3 = 1{,}356$$
$$3x = 1{,}353$$
$$x = 451,\ x + 1 = 452, \text{ and } x + 2 = 453$$

   Check: $451 + 452 + 453 = 1{,}356$ (correct).

## Chapter 3 Test: Expressions, Equations, and Functions Test (pages 192–196)

1. **C.** 2. **D.** 3. **D.** 4. **A.** 5. **C.** 6. **B.**

7. Solution: $x = -3$

$$\frac{1}{3}(3x-11)=12x$$

$$3x - 11 = 36x$$

$$-11 = 33x$$

$$\frac{-11}{33} = x$$

$$-\frac{1}{3} = x$$

8. **A.** $y = \frac{7}{5}x + 1$ The slope is $\frac{7}{5}$ and the $y$-intercept is (0, 1).

9. **C.**

10. $2x + \frac{1}{4}(x-24) = 3$

| | |
|---|---|
| $2x + \frac{x}{4} - \frac{24}{4} = 3$ | Distribute the $\frac{1}{4}$ |
| $2x + \frac{x}{4} - 6 = 3$ | Simplify the fraction $-\frac{24}{4} = -6$ |
| $2x + \frac{x}{4} = 9$ | Add 6 to both sides |
| $(\mathbf{4})2x + (\mathbf{4})\frac{x}{4} = (\mathbf{4})(9)$ | Multiply all terms by **4** |
| $8x + x = 36$ | |
| $9x = 36$ | Combine like terms |
| $x = 4$ | Divide both sides by 9; solve for $x$ |

11. **C.**

12. **Part A: B.** They are both decreasing.

**Part B: B.** It shows a negative rate of change. As $x$ decreases, $y$ increases.
Rate of change in the table is:

$$\underline{-2} \quad \frac{60-58}{0-1} = \frac{2}{-1} = -2 \text{ or } \frac{58-56}{1-2} = \frac{2}{-1} = -2$$

13. One side (width) = $x = 30$
    One side (length) = $2x + 10 = 60 + 10 = 70$

**Performance-Based Assessment Questions**

14. **Part A:** Equation of Perimeter: $x + 2x + 4 + 3x = 100$ cm or $6x + 4 = 100$ cm
    **Part B:** $x = 16$ — $6x + 4 = 100$, $6x = 96$, $x = 16$
    **Part C:** *Side a* = 16 cm — Side $a = x$, $x = 16$ cm
    *Side b* = 36 cm — Side $b = 2x + 4 = 2(16) + 4 = 36$
    *Side c* = 48 cm — Side $c = 3x = 3(16) = 48$
    **Part D:** Area = 192 sq. cm — Area of triangle $= \frac{1}{2}$(Base)(Height)
    Area $= \frac{1}{2}(8)(48) = 192$ sq. cm

15. **Part A:** $24(x + 4) = 216$
    **Part B:** \$5 — $24x + 96 = 216$
    $24x = 120$; $x = 5$

16. **Part A:** $12(e + 3) = 132$
    **Part B:** \$8 — $12e + 36 = 132$
    $12e = 96$; $e = 8$

# CHAPTER 4: PROBABILITY AND STATISTICS (8.SP)

## Probability (pages 201–203)

1. **A.** 0 — There are no cards numbered 12 in a regular deck.
2. **C.** 1 — Any card you pick would be correct since all are even numbers.
3. **C.** 1 — Yes, there is a 100% chance that it will snow in Alaska each year.
4. **A.** 0 — There is no number less than 2; zero possibility of selecting that.
5. **A.** No
   **B.** Yes — As students spent more time studying, their test score increased.
   **C.** Yes — Arnot's graph shows 10 minutes = a score of 35
   Erik's graph shows 10 minutes = between 35 and 40 if you look at the line of best fit.
   **D.** Yes — This is true in both classes.
   **E.** Yes — 10 students in Erik's class scored greater than or equal to 60 points. In Arnot's class only 9 students scored greater than or equal to 60 points.

6. **A.** Yes As popcorn sales increase, beverage sales increase, too.

   **E.** Yes If you extend the *line of best fit* this seems to be a reasonable estimate.

## Experimental Probability (pages 204–205)

1. **B.** 30% $\frac{3\text{ samples}}{10\text{ total}} = \frac{3}{10} = 0.30 = 30\%$

2. **D.** 50% $\frac{6}{12} = \frac{1}{2} = 0.50 = 50\%$

3. **B.** 50% $\frac{4}{8} = \frac{1}{2} = 0.50 = 50\%$

4. **B.** 14.29% $\frac{1\text{ Monday}}{7\text{ day in week}} = \frac{1}{7} = 0.1428571 \text{ or } 0.1429 = 14.29\%$

5. **C.** in May $\frac{1\text{ May}}{12\text{ months in year}} = \frac{1}{12} = 0.083 \text{ or } {\sim}8\%$

   $\text{Spring} = \frac{1\text{ spring}}{4\text{ seasons}} = \frac{1}{2} = 0.50 = 50\%$

   $\text{Wednesday} = \frac{1\text{ Wednesday}}{7\text{ days}} = \frac{1}{7} = 0.1428 = 14.28\%$

   $\text{Even numbers} = \frac{1\text{ even}}{2\text{ choices}} = \frac{1}{2} = 0.50 = 50\%$

6. **A.** $\frac{5}{8} = 62.5\%$ $\frac{\text{NOT odd numbers}}{\text{All choices}} = \frac{5}{8} = 0.625 = 62.5\%$

## Probability of Events (pages 207–208)

1. **B.** 0.6, 0.4 If 60% win, then 40% will lose; if 0.60, then 0.40.

2. **B.** 55% Probability 1 run = 0.30, probability 2 runs = 0.25; since this is an **or** situation, you add the two probabilities: 0.30 + 0.25 = 0.55 **or** 55%

3. **Part A:** **A.** 25% Since they are 50% together, then Zak + Kyle = 50%. Since Zak and Kyle are equal in their abilities, Zak = 25%, Kyle = 25%.

   **Part B:** **A.** 6-7% Each has 0.25; since this is an **and** situation, you multiply the two (0.25)(0.25) = 0.0625 or 6.25%, which is between 6 and 7%.

4. **D.** 10% This is an **and** situation, so you multiply: (0.50)(0.20) = 0.10 = 10%.

## Mean, Median, Mode, and Range (pages 209–210)

1(a) **B.** 12 mean (Add all the numbers, then divide by 9.)

$$\frac{7+8+9+11+12+12+13+15+21}{9}=\frac{108}{9}=12$$

(b) **B.** 12 median (Put the numbers in numerical order; this is the one in the middle.)

(c) **A.** 12 mode (the one that appears most often); 12 appears two times

2. **D.** 20 The range is $53 - 33 = 20$.

3. **Part A:** **D.** $72.10

   **Part B:** **B.** $60

   **Part C:** **A.** $152.10

4. **D.** nothing would change

5. **B.** 37 $100 - 63 = 37$

## Scatter Plots (pages 212–214)

1. **Part A:** **B.** (–8, 2)

   **Part B:** **A.** Yes, the coordinate (5, 5) seems to be on the trend line.

2. **C.** $210 She saved $230 but also withdrew $20, which means she has 230 – 20 = $210.

3. **Part A:** **A.** positive

   **Part B:** **B.** yes $\left(\frac{1}{2}\text{ hr and }95\%\right)$

   **Part C:** **C.** $\left(\frac{1}{2}, 95\right)$ $\frac{1}{2}$ hr studied and 95% on the test

   **Part D:** rising (It increases from left to right.)

4. **B.** and **F.** negative As the price of gasoline increases, the number of hours for Sunday drives decreases. This can also be described as "falling."

## Circle Graphs (pages 215–216)

1. **C.** 65%  $35 + 30 = 65$
2. **D.** 45  11% of 410 students = $(0.11)(410) = 45.1$
3. **A.**  $(410)(0.35) - (410)(0.07)$ = hamburger – chili
4. **A.**  chili, pizza, or hot dog
   chili + pizza + hot dog = 11% + 30% + 17% = 58%
5. **Part A:** **A.** 19%  Total number of students in all 315; $\frac{60}{315} = 0.19$ or 19%

   **Part B:** **C.** 56%  90 + 85 = 175 students had birthdays in April or May;
   $\frac{175}{315} = 0.56$ or 56%

   **Part C:** **A.** 0%  No students surveyed had birthdays in the fall. The fall months are September, October, and November.

## SCR Non-calculator Questions (pages 217–218)

1. There is no change.  All remain the same. 75, 80, 80, 80, 85
   Mean = 80, mode = 80, and median = 80
2. There is positive correlation. As they use more KWH, their cost increases.
3. 9 combinations  (3 flavors) × (3 kinds of sprinkles) = $3 \times 3 = 9$
4. $\frac{1}{2}$  $\frac{\text{Favorable outcomes}}{\text{Total possible outcomes}} = \frac{1\text{(head)}}{2\text{(heads or tails)}} = \frac{1}{2}$
5. **A.** Yes  $\frac{5}{16}$ is a larger number than $\frac{2}{14}$
   $\frac{5}{16}$ is more than $\frac{1}{3}$; while $\frac{2}{14}$ is only $\frac{1}{7}$
   Bonnie has about a 1 out of 3 chance of picking a red marble.

   **B.** Yes  Blue + Green = 4 + 4 = 8  and 8 is 50% of her total 16.

   **C.** Yes  Red + Blue = 2 + 5 = 7  and 7 is 50% of her total 14.

   **E.** Yes  Bonnie has the marbles and her sister 14.
6. **B.** 30 eighth-grade students as they leave the auditorium.

## PBA Questions (pages 219–221)

1. **Part A:**

| | Regular Tomato Plant | | Cherry Tomato Plant | |
|---|---|---|---|---|
| | Show work! | Height of Plant (inches) | Show work! | Height of Plant (inches) |
| Now | 12 | 12 | 6 | 6 |
| Week 1 | 12 + 1.5 = | 13.5 | 6 + 2 = | 8 |
| Week 2 | 13.5 + 1.5 = | 15 | 8 + 2 = | 10 |
| Week 3 | 15 + 1.5 = | 16.5 | 10 + 2 = | 12 |
| Week 4 | 16.5 + 1.5 = | 18 | 12 + 2 = | 14 |
| Week 5 | 18 + 1.5 = | 19.5 | 14 + 2 = | 16 |
| Week 6 | 19.5 + 1.5 = | 21 | 16 + 2 = | 18 |
| Week 7 | 21 + 1.5 = | 22.5 | 18 + 2 = | 20 |
| Week 8 | 22.5 + 1.5 = | 24 | 20 + 2 = | 22 |
| Week 9 | 24 + 1.5 = | 25.5 | 22 + 2 = | 24 |
| Week 10 | 25.5 + 1.5 = | 27 | 24 + 2 = | 26 |
| Week 11 | 27 + 1.5 = | 28.5 | 26 + 2 = | 28 |
| Week 12 | 28.5 + 1.5 = | 30 | 28 + 2 = | 30 |
| Week 13 | 30 + 1.5 = | 31.5 | 30 + 2 = | 32 |
| Week 14 | 31.5 + 1.5 = | 33 | 32 + 2 = | 34 |

**Part B:**

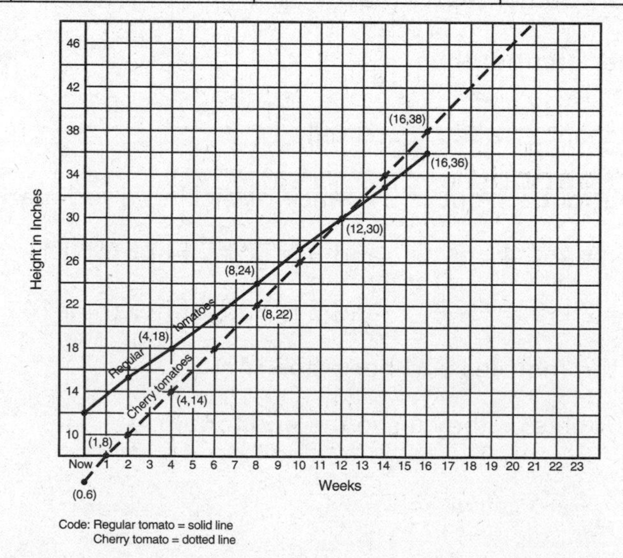


**Part C:** At week #12 both plants will be 30 inches tall.

**Part D:** The two plants are the same height where the lines intersect.

**Part E:** The cherry tomato plant begins to grow taller than the regular tomato. At week #13 the cherry tomato is 32 inches tall and the regular tomato is only 31.5 inches tall.

2. **Part A:** 41% of his sales came from bread and bagels.

   (% bread) + (% bagels) = 16% + 25% = 41%
   Then (0.41) ($23,769) = $9745.29

   **Part B:** He could expect to get $2,852.28 next year on cookie sales.

   This year, cookies are 8% of the total: (.08)(23,769), which equals $1,901.52.

   Next year, cookies would be 8% + 4% (50% more), or 12% of the total.

   (0.12)($23,769) = $2,852.28

3. **Part A:** $\frac{1}{13}$ There are four 10s in a deck of 52 cards. The probability would be

   $\frac{4}{52}=\frac{2}{26}=\frac{1}{13}$

   **Part B:** $\frac{1}{26}$ There are only two black Kings: the King of spades and the King of clubs; so, the probability of selecting a black King is $\frac{2}{52}=\frac{1}{26}$.

   **Part C:** $\frac{4}{13}$ The following are the numbered cards that are prime numbers: **2, 3, 5, 7**.

   There are four of each one, so there are (4)(4) = 16 cards that are prime numbers. The probability of selecting 16 cards out of the total of 52 is $\frac{16}{52}=\frac{4}{13}$.

## Chapter 4 Test: Probability and Statistics (pages 222-228)

1. **B.** 2. **B.** 3. 25% 4. **B**

5. **Part A:** 120 students are taller than 65 inches.

   80 are between 66-68

   30 are between 69-71

   10 are between 72-74

**Part B:** No. This type of graph just gives you information about a range of data. You can tell how many students are at least 57 inches and not taller than 59 inches. But you cannot tell how many are exactly 59 inches.

6. **Part A:** The *mean* is 68.03 years. $\frac{680.3}{10} = 68.03$

**Part B:** The *median* is 74.3 years.

Arrange the numbers in numerical order and take the middle number, or, in this case, take the average of the two middle numbers. $\frac{77.1+71.5}{2} = \frac{148.8}{6} = 74.3$

**Part C:** ***C.*** The *median will decrease.* It will go from 74.3 to 71.5 years. With the new eleven numbers arranged numerically, 71.5 is now in the middle.

45.9 49.1 55.0 57.5 62.9 **71.5** 77.1 78.8 79.4 79.8 80.7

7. **Parts A and B:** See the graph shown below.

There is a positive correlation. The taller the student the more he weighs.

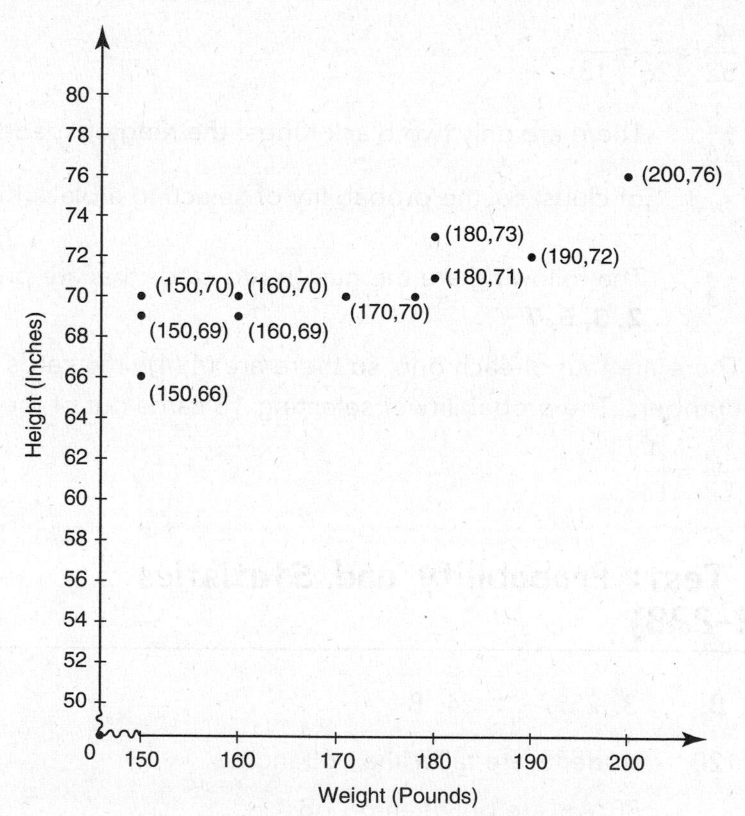

**Part C:** The slope is about $\frac{1}{10}$; I chose the coordinates (160, 69) and (170, 70)

$\frac{y_1 - y_2}{x_1 - x_2} = \frac{70-69}{170-160} = \frac{1}{10}$ (Note: Since the *x* and *y* scales are not the same, you cannot just count the boxes you must use the coordinate points and the formula for the slope.)

**Part D:** I would expect a player who is about 65 inches tall to weigh about 165 pounds.

8. **D.**
9. **A.** There is a positive correlation. As the years increase, the minimum wage also increases.
10. **C.**
11. **D.**
12. **A.**

**Performance-Based Assessment Questions**

13. **Part A:** **27.5% or 27½%** $\frac{55}{200} = 27.5 \text{ or } 27\frac{1}{2}\% = \frac{\textit{boys selected Italian}}{\textit{total boys}}$

**Part B:** **15.4%** $\frac{60}{390} = \frac{6}{39} = 15.38\% \sim 15.4\% = \frac{\textit{students selected French}}{\textit{total students}}$

**Part C:** **47.4%** $\frac{50+40}{190} = \frac{90}{190} = 47.4\%$

$\frac{\textit{girls selected Spanish} + \textit{girls selected Mandarin}}{\textit{total girls}}$

**Part D:**

| | | | |
|---|---|---|---|
| Spanish: | **5** classes | 120 | (four classes of 25 + one class of 20) |
| French: | **3** classes | 60 | (three classes of 20) |
| Italian: | **5** classes | 105 | (four classes of 20 and one of 25) |
| German: | **1** class | 25 | (one class of 25) |
| Mandarin: | **4** classes | 80 | (four classes of 20) |

14. Using the circle graph (Total students surveyed = 520; 20% basketball, 12% bowling, 32% indoor soccer, 25% ice hockey, 11% ice skating)

**Part A:** 130 students selected ice hockey (0.25)(520) = 130

**Part B:** 166 students selected basketball or bowling

Basketball: (0.20)(520) = 104 students

Bowling: (0.12)(520) = 62.4 students

Basketball + Bowling = 104 + 62.4 or appoximately 166 students

**Part C:** Approximately 109 more students selected indoor soccer over ice skating.

Indoor soccer: (0.32)(520) = 166.4 students

Ice skating (0.11)(520) = 57.2 students

166 – 57 = 109 students

**Part D:** If the percent for ice skating was increased, the total percent would be greater than 100% and that is not possible if all the other information is correct.

20% + 12% + 32% + 25% + 11% = 100% now

15. **B.**

# Practice Test Performance-Based Assessment Session I

## (No Calculator Permitted.)

Today you will be taking the Grade 8 Mathematics Performance-Based Assessment (PBA).

Read the directions carefully in each question. If you do not know the answer to a question, skip it and go on. If time permits, you may return to earlier questions. **Do your best to answer every question!** Read each question carefully. You may circle, highlight or underline important words. You may also use the Grade 8 Reference sheet for some formulas.

This Performance-Based Assessment is divided into two sessions. Session I, no calculator permitted and Session II, scientific calculator permitted.

- Some are **multiple-choice** questions with only **one-right answer**. Here you fill in the correct choice. (Fill in one letter only.)

  *Example: What is the area of a square with a side measuring 4.5 inches?*

  ○ **A.** *8.10 sq. in.* ○ **B.** *9 sq. in.* ○ **C.** *18 sq. in.* ○ **D.** *20.25 sq. in.*

  *The correct answer is* **D**. *20.25 sq. in.* *Area of a square = (4.5)(4.5)*

- Some are **new multiple-choice** type questions with **more than one right answer**. Here you should check all that are correct (all that apply).

  *Example: Which statements show a value that is equivalent to* $4^{-2}$*?*

  ☐ **A.** 8 ☐ **B.** $\sqrt{8}$ ☐ **C.** $\frac{1}{16}$ ☐ **D.** $\frac{-2}{-32}$

  *The two correct answers are:*

  ☑ **C.** $\frac{1}{16}$ and ☑ **D.** $\frac{-2}{-32}$ *since both are equivalent to* $4^{-2}$ *or* $\frac{1}{4^2}$*?*

- Other new **multiple-choice** examples are **Yes/No** statements. Here you should check Yes √ or No √ for each statement.

  *Example: Used the equation* $\boldsymbol{y = 2x + 4}$ *to answer the following:*

  **A.** *This equation represents a curved line.* *Yes ------ No ------*

  **B.** *This equation has a slope of 2.* *Yes ------ No ------*

  **C.** *The line* $y = 1/2x + 4$ *is parallel to this line.* *Yes ------ No ------*

  **D.** *The point (8, 20) falls on this line.* *Yes ------ No ------*

  The correct answers are:

  **A.** No √ **B.** Yes √ **C.** No √ **D.** Yes √

- Some questions ask that you ***show your work*** or ***label a diagram***.
- Other questions ask that you ***explain***, or ***justify*** your answer. Here you can use sentences, bulleted points, or sometimes a graph or diagram to explain. Answer ALL parts of these questions.

## SESSION I NO CALCULATOR (50 MINUTES)

**Name:** ..................................... **Date:** .....................................

**1.** Which expression below is equivalent to $\frac{1}{81}$?

○ **A.** 1/9
○ **B.** $9^2$
○ **C.** $3^4$
○ **D.** $3^{-4}$

**2.** The average distance of the planet Mercury from the sun is $5.8 \times 10^9$ km.

The average distance of the planet Jupiter from the sun is $7.78 \times 10^{11}$ km.

About how many times farther from the sun is Jupiter than Mercury?

○ **A.** About 10 times greater
○ **B.** About 50 times greater
○ **C.** About 130 times greater
○ **D.** More than 200 times greater

**3.** The number $2.86 \times 10^{-4}$ is written in *scientific notation form.*

Which numbers below are equivalent to $2.86 \times 10^{-4}$?

**(There is more than one right answer. Select all that are correct.)**

☐ **A.** 28,600
☐ **B.** 2,860,000
☐ **C.** $2.86 \times \frac{1}{10^4}$
☐ **D.** $\frac{1}{286 \times 10^4}$
☐ **E.** $2.86 \div 10^4$

**4.** Solve for $x$ in the linear equation given. Show your work in the box below.

$$2x + 4 - 3(x + 5) = 8x - 20$$

Final answer: $x =$ ____________

**5.** Solve for $a$ in the linear equation given. Show your work in the box below.

$$14 - 4a = -8a - 4 - 2(a - 3)$$

Final answer: $a =$ ____________

**6.** Filipo needs to solve this problem about systems of equations.

He is given two lines.

*Line p* goes through the points (3, 3) and (0, 6).

*Line q* goes through the points (0, 0) and (–2, –2).

**Part A:** Does *line p* intersect with *line q*? Yes ______ No ______

**Part B:** Explain your answer in the box below, or use the graph below to model your answer.

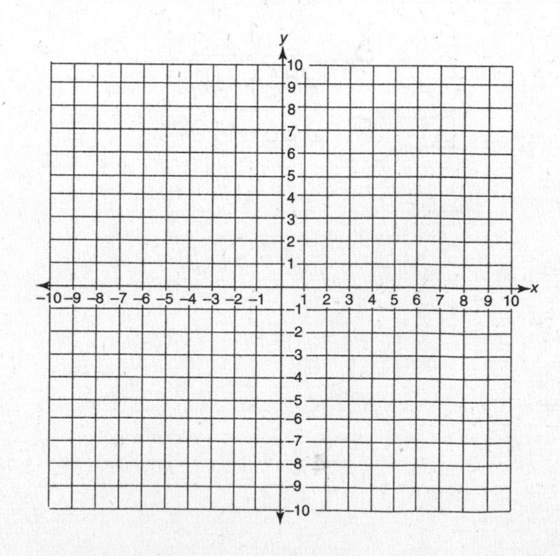

**7.** Which of the following is or is not a function? Select each one.

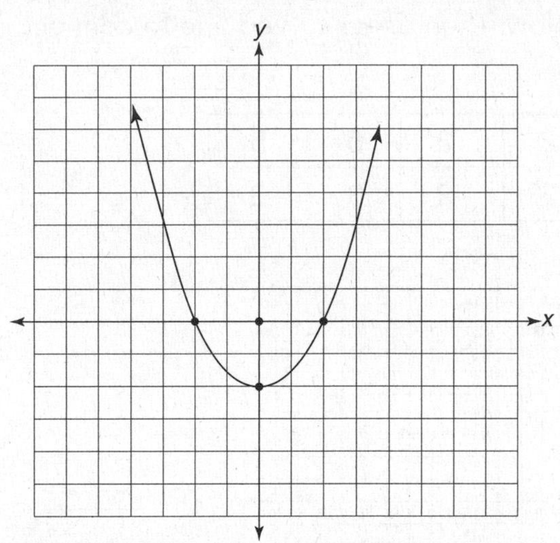


**A.** The diagram above is — a function ☐ — not a function ☐

**B.** the line described by $y = 4$ — a function ☐ — not a function ☐

**C.** a circle — a function ☐ — not a function ☐

**D.** $x = -2$ — a function ☐ — not a function ☐

**E.** $y = 3x + 5$ — a function ☐ — not a function ☐

**F.** Use the graph below. This is — a function ☐ — not a function ☐

y

x

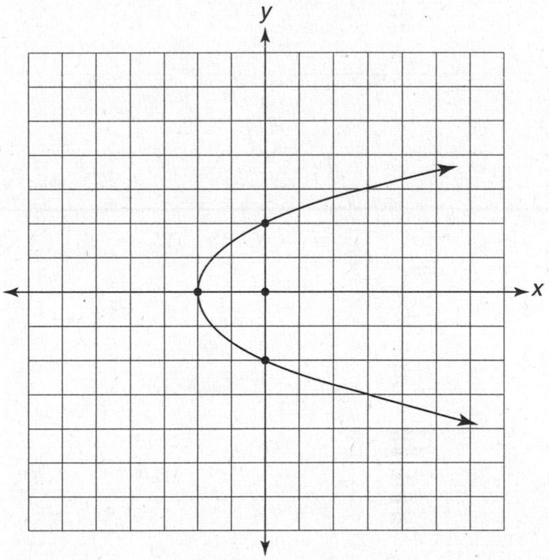

**8. Part A:** Use the information in the data in the table shown and plot the five coordinates on the grid below. Then use a curved line to connect the points.

| *x* | 0 | 2 | 2 | 5 | 5 |
|---|---|---|---|---|---|
| *y* | 0 | 2 | −2 | 3 | −3 |

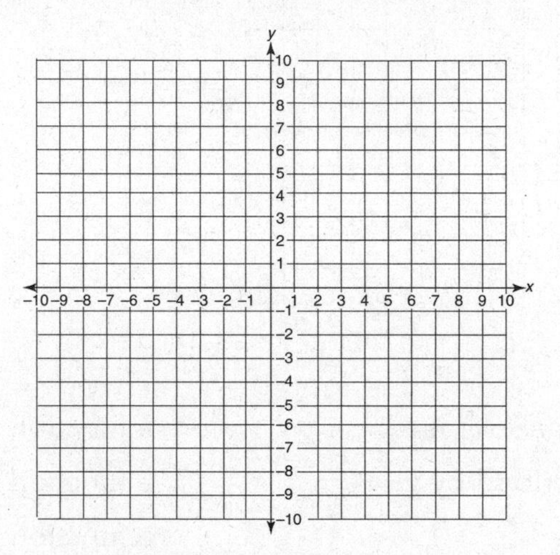


**Part B:** Does this represent a function? Yes ______ No ______

**Part C:** Explain your answer in the box below.

**9. Part A:** Use the figure to determine your answers:

Which explanations of transformations proves triangle *ABC* is congruent to triangle *A′B′C′*?

**(Note: there is more than one correct choice. Select all that are correct.)**

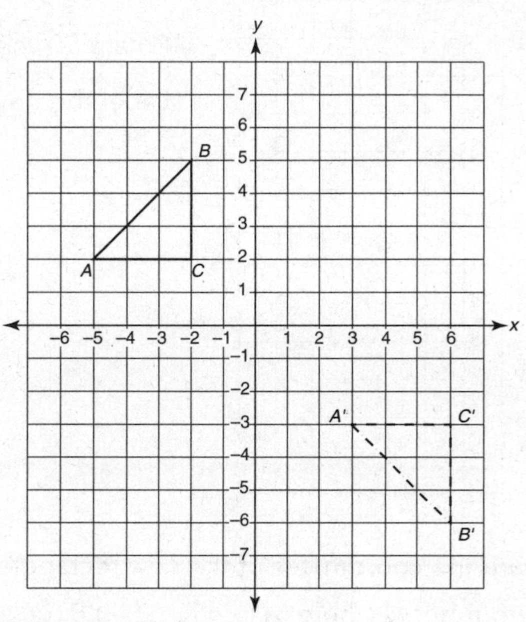


☐ **A.** translate △*ABC* over the *y-axis* (move right 8 units), flip, then move down 4 units

☐ **B.** translate △*ABC* over the *y-axis,* (move right 8 units), flip, then move down 5 units

☐ **C.** flip △*ABC*, then translate over the *x-axis* (move down 5 units), then move 8 units to the right

☐ **D.** flip △*ABC*, translate over the *x-axis,* move down 4 units move 8 units to the right

☐ **E.** translate △*ABC* over the *y-axis,* then translate over the *x-axis*.

**Part B:** What are the coordinates of the translated triangle *A′B′C′*?

Answer: *A′*(______, ______) *B′*(______, ______) and *C′*(______, ______)

**10.** **Part A:** Follow the translation directions to move rectangle *ABCD* *A*(3, 6) *B*(5, 6) *C*(5, 2) *D*(3, 2) from quadrant I to quadrant III. Sketch the new rectangle *A′B′C′D*.

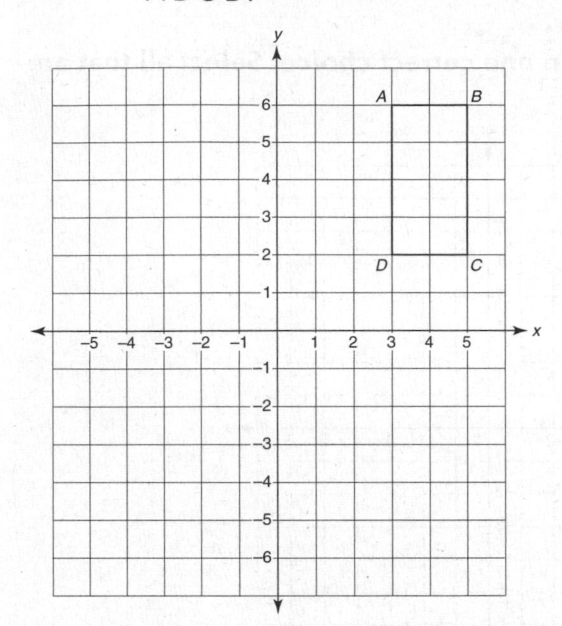


Directions:

Translate over the *x-axis.*

Rotate 90° clockwise **around point "A"**

Label the new rectangle *A′B′C′D′*

**Part B:** What are the coordinates of the new rectangle *A′B′C′D′*?

Write your answer here: *A′*(______, ______) *B′*(______, ______) *C′*(______, ______) *D′*(______, ______)

**11.** Answer the following statements about these two similar triangles below.

Triangle *ABC* is similar to Triangle *EFG*

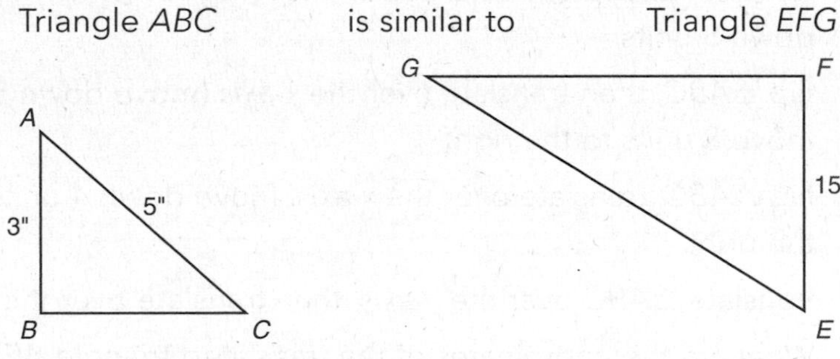


**A.** Side $\overline{EG}$ measures 20". Yes ______ No ______

**B.** The measure of angle *A* = the measure of angle *E*. Yes ______ No ______

**C.** If angle *B* measures 90°, then the sum of $\angle G + \angle E = 90°$. Yes ______ No ______

**D.** If side $\overline{BC}$ measures 4", then length of side $\overline{GF}$ measures 8". Yes ______ No ______

**E.** If one △ is a right triangle, then the other △ is also a right triangle. Yes ______ No ______

**12.** Look at the triangles *WXY* and *ABC* and *A′B′C′*.

**Part A:** Explain using a sequence of translations, reflections, rotations, and/or dilations to show that $\triangle A'B'C'$ is similar to $\triangle WXY$.

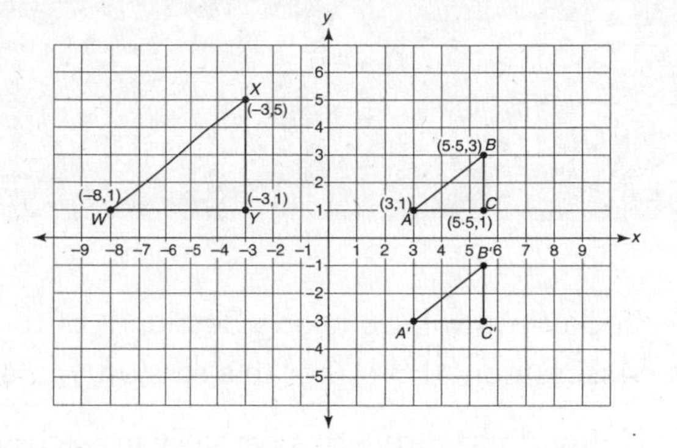


Write or model your explanation in the box below.

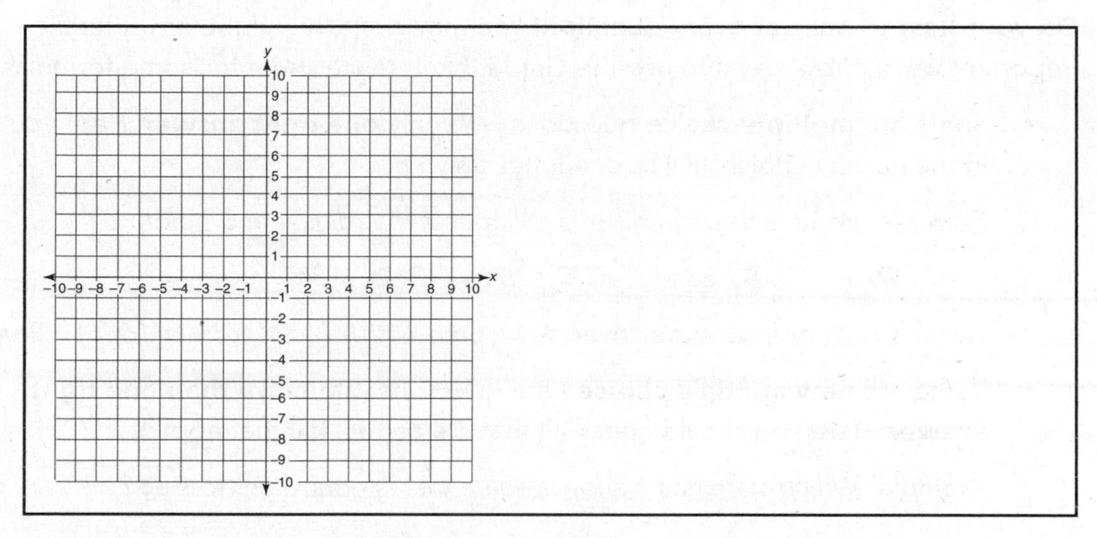

**Part B:** What are the coordinates of $\triangle A'B'C'$?

$A'$(______, ______) $B'$(______, ______) $C'$(______, ______)

**Part C:** What is the area of $\triangle WXY$ and area of $\triangle A'B'C'$?

Area of $\triangle WXY$ is ______ square units

Area of $\triangle A'B'C'$ is ______ square units

**Part D:** Are the areas of $\triangle WXY$ and $\triangle ABC$ in the same proportion as their sides?

Yes ______ No ______

# Practice Test Performance-Based Assessment Session II

## (Calculator Permitted.)

Today you will be taking Session II of the Grade 8 Mathematics Performance-Based Assessment (PBA). For this session you are permitted to use a scientific calculator.

Read the directions carefully in each question. If you do not know the answer to a question, skip it and go on. If time permits, you may return to earlier questions. **Do your best to answer every question!** You may circle, highlight, or underline important words. You may also use the Grade 8 Reference sheet for some formulas.

- Some are **multiple-choice** questions with only **one-right answer**. Here you fill in the correct choice. (Fill in one letter only.)

  *Example: What is the volume of a cylinder with radius 4 and height 6?*

  ○ **A.** $96\pi$ ○ **B.** $36\pi$ ○ **C.** $28\pi$ ○ **D.** $12\pi$

  *There is only one correct answer,* **A**. *Volume* $= \pi r^2h = \pi(4^2)(6) = \pi(16)(6) = 96\pi$

- Some are **new multiple-choice** type questions with **more than one right answer**. Here you should check all that are correct (all that apply).

  *Example: Which statements show a value that is equivalent to* $\sqrt{36}$*?*

  ☐ **A.** 9 ☐ **B.** $2^3 - 2$ ☐ **C.** $\frac{1}{6}$ ☐ **D.** $6^{3-2}$

  *The two correct answers are:*

  ☑ **B.** $2^3 - 2 = 8 - 2 = 6$ *and* ☑ **D.** $6^{3-2} = 6^1$ *since both are equivalent to 6.*

- Other **new multiple-choice** examples are **Yes/No** statements. Here you should check Yes √ or No √ for each statement.

  *Example: Used the equation* $\mathbf{y = 2x + 4}$ *to answer the following.*

  **A.** *This equation represents a curved line.* Yes ------ No ------

  **B.** *This equation has a slope of 2.* Yes ------ No ------

  **C.** *The line* $y = 1/2x + 4$ *is parallel to this line.* Yes ------ No ------

  **D.** *The point (8, 20) falls on this line.* Yes ------ No ------

The correct answers are:

**A.** No √ **B.** Yes √ **C.** No √ **D.** Yes √

- Some questions ask that you ***show your work*** or ***label a diagram***.
- Other questions ask that you ***explain***, or ***justify*** your answer. Here you can use sentences, bulleted points, or sometimes a graph or diagram to explain. Answer ALL parts of these questions.

## SESSION II WITH CALCULATOR (50 MINUTES)

**Name:** ................................ **Date:** ................................

1. Complete the following multiplication, division, addition, and subtraction *scientific notation* problems. Show your work.

   **Part A:** multiply $(3.2 \times 10^5)\,(4.89 \times 10^2)$

   Answer: ................

   **Part B:** Evaluate the expression below. Write your answer in proper *scientific notation* format. $\dfrac{5.9 \times 10^6}{2.3 \times 10^4}$

   Answer: ................

   **Part C:** Add these two expressions. Write your answer in proper *scientific notation* format. $4.6 \times 10^2 + 3.8 \times 10^2$

   Answer: ................

   **Part D:** Subtract these two expressions. Write your answer in proper *scientific notation* format. $9.2 \times 10^5 - 4.5 \times 10^2$

   Answer: ................

2. Compare the two types of information given below. Each describes a line.

   **line *p***

   $y = \dfrac{1}{5}x$

   **line *r***

   $y = -\frac{1}{2}x + 2$

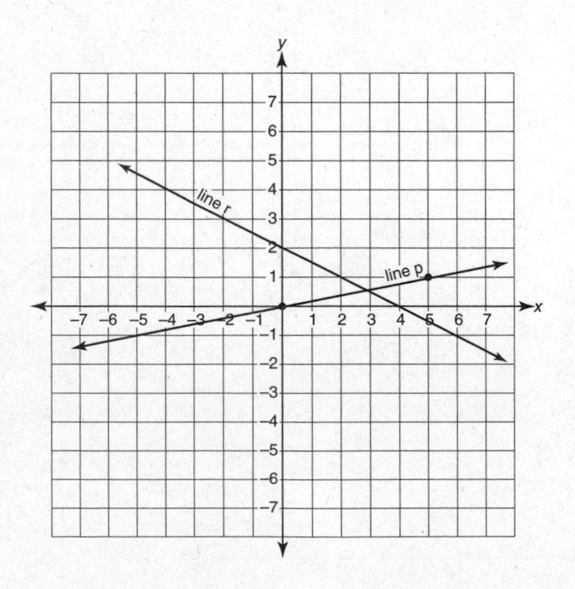

**Part A:** What is the slope of each line?

The slope of line $p$ is ______. The slope of line $r$ is ______.

**Part B:** Which lines shows a positive *rate of change* and which shows a *negative rate of change?*

Line $p$ *shows a* ______________________________ *rate of change.*

Line $r$ *shows a* ______________________________ *rate of change.*

**Part C:** Aarte has been collecting information about the amount of time she spends on schoolwork each week. She sees that when she spends more time on homework and studying, her grades get higher. She makes a graph to show this.

Aarte's dad has been collecting information about their monthly heating costs for their house. His graph tells him that when the weather is warmer their heating bills are lower.

Which data (line $p$ or line $r$) could represent Aarte's information, and which her dad's?

Line: ______ *could represent Aarte's information about homework.*

Line: ______ *could represent her dad's information about heating costs.*

**Part D:** Explain how you know. (Write your answer in the box below.)

3. Compare the two types of information below. Each describes a line.

Note: the *x-axis* represents time (hours); the *y-axis* represents the height (feet).

**line *a***

| x | y |
|---|---|
| 2 | 6 |
| 1 | 5 |
| 0 | 4 |
| –1 | 3 |
| –2 | 2 |

**line *b***

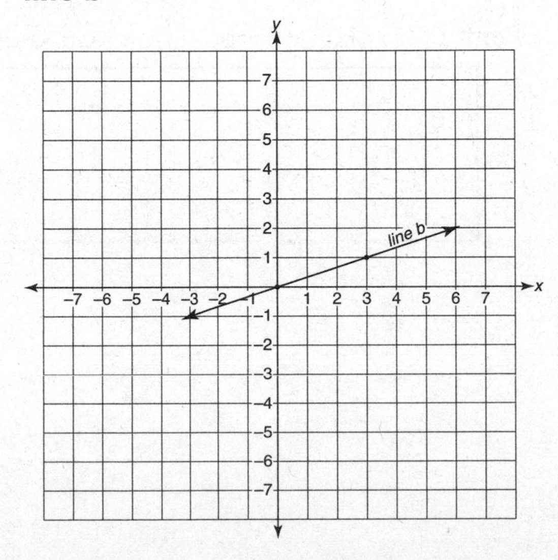

**Part A:** Which line has a steeper slope, *line a* or *line b*?

*Line* ------ has a steeper slope.

**Part B:** Line *a* and *b* each represent the amount of water being poured into a swimming pool each hour. Which pool was being filled the fastest?

*Line* ------ shows that pool was being filled the fastest.

**Part C:** Explain why you selected your answer to Part B above.
(Write your explanation in the box below.)

4. **Part A:** On the coordinate grid provided plot the points and connect them to create the two triangles described. Label each point as noted below.

- Triangle *ABC*: *A*(–1,1) *B*(–4,1) *C*(–4,9)
- Triangle *A′B′C′*: *A′*(4,–10) *B′*(1,–10) *C′*(1,–2)]

y
10 9 8 7 6 5 4 3 2 1
–10 –9 –8 –7 –6 –5 –4 –3 –2 –1 1 2 3 4 5 6 7 8 9 10 x
–1 –2 –3 –4 –5 –6 –7 –8 –9 –10

**Part B:** Are these two triangles similar? Yes ------ No ------

**Part C:** Explain or model how you know.

5. Shelila is an artist working for an advertising company, and for the company's new contract she is designing a logo for a major airline company; she needs to know if these two triangles are similar. One way is to first find the slope of line segments *FH* and *GH*.

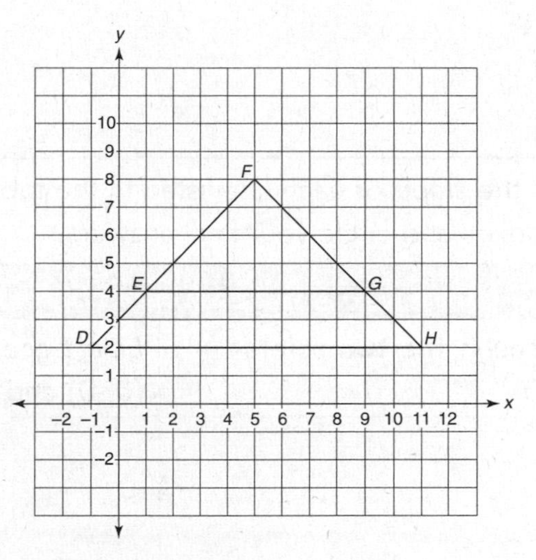


**Part A:** • What is the slope of line segment $\overline{FH}$ (the side of the larger triangle)? ------

• What is the slope of line segment $\overline{FG}$ (the side of the smaller triangle)? ------

**Part B:** Model **or** explain how you found the slope of each line. (Show you work in the box below.)

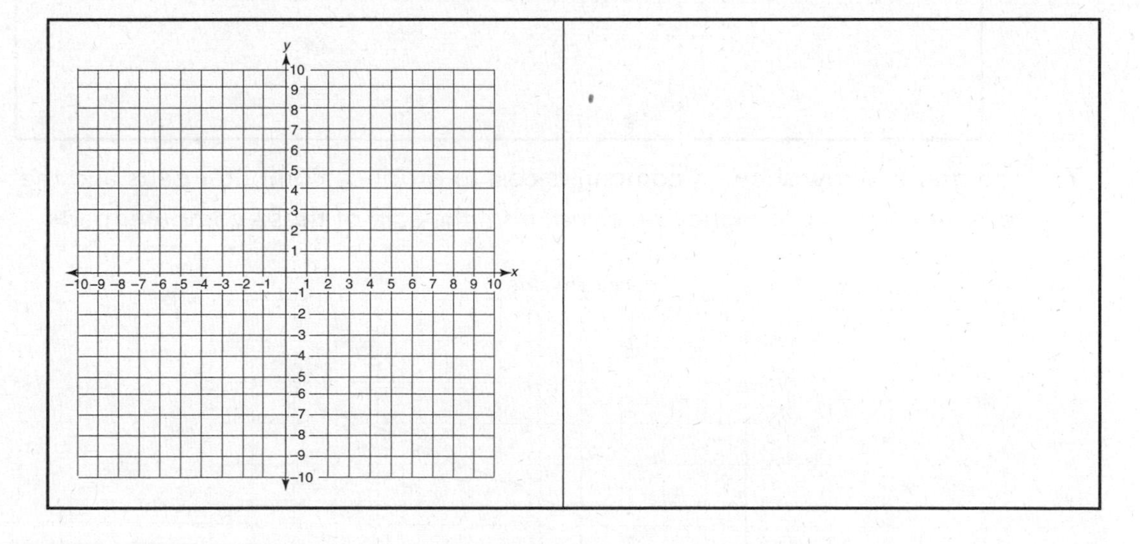

6. **Part A:** Write the equation of the line in $y = mx + b$ form for a line that goes through the point (4, 8) and has a *y-intercept* of 4.

Show your work in the box below and circle your final answer.

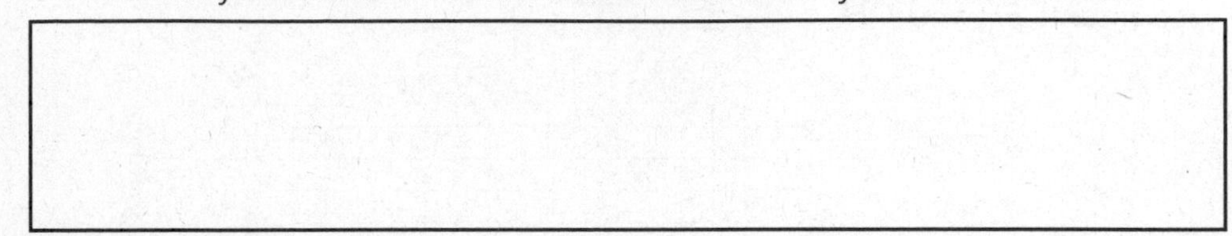

**Part B:** What is the *slope* of each line listed in the table below? Show your work in the boxes and circle your final answers.

| *Line s* | *Line t* |
|---|---|
| *Line s* goes through the two points (0, 6) and (4, 0). | *Line t* goes through the two points (0, 0) and (5, 8). |
| | |

**Part C:** Which line has a negative slope? *Line* ______ has a negative slope.

**Part D:** Are these lines parallel? Yes ______ No ______

**Part E:** Explain or model how you know whether the two lines are parallel or not. (Write your answer in the box below.)

7. The graph below shows a company's cost to produce computer parts and the revenue (amount of money received) from the sale of those computer parts.

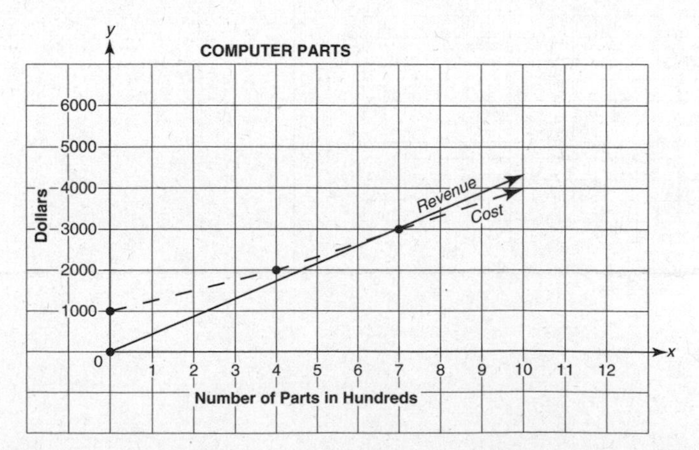

**Part A:** At approximately what point (level of production) does the cost of the computer parts equal the revenue (the amount of money received from selling the parts)?

Write your answer here: ☐

**Part B:** The company will make a profit after it receives more money than it spends to produce the computer parts. Use the graph to estimate the profit on the sale of 1,100 computer parts.

- ○ **A.** \$4,000
- ○ **B.** \$3,000
- ○ **C.** \$1,000 (5,000 – 4,000 = 1,000)
- ○ **D.** \$700 (4,700 – 4,000) = 700

**8.** **Part A:** Solve for *x* in the equation below.

$4x + 2 - 2(x + 4) = 3 + 15 + 2(x - 4)$

Write your final answer here: ------

**Part B:** Does this equation have one solution, infinitely many solutions, or no solutions? It has ---------------------------- solution(s).

**Part C:** Explain your answer.

☐

**9.** For each linear equation in the table below, select whether the equation has *no solution, one solution,* or *infinitely many solutions.* Put a check in the appropriate box.

| | Equation | No solution | One solution | Infinitely many solutions |
|---|---|---|---|---|
| A. | $16x - 2(2x) - 6 = 2(5x) - 6 + 2x$ | | | |
| B. | $x = x + 1$ | | | |
| C. | $12(x + 2) = -4x + 2$ | | | |

**10.** Explain or model how you know the graph containing the points (–1, –2), (2, 4), and (6, 6) does **not** represent a straight line. (Write your answer in the box below.)

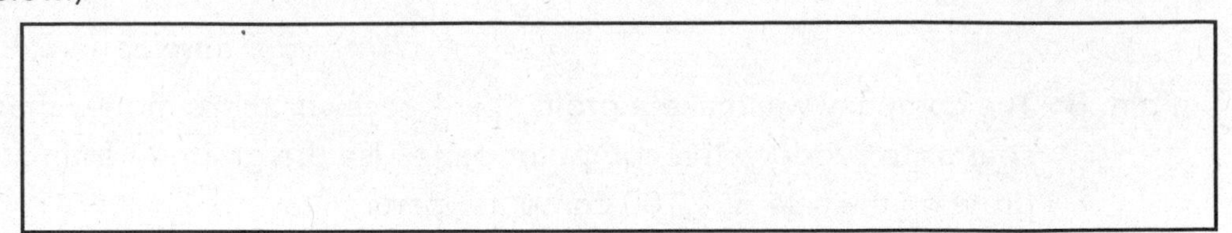

**11.** **Part A:** Which set of translations will move the two-dimensional rectangle *ABCD* in quadrant IV to be the larger rectangle *A′B′C′D′* in quadrant II?

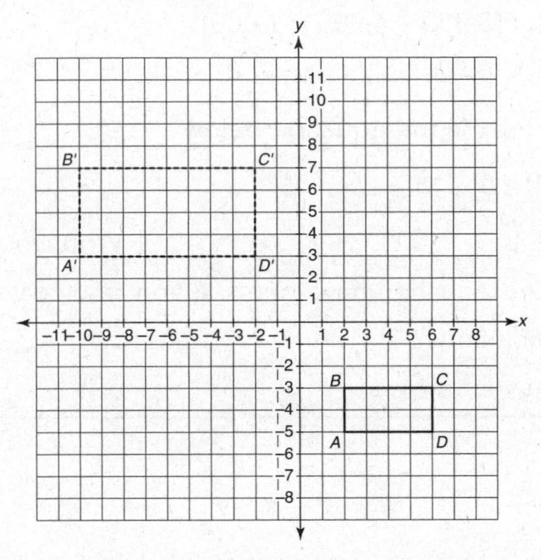


○ **A.** Translate rectangle *ABCD* 8 units left, then 10 units up, then dilate by a factor of 3.

○ **B.** Translate rectangle *ABCD* 8 units left, then 10 units up, then dilate by a factor of 2.

○ **C.** Translate rectangle *ABCD* 10 units up, then double each side, and reflect over the *y-axis*.

○ **D.** Translate rectangle *ABCD* 9 units up and then 9 units left over the *y-axis*.

**Part B:**
- What are the dimensions of rectangle *ABCD*?

  Length = ______ Width = ______

- What are the dimensions of rectangle *A′B′C′D′*?

  Length = ______ Width = ______

**Part C:** Is rectangle *ABCD* in quadrant IV *similar* to rectangle *A′B′C′D′* in quadrant II?

Check off your answer here: Yes ______ No ______

**Part D:** Explain why the rectangles **are** or **are not** similar. Write your explanation in the box below.

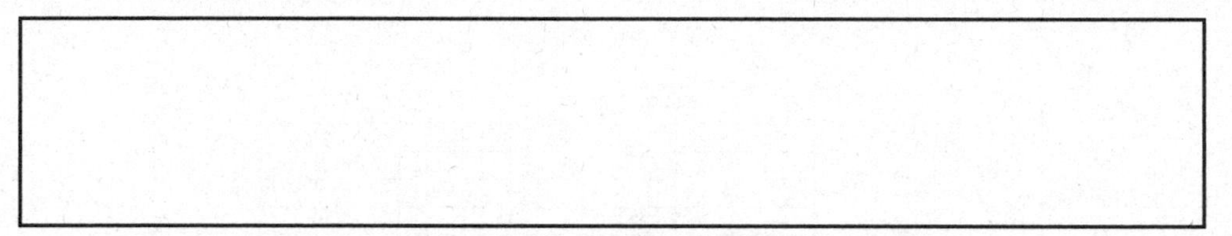

**12. Part A:** What is the sum of the interior angles of any triangle? ______°

**Part B:** What is the measure of any *straight angle*? ______°

**Part C:** If the sum of the interior angles of any triangle is what you have said in Part A above, fill in the missing angle measures in the triangles below.

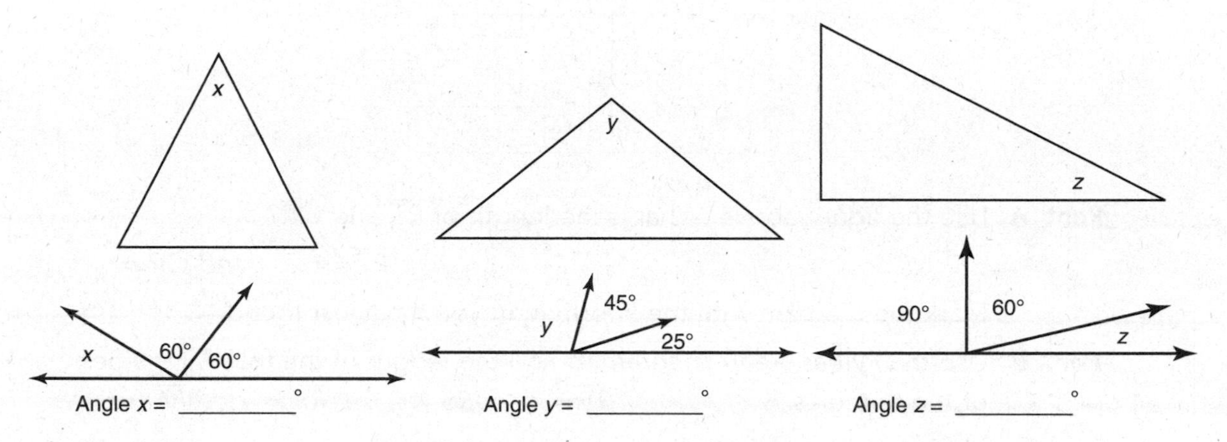


**13. Part A:** Solve for $w$ in the equation below. Show your work and write your answer in simplest form. For **Parts A and B,** you may need scrap paper.

$$4w - 3w(w - 2) = -3w^2 - 4(3)$$

**Part B:** Who is correct and why?

<u>Awni's work</u>

$6a + 7 - 3(a - 2) = 16 - 6a$

$6a + 7 - 3a + 6 = 16 - 6a$

$6a - 3a + 7 + 6 = 16 - 6a$

$3a + 13 = 16 - 6a$

$9a = 3$

$a = 3/9 = 1/3$

<u>Muhammad's work</u>

$6a + 7 - 3(a - 2) = 16 - 6a$

$6a + 7 - 3a - 6 = 16 - 6a$

$6a - 3a + 7 - 6 = 16 - 6a$

$3a + 1 = 16 - 6a$

$9a = 15$

$a = 15/9 = 5/3$

**Check the name of the student who is correct: Awni ______ Muhammad ______ Where was the mistake made? Explain or model your answer.**

**14.** Dyane and Pat are working together on their math homework. They need to find out if these two triangles are congruent.

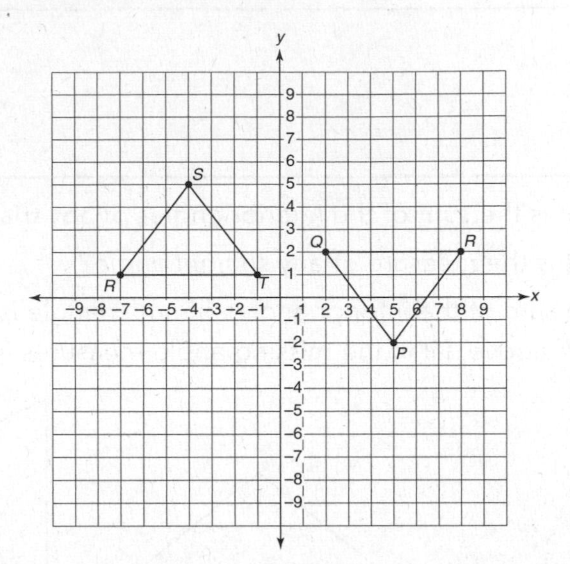


**Part A:** Use the figure above. What is the length of $\overline{RT}$ *and* $\overline{QR}$?

$\overline{RT}$ = ______ *and* $\overline{QR}$ = ______

Mark the diagram with the information you have just found.

**Part B:** Use the *Pythagorean Theorem* to find the length of the following sides of the triangles. $\overline{ST}$ = ______ $\overline{RS}$ = ______; $\overline{RP}$ = ______ $\overline{PQ}$ = ______

Mark the diagram with this new information.

**Part C:** In the box below, describe the translations needed in the correct sequence to show that $\triangle RST \cong \triangle QPR$. Use *translate (slide up, down, or sideways), rotate, and/or reflect (flip).* (Hint: You should need only three steps.)

**Part D:** Is $\triangle QRP$ congruent to $\triangle RST$? Yes ______ No ______

How do you know? (Write your answer in the box below.)

**15.** Use the diagram below of an *isosceles trapezoid* to answer the following questions.

Label the diagram as you discover more information.

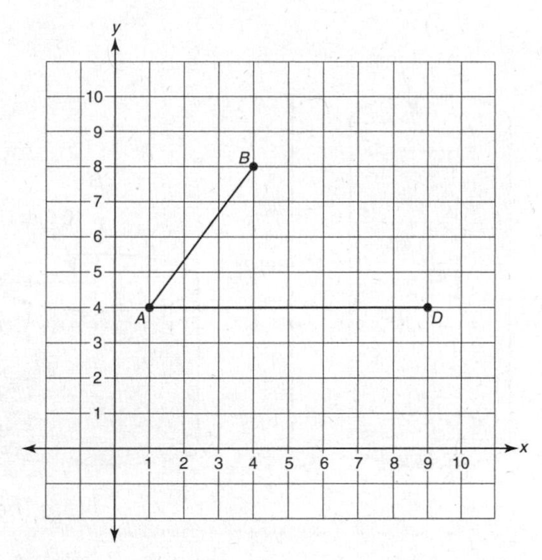


**Part A:** What are the coordinates of points *C* and *D* of this *isosceles trapezoid?*

Answer: *C* (------, ------) and D (------, ------)

**Part B:** What is the height of the trapezoid? Answer: height = ------ units

Use the *Pythagorean Theorem* to find the length of sides $\overline{AB}$ and $\overline{CD}$.

Answer: $\overline{AB}$ = ------ *units long and* $\overline{CD}$ = ------ *units long*

**Part C:** What is the perimeter of trapezoid *ABCD*? Answer: ------ units

What is the area of trapezoid *ABCD*? Answer: ------ sq. units

**16. Part A:** What is the approximate probability that a spinning penny will land heads up?

○ **A.** 25% ○ **B.** 40% ○ **C.** 50% ○ **D.** 75%

**Part B:** What is the probability that the spinner in the figure below will land on a perfect-square number?

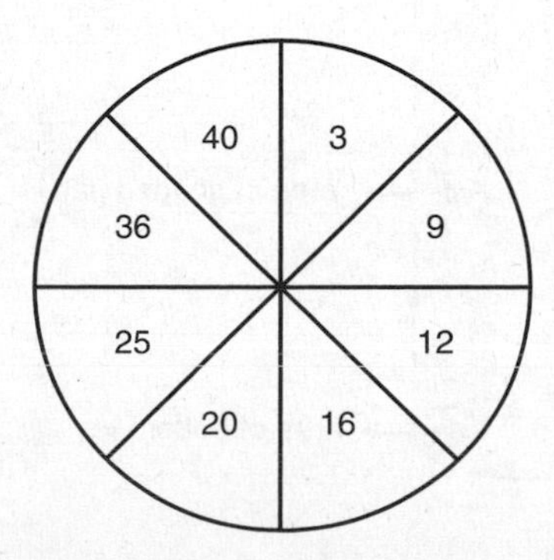


○ **A.** 25% ○ **B.** 40% ○ **C.** 50% ○ **D.** 75%

**17. Part A:** What is the volume of each cylinder shown below? Round your final answer up to the nearest whole number. (The formula for volume of a cylinder is $\pi r^2 h$, or think of it as the area of the base times the height.)

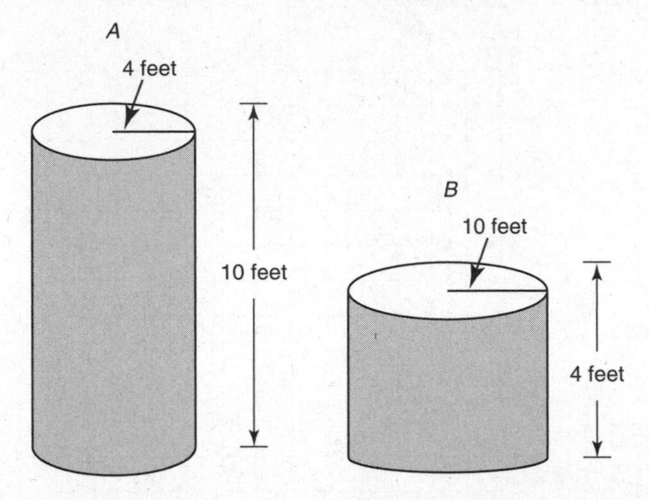


*Note, figures not drawn to scale.*

The volume of the tall cylinder *A* is about ______ cubic feet.

The volume of the shorter cylinder *B* is about ______ cubic feet.

**Part B:** Check off the shapes that make up cylinder *A*. (*Shapes are not drawn to scale.*)

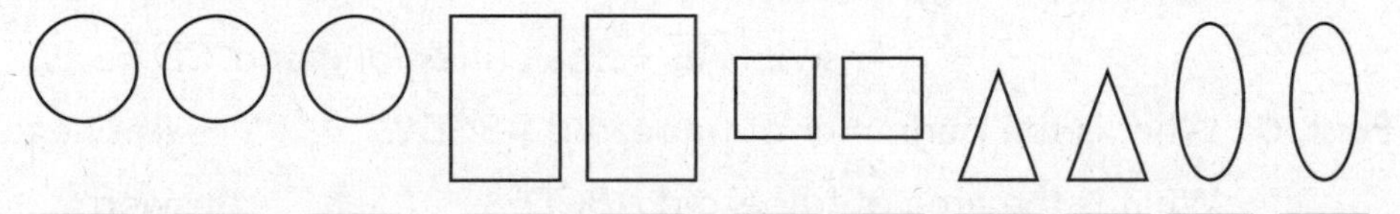

**Part C:** Jeremiah needs to find the *surface area* of cylinder *A* to estimate the total gallons of paint he will need to paint 100 such cylinders in his dad's factory. The following information will help him. **Mark up the diagram below as you calculate the dimensions needed.**

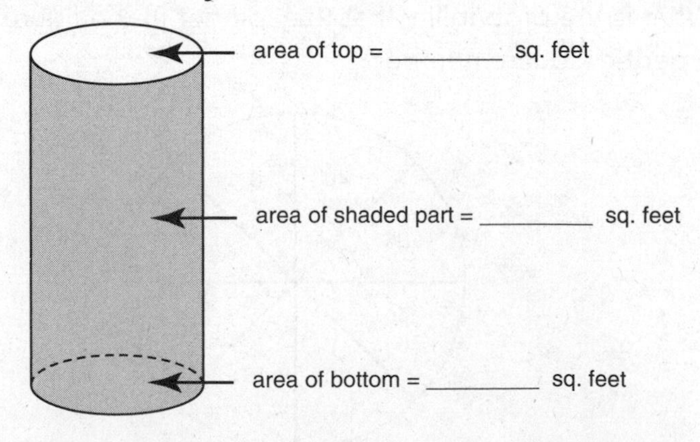

Step 1: What is the *area of the top* of cylinder *A*? (Note: $\pi r^2$ = area of a circle; use 3.14 for $\pi$ and round your answer up to the nearest whole number.)

The *area* of the top of cylinder *A* = ______ sq. feet

Step 2: What is the *area of the shaded part* of cylinder *A*?

Step 3: First find the *circumference of the top* of cylinder *A*? The *circumference* is actually the *length* of the *rectangle*. (Note: $2\pi r$ = circumference; use 3.14 for $\pi$ and round your answer up to the nearest whole number.)

The *circumference* of the top of cylinder *A* = ______ feet

The *area* of the *shaded part* of cylinder *A* = ______ sq. feet

Step 4: What is the *entire surface area* of cylinder *A*. (Round your answer up to the nearest whole number.)

The entire *surface area* of cylinder *A* = about ______ sq. feet

**18.** There are five grade 8 classes in Stephanie's middle school. Stephanie and her group decided to survey all the grade 8 students as part of a science project. The table below shows the results of their survey. Use these results to answer Parts A-G.

| Students | Right-Handed | Left-Handed | Total |
|---|---|---|---|
| **Boys** | 52 | 11 | 63 |
| **Girls** | 46 | 5 | 51 |
| **Total** | 98 | 16 | 114 |

**Part A:** How many right-handed boys were surveyed?

Answer: ____________

**Part B:** How many left-handed students were surveyed?

Answer: ____________

**Part C:** How many students were surveyed in all?

Answer: ____________

**Part D:** What percent of students surveyed were right-handed?
(Round your answer to 2 decimal places.)

Answer: __________%

**Part E:** What percent of children surveyed were left-handed?

Answer: __________%

**Part F:** What percent of boys are right-handed?
(Round your answer to 2 decimal places.)

Answer: __________%

**Part G:** What percent of right-handed students are boys?

Answer: __________%

**19.** Another group of students decided to survey students in their class to compare two different questions. They collected data about whether or not students have assigned chores at home and whether or not they had a school-night curfew (a time they must be home on school nights).

| | Have a curfew | No curfew | Totals |
|---|---|---|---|
| Have assigned chores | 10 | 1 | **11** |
| No chores | 2 | 8 | **10** |
| **Totals** | **12** | **9** | **21** |

**Part A:** What percentage of all students who have a curfew also have chores? (Round your answer to the nearest hundredth.) ------%

**Part B:** What percentage of all students have no chores? ------% (Round your answer to the nearest percent.)

# PERFORMANCE-BASED ASSESSMENT ANSWERS

## Session I No Calculator Permitted (pages 281–287)

**8.EE.1, 8.EE.A.1**

1. **D.** $3^{-4}$ $\quad \frac{1}{3^4}$ *which* $= \frac{1}{3} \times \frac{1}{3} \times \frac{1}{3} \times \frac{1}{3} = \frac{1}{9} \times \frac{1}{9} = \frac{1}{81}$

**8.EE.A3**

2. ***C.*** About 130 times greater

   $5.8 \times 10^9 = 5{,}800{,}000{,}000$ written in standard form

   $7.78 \times 10^{11} = 778{,}000{,}000{,}000$ written in standard form

   For a rough estimate try 7,800 ÷ 60 and you'll get **130** or use 800 ÷ 6 and you'll get approximately 133.

   For a more accurate figure $\frac{778{,}0\cancel{00{,}000{,}000}}{5{,}8\cancel{00{,}000{,}000}} = \frac{7{,}780}{58} \sim 134.1$

   Divide and simplify and write the fraction as a decimal approximation.

**8.EE.A4**

3. ***C.*** $2.86 \times \frac{1}{10^4}$ $\quad$ Remember, $10^{-4}$ really means $\frac{1}{10^4}$

   **E.** $2.86 \div 10^4$

**8.EE.C7, 8.EE.7b**

4. $x = 1$

$$\begin{aligned}
2x + 4 - \mathbf{3(x + 5)} &= 8x - 20 \\
\mathbf{-3x - 15} & \qquad \textit{distribute the } -3 \\
\mathbf{2x} + 4 \mathbf{-3x} - 15 &= 8x - 20 \\
+ 4 \mathbf{-x} - 15 &= 8x - 20 \qquad \textit{combine like terms (the x's)} \\
-x - 11 &= 8x - 20 \\
\underline{+x} \qquad & \underline{+x} \qquad \textit{add "x" to both sides} \\
-11 &= 9x - 20 \\
\underline{+20} \qquad & \underline{+20} \qquad \textit{add +20 to both sides} \\
9 &= 9x \\
1 &= x \qquad \textit{divide both sides by 9}
\end{aligned}$$

5. $a = -2$

| | | |
|---|---|---|
| $14 - 4a - = -8a - 4 - \mathbf{2(a - 3)}$ | | *distribute the −2* |
| $= -8a - 4 \mathbf{- 2a + 6}$ | | *combine like terms (the a's)* |
| $= -10a - 4 + 6$ | | *combine like terms (−4 + 6)* |
| $14 - 4a$ | $= -10a + 2$ | *combine like terms; add 10a to both sides)* |
| $\underline{+10a}$ | $\underline{+10a}$ | |
| $14 + 6a$ | $= 2$ | |
| $\underline{-14}$ | $\underline{-14}$ | *subtract 14 from both sides* |
| $6a = -12$ | | *divide both sides by 6* |
| $a = -2$ | | |

**8.EE.C8a**

6.

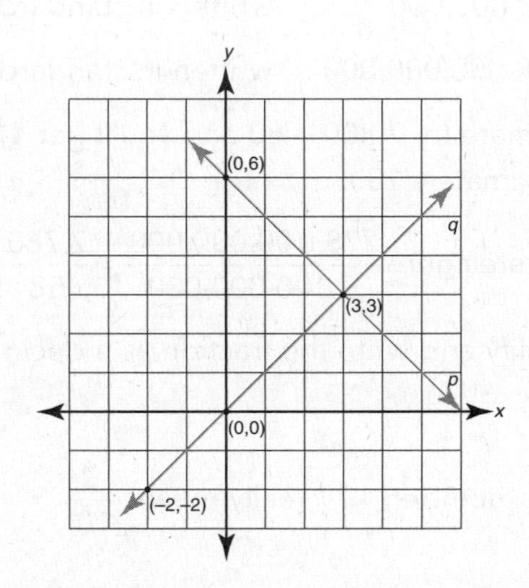


**Part A:** Yes, line *p* intersect with line *q*.

**Part B:** Explanation: When I found the slope of each line they were NOT the same, so the lines are NOT parallel. The lines will intersect.

$$\textit{Line } p\text{: slope} = \frac{y_1 - y_2}{x_1 - x_2} = \frac{6-3}{0-3} = \frac{3}{-3} = \frac{1}{-1} = -1$$

$$\textit{Line } q\text{: slope} = \frac{0-(-2)}{0-(-2)} = \frac{+2}{+2} = 1$$

Model: See the graph after you've plotted the points and drawn the lines. You now can count to find the $\frac{rise}{run}$ of each line and will see the slope of one line is $\frac{1}{1}$ and the slope of the other is $-\frac{1}{1}$. The slopes are not the same; the lines are not parallel. In fact, these lines are perpendicular to each other. You also can see that lines *p* and *q* meet at (3, 3).

**8.F.A.1**

7. **A.** a function **B.** a function **C.** not a function

   **D.** not a function **E.** a function **F.** not a function

8. **Part A:**

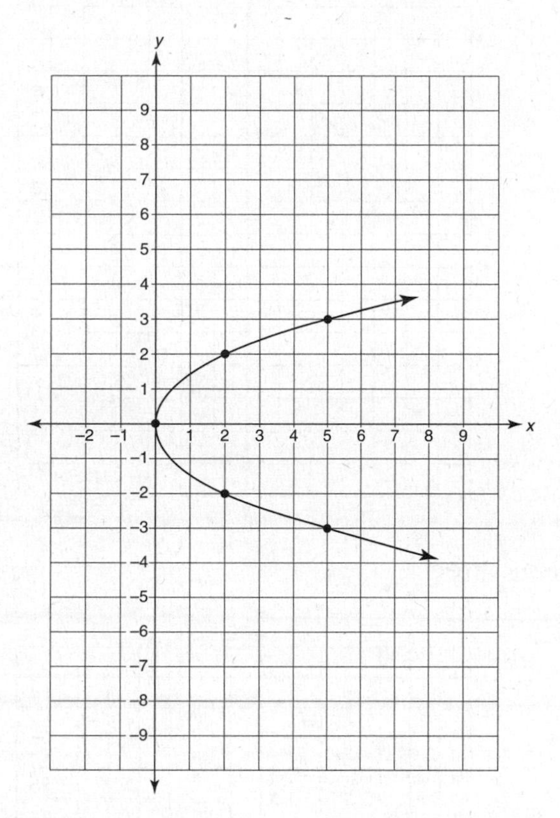


**Part B:** No

**Part C:** If this were a function there would be ONLY one value for *y* for each value of *x*.

In this equation when $x = 2$, *y could* $= 2$ *or* $-2$. *When* $x = 5$, *y could* $= 3$ *or* $-3$. This is not a function.

**8.G.A.1, 8.G.A.1a**

9. **Part A: B** is correct

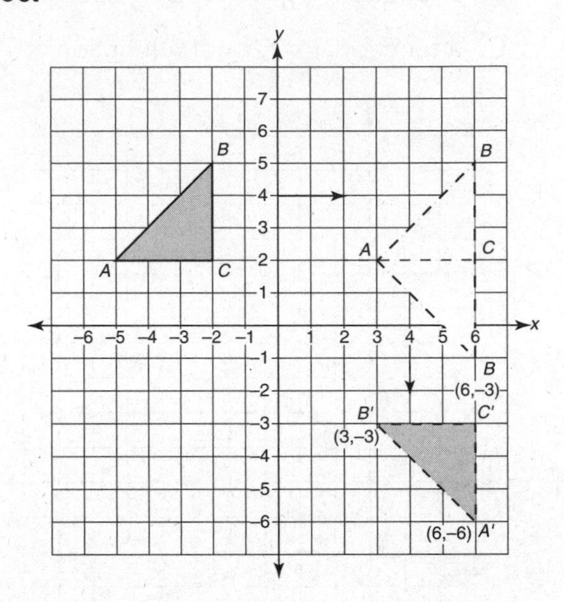


*C* is also correct

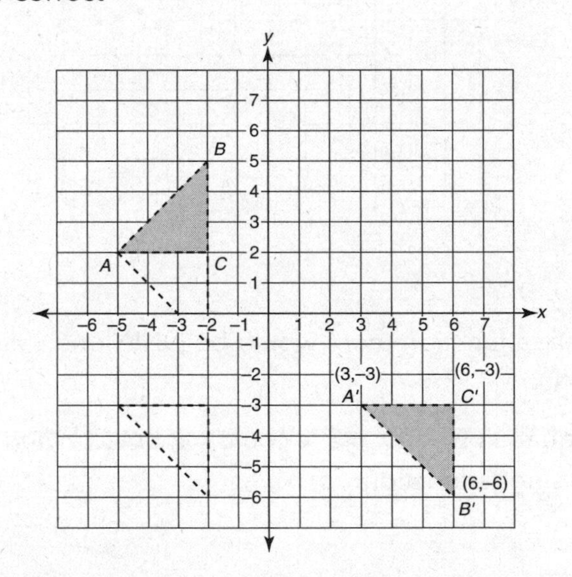


**Part B:** The new coordinates of the translated and congruent triangle are *A*′(3, –3), *B*′(6, –6), and *C*′(6, –3).

**8.G.A.1, 8.G.A.1b**

10.

**Part A:**

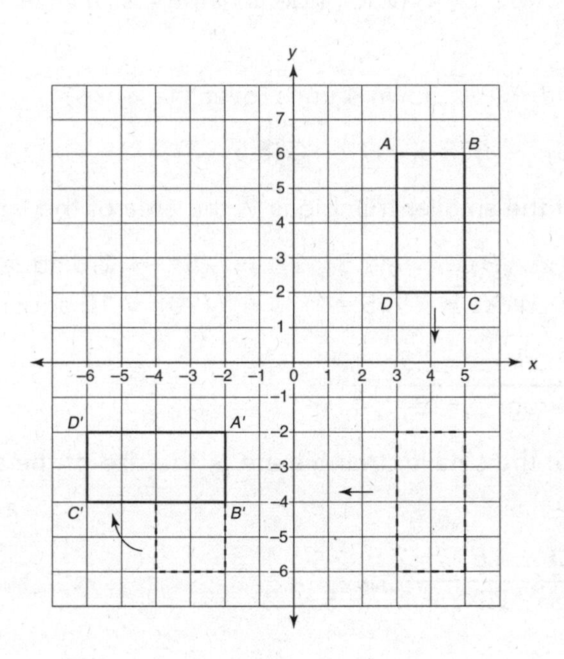


**Part B:** $A'(-2, -2)$ $B'(-2, -4)$ $C'(-6, -4)$ $D'(-6, -2)$ See diagram.

**8.G.A.2, 8.G.A.3, 8.G.A4**

11. **A.** No — Side $\overline{EG}$ measures 25"

In similar triangles the sides are in proportion. In this case the sides are in the proportion of $\frac{\overline{AB}}{\overline{EF}} = \frac{3}{15}$ *which* $= \frac{1}{5}$. $\frac{5}{x} = \frac{1}{5}$ therefore $x = (5)(5) = 25$.

**B.** Yes — In similar triangles corresponding angles are equal.

**C.** Yes — If angle $B$ measures 90° then angle $F$ also measures 90° since they are corresponding angles of similar triangles. Therefore, the sum of $\angle G + \angle E = 90°$ since the three angles of any triangle add to 180°.

**D.** No — The sides of the larger triangle are each 3 times the size of the smaller triangle.

If side $\overline{BC}$ measures 4" then the length of side $\overline{GF}$ (its corresponding side) would measure $4 \times 3$ or 12" (which means 12 inches).

**E.** Yes — Corresponding angles of similar $\Delta$'s have the same measure.

12. **Part A:** Translate $\triangle WXY$ 8.5 units to the right (over the *y-axis*)

Dilate $\triangle ABC$ by $\frac{1}{2}$ (Each side is half the size of its corresponding side in $\triangle WXY$.)

Translate $\triangle ABC$ down 4 units (over the *x-axis*)

**Part B:** $A'(3, -3)$ $B'(5.5, -1)$ $C'(5.5, -3)$

**Part C:** **Area** of the smaller triangle is $\frac{1}{4}$ the area of the larger triangle.

Area of $\triangle A'B'C' = \frac{1}{2}(2.5 \times 2) = \frac{1}{2}(5) = \mathbf{2.5}$ square units
Area of $\triangle WXY = \frac{1}{2}(5 \times 4) = \frac{1}{2}(20) = \mathbf{10}$ square units

$$\frac{\textit{area smaller}\ \triangle = 2.5}{\textit{area larger}\ \triangle = 10} = \frac{1}{4}$$

**Sides** of the smaller triangle are $\frac{1}{2}$ the size of the sides in the larger triangle.

$$\frac{\textit{sides of}\ \triangle A'B'C' = 2}{\textit{sides of}\ \triangle WXY = 4}\ \textit{or}\ \frac{2.5}{5} = \frac{1}{2}$$

**Part D:** No

# PERFORMANCE-BASED ASSESSMENT ANSWERS

## Session II Calculator Permitted (pages 290–302)

**(8.EE.A.4)**

1. **Part A:** $1.5648 \times 10^8$ $(3.2 \times 10^5)(4.89 \times 10^2) = (3.2)(4.89) \times (10^{5+2})$ $= 15.648 \times 10^7 = 1.5648 \times 10^8$

   **Part B:** $2.565217391 \times 10^2$ $\frac{5.9 \times 10^6}{2.3 \times 10^4} = \frac{5.9}{2.3} \times 10^{6-4} = 2.565217391 \times 10^2$

   **Part C:** $8.4 \times 10^2$ $4.6 \times 10^2 + 3.8 \times 10^2 = (4.6 + 3.8) \times 10^2 = 8.4 \times 10^2$

   **Part D:** $9.1955 \times 10^5$ $9.2 \times 10^5 - 4.5 \times 10^2$

   Step 1: Rewrite with the same exponents $9.2 \times 10^5 - 0.0045 \times 10^5$

   Step 2: Subtract the decimal numbers $(9.2 - 0.0045) \times 10^5 = 9.1955 \times 10^5$

This is already in the correct scientific notation form.

**(8.EE.B5)(8.EE.5-2)**

2. **Part A:** The slope of line *p* is $\frac{1}{5}$.

   The slope of line *r* is $\frac{-1}{2}$.

   **Part B:** Line *p* shows a **positive** rate of change. As the *x* values increase, the *y* values also increase. Line *p* has a positive slope.

   Line *r* shows a **negative** rate of change. As the *x* values increase, the *y* values do the opposite, they decrease. Line *r* has a negative slope.

   **Part C:** Line *p* could represent Aarte's graph about her homework.

   Line *r* could represent her father's information about their heating bills.

   **Part D:** The more time Aarte spends on her homework the higher her grades are. As her homework time increases, her grades also increase. This is a positive rate of change. Line *p* has a positive rate of change.

   As the weather gets warmer (the temperature gets higher) then their heating bills get lower. As one increases the other decreases; this is a negative rate of change. Line *r* has a negative rate of change and could represent her dad's information.

3. **Part A:** *Line a* has a steeper slope

Slope of line $a = 1$ $\quad \dfrac{6-5}{2-1}=\dfrac{1}{1}=1$ *or* $\dfrac{3-2}{-1--2}=\dfrac{1}{-1+2}=\dfrac{1}{1}=1$

The slope of *line b* is only $\dfrac{1}{3}$. See diagram below.

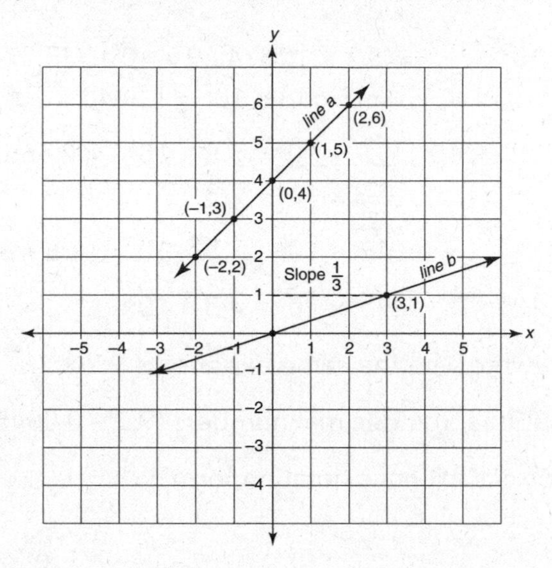


**Part B:** Pool *a*, Line *b*

**Part C:** Answers will vary.

The *x-axis* represents time and the *y-axis* represents the height of water in the pool. Just by looking at the graph I can see that it takes 3 (hours) for pool *b* to reach 1 foot high. In 3 hours pool *a* reaches 7 feet high. A steeper slope means a faster rate of change; this pool is being filled much faster.

**(8.G.A.4)**

4. **Part A:**

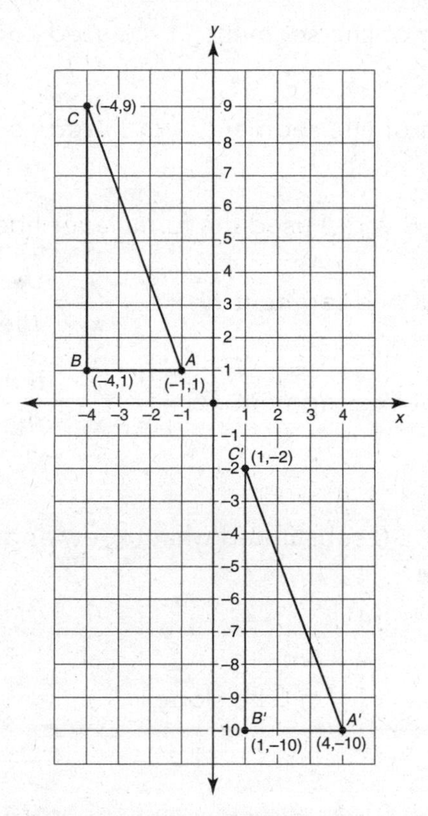


**Part B:** Yes, they are similar.

**Part C:** You can translate triangle *ABC* using the following steps, and your new triangle will be directly on top of triangle *A′B′C′*. The vertices match up exactly and the sides are the same length.

Step 1: Translate (slide) 5 units to the right over the *y*-axis into quadrant I.

Step 2: Translate (slide) 11 units down over the *x*-axis into quadrant IV.

or

Step 1: Translate (slide) 11 units down over the *x*-axis into quadrant III.

Step 2: Translate (slide) 5 units to the right over the *y*-axis into quadrant IV.

**(8.C.11, 8.EE.B.6)**

5. **Part A:** • The slope of line segment $\overline{FH}$ is –1 — I used coordinates $F(5, 8)$ and $H(11, 2)$.

   • The slope of line segment $\overline{FG}$ is also –1 — I used coordinates $F(5, 8)$ and $G(9, 4)$.

   **Part B:** Answers may vary. I used the formula for finding the slope of a line.

   The slope of line segment $\overline{FH} = \frac{y_1 - y_2}{x_1 - x_2} = \frac{\textit{the change in } y}{\textit{the change in } x} = \frac{2-8}{11-5} = \frac{-6}{6} = 1$

   The slope of line segment $\overline{FG} = \frac{y_1 - y_2}{x_1 - x_2} = \frac{\textit{the change in } y}{\textit{the change in } x} = \frac{4-8}{9-5} = \frac{-4}{4} = -1$

6. **Part A:** $y = x + 4$

   $\mathbf{y} = m\mathbf{x} + b$ (I substituted what I knew and solved for $m$.)

   $\mathbf{8} = m(\mathbf{4}) + 4$ $\quad 8 = 4m + 4$

   $\underline{-4} \quad \underline{-4}$

   $4 = 4m$

   $1 = m$ (The slope is 1.)

   **Part B:**

   Slope of line $s = \frac{3}{-2}$ — slope of line $t = \frac{8}{5}$

   Work shown: $\frac{6-0}{0-4} = \frac{6}{-4} = \frac{3}{-2}$ — work shown: $\frac{8-0}{5-0} = \frac{8}{5}$

   **Part C:** Line **s** has a negative slope.

   **Part D:** No

   **Part E:** Answers may vary. See my "work shown" in example B above.

   or

   I sketched the two lines on a coordinate grid and counted the boxes to determine the rise over run. (Student includes sketch.)

**(8.C.1.2, 8.EE.C.8)**

7. **Part A:** (7, 2,700 to 3,000) Note: A $y$-value between 2,700–3,000 is acceptable.

   **Part B:** **D.** $700 would be their profit

   COST: By reading the graph you can see that the cost of producing 1,100 computer parts (the dotted line) is at the coordinate point (**11**,4,000); it costs $4,000 to produce **1,100** parts. *Remember, the 11 represents 1,100 in this graph.*

RECEIVED: The revenue, the amount of money received after selling 1,100 computer parts (the solid line), is at the point of approximately (11, **4,700**); The company would make $4,700 if it sold 1,100 parts.

PROFIT: The company's profit would be the amount of money it received less the cost: $4,700 – $4,000 = $700 is the profit.

**(8.EE.C.7, 8.EE.C.8b)**

8. **Part A:** $4x + 2 - 2(x + 4) = 3 + 15 + 2(x - 4)$

$$4x + 2 \mathbf{- 2x - 8} = \mathbf{18} \mathbf{+ 2x - 8}$$
$$+2x - 6 = 10 + 2x$$
$$\underline{-2x} \qquad \underline{-2x}$$
$$-6 = 10$$

**Part B:** This is not possible. No solution.

**Part C:** –6 will never equal 10.

9.

| | *Equation* | *No solution* | *One solution* | *Infinitely many solutions* |
|---|---|---|---|---|
| A. | $16x - 2(2x) - 6 = 2(5x) - 6 + 2x$ | | | √ |
| B. | $x = x + 1$ | √ | | |
| C. | $12(x + 2) = -4x + 2$ | | √ | |

**A.** *Infinitely many solutions* $16x - 2(2x) - 6 = 2(5x) - 6 + 2x$

$$16x - 4x - 6 = 10x - 6 + 2x$$
$$12x - 6 = 12x - 6$$

**B.** *No solution* no number "x" can equal "x + 1"

**C.** *One solution*

$$12(x + 2) = -4x + 2$$
$$12x + 24 = -4x + 2$$
$$\underline{+4x} \qquad \underline{+4x}$$
$$16x + 24 = 2$$
$$\underline{-24} \qquad \underline{-24}$$
$$16x = -22$$
$$x = \frac{-22}{16} = \frac{-11}{8}$$

**(8.C.3.1)(8.F.A.3)**

10. Explanations may vary but should discuss slope, (or (–1, –2) to (2, 4) would be the same as the slope from (2, 4) to (6, 6) or from (1, 1) to (6, 6). Their slopes are not equal.

Slope from (–1, –2) to (2, 4) $=\dfrac{4-(-2)}{2-(-1)}=\dfrac{4+2}{2+1}=\dfrac{6}{3}\,or\,2$

Slope from (–1, –2) to (6, 6) $=\dfrac{6-(-2)}{6-(-1)}=\dfrac{6+2}{6+1}=\dfrac{8}{7}$

Slope from (2, 4) to (6, 6) $=\dfrac{6-4}{6-2}=\dfrac{2}{4}=\dfrac{1}{2}$

**(8.C.3.2, 8.G.A.2, 8.G.A.4)**

11. **Part A: B.** Translate rectangle *ABCD* 8 units left then 10 units up, then dilate by a factor of 2. (This means you doubled the width of rectangle *ABCD* from 2 to 4, and doubled the length from 4 to 8.)

**Part B:** • The dimensions of rectangle *ABCD* are Length = 4, width = 2 units

• The dimensions of rectangle *A′B′C′D′* are Length = 8, width = 4 units

**Part C:** Yes, rectangle *ABCD* is similar to rectangle *A′B′C′D′*.

**Part D:** Since they are both rectangles all angles are 90° (**corresponding angles are equal**) and their **corresponding sides are in proportion** $\dfrac{\overline{AB}}{\overline{BC}}=\dfrac{2}{4}=\dfrac{1}{2}$

$\dfrac{\overline{A'B'}}{\overline{B'C'}}=\dfrac{4}{8}=\dfrac{1}{2}$

**(8.C.3.3, 8.G.A.5)**

12. **Part A:** 180°

**Part B:** 180°

**Part C:** $x = 60°$ (60 + 60 = 120; 180 – 120 = **60°**)

$y = 110°$ (45 + 25 = 70; 180 – 70 = **110°**)

$z = 30°$ (90 + 60 = 150; 180 – 150 = **30°**)

**(8.C.4.1, 8.EE.8C)**

13. **Part A:**

| | |
|---|---|
| $4w - \mathbf{3w}(w-2) = -3w^2 - 4(3)$ | distribute the $-3w$ |
| $4w - 3w^2 + 6w = -3w^2 - 12$ | |
| $\underline{+\mathbf{3w^2}} \quad \underline{+\mathbf{3w^2}}$ | add $\mathbf{3w^2}$ to both sides |
| $4w + 6w = -12$ | combine like terms (the $w$'s) |
| $\dfrac{10w}{10} = \dfrac{-12}{10}$ | divide both sides by 10 |
| $w=\dfrac{-12}{10}=\dfrac{-6}{5}\,or\,-1\dfrac{1}{5}$ | simplify the fraction |

**Part B:** Awni is correct.

Muhammad incorrectly distributed the –3.

$6a + 7 - 3(a - 2) = 16 - 6a$

$6a + 7 - 3a \mathbf{-6} = 16 - 6a$ It should have been $6a + 7 - 3a \mathbf{+6} = 16 - 6a$

**(8.C.5.1, 8.EE.B.6)**

14.

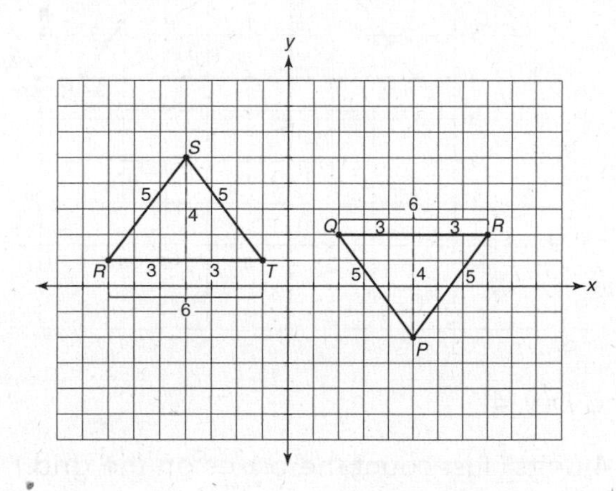


**Part A:** $RT = 6$ units long $QR = 6$ units long

**Part B:** $\overline{ST} = 5$ $\overline{RS} = 5$ $\overline{RP} = 5$ $\overline{PQ} = 5$

Use the *Pythagorean Theorem* $3^2 + 4^2 = x^2$ (the hypotenuse)

$9 + 16 = x^2$

$25 = x^2$; therefore $5 = x$

**Part C:** Translate (slide) $\triangle QRP$ 1 unit down

Translate (slide) $\triangle QRP$ 9 units to the left (into quadrant III)

Flip $\triangle QRP$ up.

**Part D:** Yes, they are congruent

When you flip $\triangle QRP$ up it will fit exactly on top of $\triangle RST$. Each side is the same length as its corresponding side, and corresponding angles are equal. The coordinates of the translated $\triangle QRP$ are now the same as the coordinates of the original $\triangle RST$.

**(8.C.5.2, 8.G.A.2, 8.G.A.4)**

15.

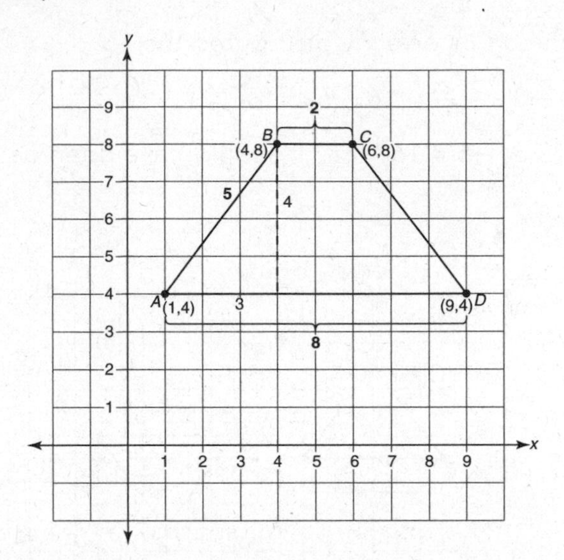


**Part A:** $C(6, 8)$ and $D(9, 4)$

**Part B:** Height = 4 units (Just count the boxes on the grid.)

Side $AB$ is 5 units long; side $CD$ is also 5 units long.

$3^2 + 4^2 = x^2$ Use the *Pythagorean Theorem*

$9 + 16 = x^2$

$25 = x^2$

$x = 5$ units

**Part C:** • The **perimeter is** 5 + 5 + 2 (the top) + 8 (the base) = ***20 units long***

• The area = (the average of the two bases)(height)

The **area** = $\left(\frac{2+8}{2}\right)(4) = (5)(4) =$ ***20 square units.***

**(8.C.6, 8.D.1, 8.D.2, 7.RP.A, 7.NS.3, 7.EE, 7.G, and 7.SP.B)**

16. **Part A:** ***C.*** 50% It will land half the time on heads and half the time on tails.

**Part B:** ***C.*** 50% There are 4 out of 8 numbers that are perfect-square numbers: 9, 16, 25, and 36. $\frac{4}{8} = \frac{1}{2} = 50\%$

17. **Part A:** The volume of cylinder $A$ is approximately **503 cubic feet.**

Vol. cylinder = $\pi r^2 h = \pi(4^2)(10) = \pi(16)(10) = 160\pi$

$160\pi = 160(3.14) = 502.4$ which rounds up to 503 cubic feet.

The volume of cylinder $B$ is **1,256** cubic feet.

Vol. cylinder = $\pi r^2 h = \pi(10^2)(4) = \pi(100)(4) = 400\pi$

$400\pi = 400(3.14) = 1,256$

**Part B:** (Two circles and one rectangle should be checked; not any squares.)

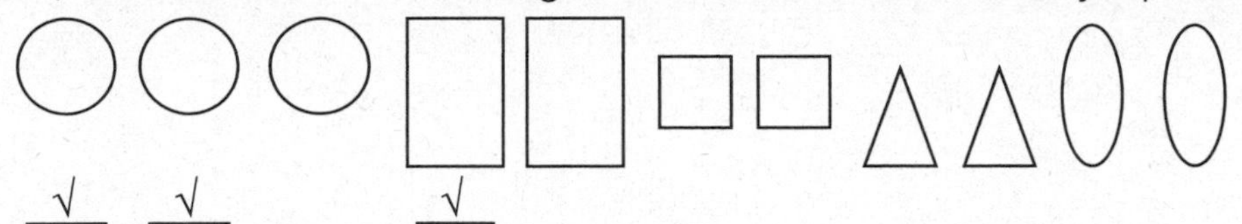

**Part C:** The *surface area* of cylinder *A* is approximately **2,613 sq. feet.**

- The area of the top of cylinder *A* (a circle) = $\pi r^2 = \pi 4^2 = 16\pi =$ **~50.24 sq. ft**
- The shaded part of the cylinder is a rectangle; its area is **~251.2 sq. ft.**

  The area of a rectangle is *length x width.*

  The width of the rectangle is the *circumference* of the circle (~**25.12**)

  (The circumference of circle is $A = 2\pi r = 2\pi 4 = 8\pi = 8(3.14) \sim 25.12$ ft)

  The *length (or height)* of the rectangle is ***10*** *feet.*

  The area of the rectangle is *lw* or *(**25.12**)(**10**) = 251.2 sq. feet*

- The total *surface area of the cylinder is Area of* top + bottom + shaded part 50.24 + 50.24 + 251.2 = about 351.68 or rounded up it would be **352 sq. feet**

**(8.SP.A.4)**

18. **Part A:** 52 right-handed boys were surveyed

**Part B:** 16 left-handed students were surveyed

**Part C:** 114 students were surveyed in all

Fraction = decimal = percent rounded to 2 decimal places

**Part D:** 85.96% $\frac{98}{114} = 0.85964$ is 85.964% which rounds to 85.96%

**Part E:** 14.04% $\frac{16}{114} = 0.14035$ is 14.035% which rounds to 14.04%

**Part F:** 82.54% $\frac{52}{63} = 0.82539$ is 82.539% which rounds to 82.54%

**Part G:** 53.06% $\frac{52}{98} = 0.53061$ is 53.061% which rounds to 53.06%

19. **Part A:** 83% Have curfew = 12 have chores = 10

$10/12 = 5/6 = 83.\overline{33}\%$

**Part B:** 48% Total number students = 21 have no chores = 10

10/21 = 47.6%

# Practice Test End-of-Year Assessment Session I (No Calculator Permitted.)

Today you will be taking the Grade 8 Mathematics End-of-Year Assessment (EOY).

Read the directions carefully in each question. If you do not know the answer to a question, skip it and go on. If time permits, you may return to earlier questions. **Do your best to answer every question!** Read each question carefully. You may circle, highlight, or underline important words. You may also use the Grade 8 Reference sheet for some formulas.

This End-of-Year Assessment is divided into two sessions. Session I, no calculator permitted, and Session II, scientific calculator permitted.

- Some are **multiple-choice** questions with only **one-right answer.** Here you fill in the correct choice. (Fill in one letter only.)

  *Example: What is the slope of the line $y = 3x - 4$?*

  ○ **A.** 1/3 ○ **B.** 3 ○ **C.** 4 ○ **D.** –4 *The correct answer is* **B**. *3*

- Some are **new multiple-choice** type questions with **more than one right answer.** Here you should check all that are correct (all that apply).

  *Example: Which statements show a value that is equivalent to $4^2$?*

  ☐ **A.** $2^3 \times 2^1$ ☐ **B.** $\sqrt{8}$ **C.** $(8)(8)$ ☐ **D.** $-2 \times -8$

  *There are two correct answers; they are:*

  ☑ **A.** $2^3 \times 2^1$ *and* ☑ **D.** $-2 \times -8$ *since both are equivalent to $4^2$ or 16.*

- Other **new multiple-choice** examples are Yes/No statements. Here you should check Yes √ or No √ for each statement.

  *Example: Use the equation* $\mathbf{y = 2x + 4}$ *to answer the following:*

  **A.** *This equation represents a curved line.* Yes ------ No ------

  **B.** *This equation has a slope of 2.* Yes ------ No ------

  **C.** *The line $y = 1/2x + 4$ is parallel to this line.* Yes ------ No ------

  **D.** *The point (8, 20) falls on this line.* Yes ------ No ------

  *The correct answers are:*

  **A.** No √ **B.** Yes √ **C.** No √ **D.** Yes √

- **Short-Constructed** questions ask you to write your answer with no multiple-choice options.

  *Example: Solve for y in the equation* $\frac{2x+y}{2}=3$ Answer: <u>$y = 6 - 2x$</u>

## SESSION I NO CALCULATOR (55 MINUTES)

**Name:** ................................ **Date:** ................................

**1.** $\frac{5}{6}$ is equivalent to which decimal number below?

○ **A.** $1.\overline{222}$
○ **B.** $0.8\overline{33}$
○ **C.** $8.\overline{3232}$
○ **D.** $7.\overline{333}$

**2.** Consider the expression $\sqrt{40}$ and answer the following statements.
Check Yes ..✔... or No ..✔... .

**A.** This represents an irrational number. Yes ...... No ......
**B.** This represents a perfect square number. Yes ...... No ......
**C.** This has the same value as 20. Yes ...... No ......
**D.** When simplified, this is a negative integer. Yes ...... No ......
**E.** The most exact way to write $\sqrt{40}$ would be in decimals. Yes ...... No ......

**3.** $\sqrt{83}$ *is about*

○ **A.** 6
○ **B.** 1
○ **C.** 7
○ **D.** 9

**4.** What letter on the number line below represents the approximate value of each of the following? (Circle your answer choices.)

**Part A:** $\sqrt{26}$ A B C D E F G H I J K
**Part B:** $\sqrt{99}$ A B C D E F G H I J K
**Part C:** $\pi$ A B C D E F G H I J K
**Part D:** $\sqrt{4}+\sqrt{9}$ A B C D E F G H I J K
**Part E:** $2\times\sqrt{16}$ A B C D E F G H I J K

| 0 | 1 | 2 | 3 | 4 | 5 | 6 | 7 | 8 | 9 | 10 |
|---|---|---|---|---|---|---|---|---|---|---|
| A | B | C | D | E | F | G | H | I | J | K |

5. Which expressions are equivalent to $\frac{2^{-6}}{2^{-4}}$? Select all that apply.

- ☐ A. $\frac{1}{4}$
- ☐ B. $\frac{1}{8}$
- ☐ C. 4
- ☐ D. 8
- ☐ E. $\frac{1}{2^2}$

6. $2(6 - 3)^2 - 3(2^2) =$

- ○ A. 30
- ○ B. 6
- ○ C. 0
- ○ D. −6

7. Select **Yes** or **No** for each statement below.

| | | |
|---|---|---|
| **A.** If $x^2 = 144$ then $x = 11$ | Yes ------ | No ------ |
| **B.** If $x^3 = 8$ then $x = 4$ | Yes ------ | No ------ |
| **C.** If $x = 3$ then $x^3 = x^2 + 3$ | Yes ------ | No ------ |
| **D.** If $x^3 = 27$, then $x^3 + x^2 = 18$ | Yes ------ | No ------ |
| **E.** If $x^3 = -8$, then $x = -2$ | Yes ------ | No ------ |

8. $\sqrt{81}$

Answer: $\sqrt{81} =$ ________________

9. If $y = -3$ then $y^3 =$

- ○ A. 27
- ○ B. −18
- ○ C. 21
- ○ D. −27

10. What is the value of $\sqrt[3]{8} + \sqrt[2]{16}$?

- ○ A. 6
- ○ B. 8
- ○ C. 8/16 = ½
- ○ D. 10

**11.** Instead of writing very large numbers in the standard way, it is easier to write them in *scientific notation* form especially when we are comparing them. How much larger is $8 \times 10^9$ *than* $2 \times 10^8$?

- O **A.** 10 times larger
- O **B.** 20 times larger
- O **C.** 40 times larger
- O **D.** 80 times larger

**12.** The estimated population of the United States is about $3 \times 10^8$ and the estimated population of the world is about $7 \times 10^9$.

Nicole says the world's population is more than 20 times larger than the U.S.'s population. Caitlyn says that is not true. Who is correct?

- O **A.** Nicole is correct.
- O **B.** Caitlyn is correct.
- O **C.** They are both incorrect.

**13.** Solve for *w*. $3(w + 4) = -18$

Answer: ________________

**14.** Solve for *a* $2(a - 3) + 4a = -24$

Answer: ________________

**15.** $100 \times 5^{-2} + 2^3 =$

- O **A.** 2,508
- O **B.** 133
- O **C.** 32
- O **D.** 12

**16.** Solve the equation of this line for *y*.

$3x + 6y = 4(3x) - 24$

Answer: $y =$ ________________

**17.** Below are the equations of two lines: Line *m* and line *q*.

At what point do these two lines intersect?

$$\begin{cases} \text{line } m \; y = 3x + 9 \\ \text{line } q \; 3y = 9x - 18 \end{cases}$$

**A.** They intersect at the point (0, 6).

**B.** They intersect at the point (–2, 6).

**C.** They intersect at the point (2, –3).

**D.** They are parallel lines and do not intersect.

**18.** The following two lines are on the same coordinate grid. How do you know that they intersect?

$y = \frac{3}{2}x - 8$ *and* $y = \frac{2}{5}x - 16$

- ○ **A.** Because –16 and –8 are both divisible by 2.
- ○ **B.** Because they have the same slope.
- ○ **C.** Because one line is *rising* while the other is *falling*.
- ○ **D.** Because they do not have the same slope.

**19.** Which graph represents the two lines created from the data given in the two tables below?

| X | Y |
|---|---|
| 1 | 1 |
| 3 | 3 |
| –1 | –1 |
| 3 | –3 |

| X | Y |
|---|---|
| 1 | 1 |
| 3 | –3 |
| –1 | 1 |
| –3 | 3 |

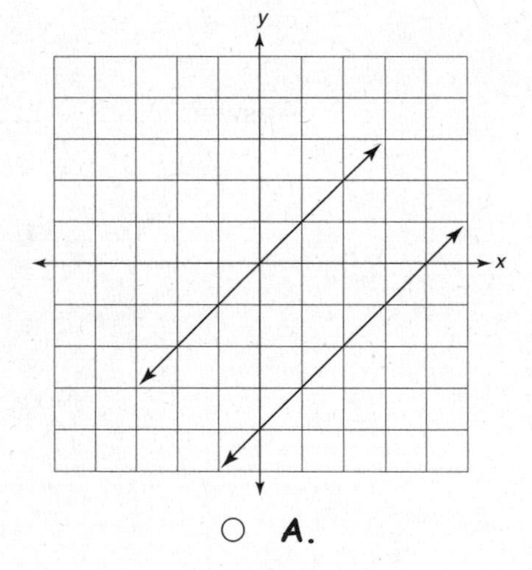


○ **A.**

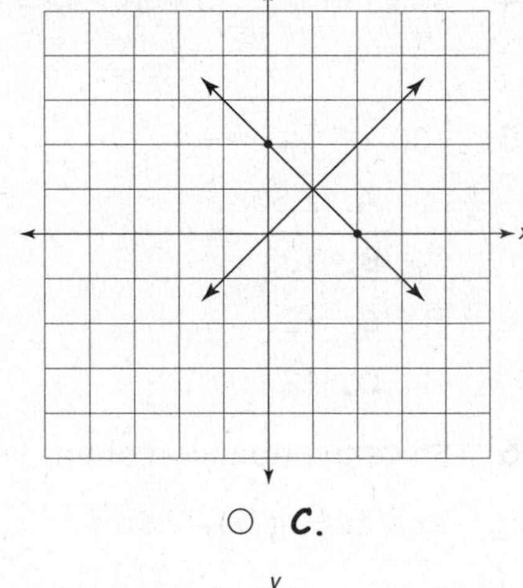


○ **C.**

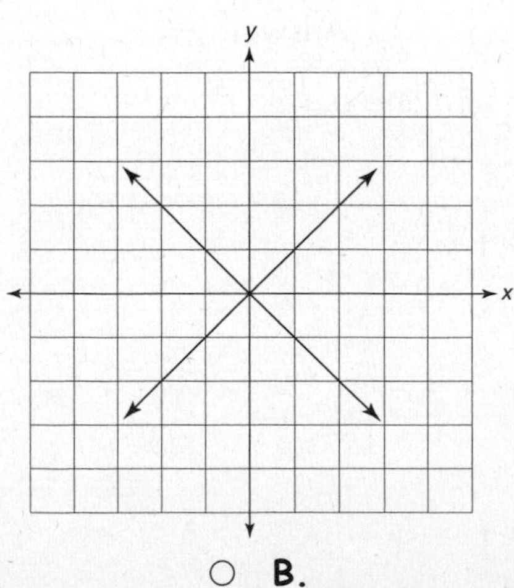


○ **B.**

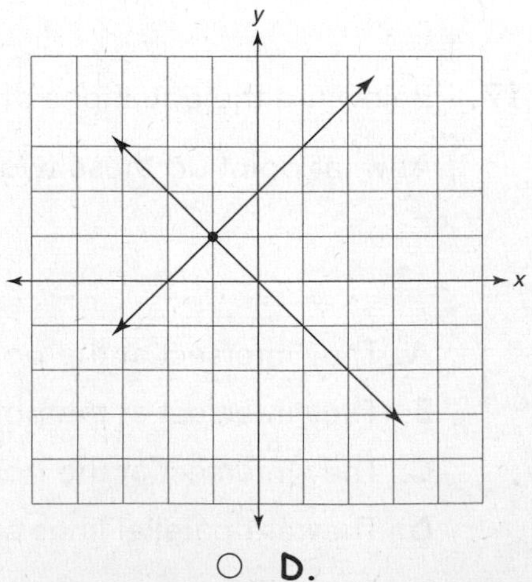


○ **D.**

**20.** Nick and Sharif are comparing the two lines shown by the equations below. Nick says these lines do not intersect. Sharif says they do; they are actually the same line. Why is Sharif correct?

$$2y = 3x + 10 \quad and \quad 4y = 6x + 20$$

- ○ **A.** Because both are rising and are not parallel.
- ○ **B.** Because 10 and 20 are both divisible by 2.
- ○ **C.** Because they intersect at the point (0,10).
- ○ **D.** Set each equation equal to $y$ and they are the same equation.

**21.** Which is an example of a function?

- ○ **A.** input = Joey Brown's name output = his birthday
- ○ **B.** input = Tuesday output = 3 options on the lunch menu
- ○ **C.** input = March 3 output = birthday of three students
- ○ **D.** input = ice cream output = 4 flavors as choices

**22.** Use the equation $f(x) = 5x - 3$ to determine if the following statements are correct.

**A.** This function represents a curved line. Yes ------ No ------

**B.** The $y$-intercept of the relationship is at (–3,0). Yes ------ No ------

**C.** The slope is negative. Yes ------ No ------

**D.** $f(x) = 10x - 6$ is equivalent to $f(x) = 5x - 3$. Yes ------ No ------

**E.** When the input is 2, then the output is 7. Yes ------ No ------

**F.** $g(x) = 5x - 8$ is a line parallel to $f(x) = 5x - 3$. Yes ------ No ------

**23.** Are the three points listed below on the same line?

$A(-2, 4)$ $B(0, 0)$ $C(2, 3)$ Answer: Yes ------ No ------

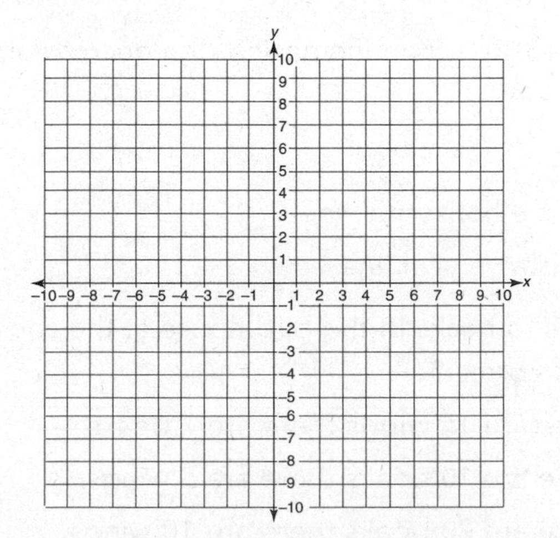

**24.** Which of the graphs below is *not* a function?

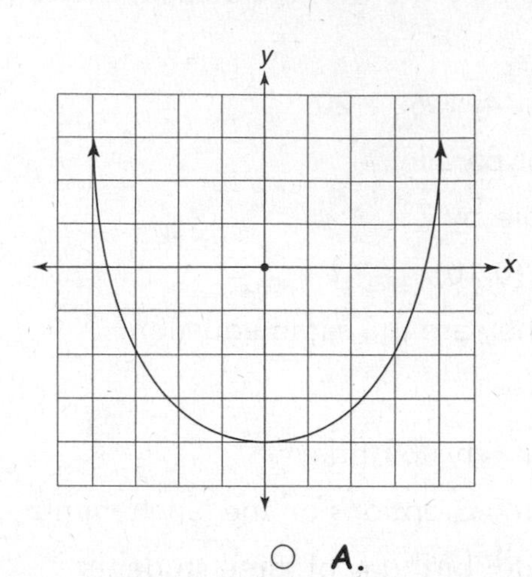


○ **A.**

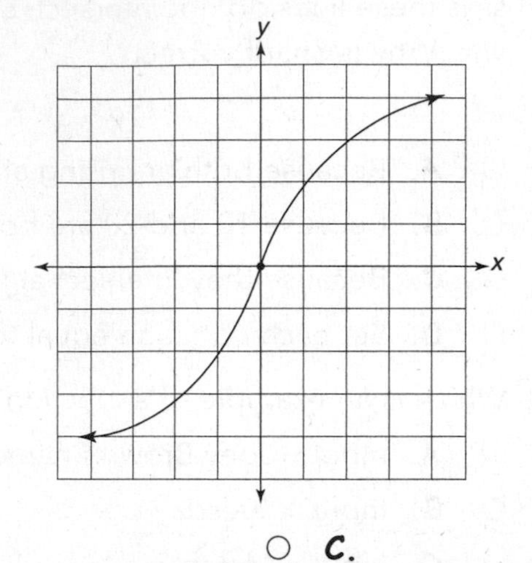


○ **C.**

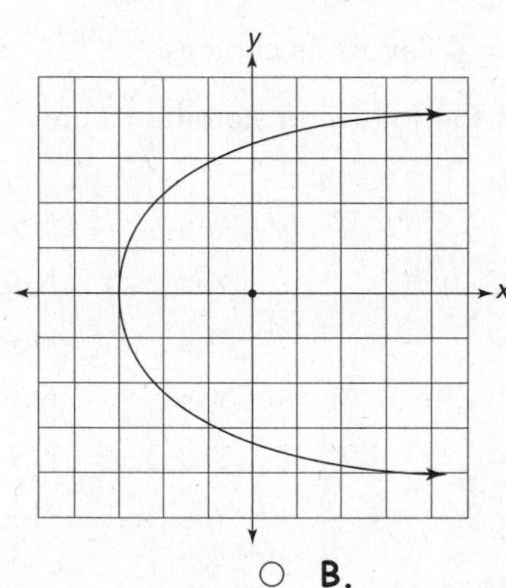


○ **B.**

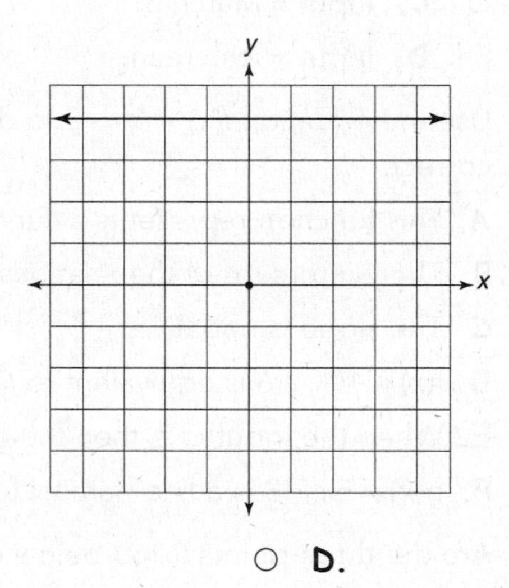


○ **D.**

**25.** Is the function $y = -3/2x + 8$ an *increasing* or a *decreasing* function?

○ **A.** increasing

○ **B.** decreasing

○ **C.** neither; it is a horizontal line

○ **D.** neither; it is a vertical line

**26.** If the ratio of wings to beaks in the bird house at the zoo was 2:1, which statement below is correct?

○ **A.** When there are 14 wings there are 7 beaks.

○ **B.** When there are 10 beaks there are 5 wings.

○ **C.** When there are 20 beaks there are 10 wings.

○ **D.** When there are 10 wings there are 4 beaks.

**27.** If a Claudine walks $\frac{1}{2}$ mile in each $\frac{1}{4}$ hr, compute her unit rate.

Answer: ---------------- miles per hour is her unit rate.

**28.** The transformation of figure ***A*** to figure ***B*** used the following translations, rotations, or reflections.

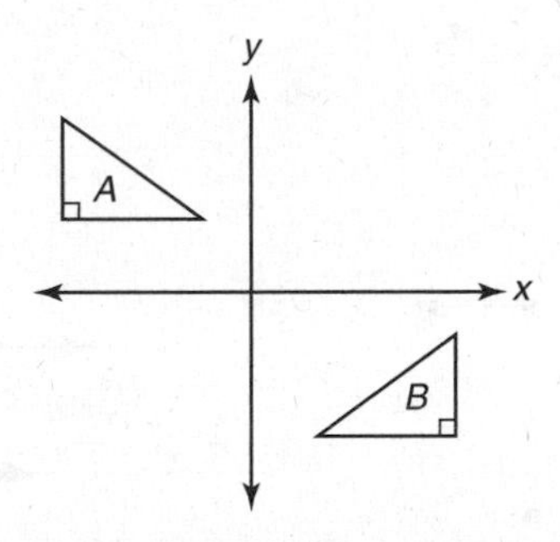


- ○ **A.** ***A*** was reflected over the *y-axis,* then translated to quadrant IV.
- ○ **B.** ***A*** was rotated 90° over the *y-axis,* then reflected over the *x-axis.*
- ○ **C.** ***A*** was reflected over the *x-axis,* then over the *y-axis*
- ○ **D.** ***A*** was translated over *x-axis, then translate it over the y-axis.*

**29.** Look at the two congruent figures drawn on the coordinate grid below. They are dilations. What are the coordinates of the *center of dilation?*

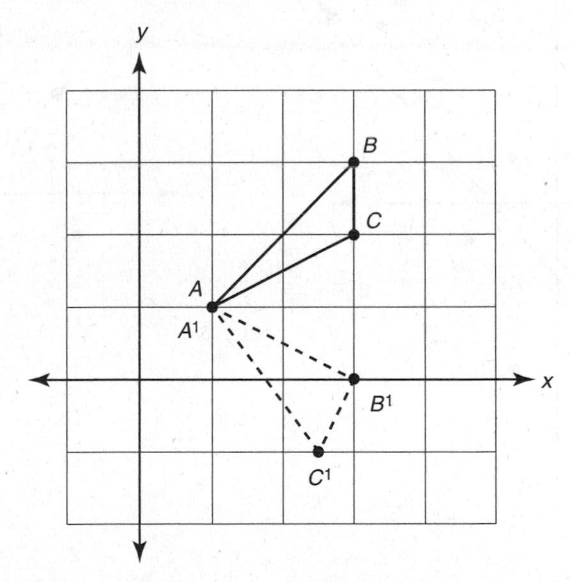


- ○ **A.** point (8, 4)
- ○ **B.** point (8, 0)
- ○ **C.** point (0, 0)
- ○ **D.** point (1, 1)

**30.** Which diagram below shows the best estimate for the *line of best fit* (the dashed line)?

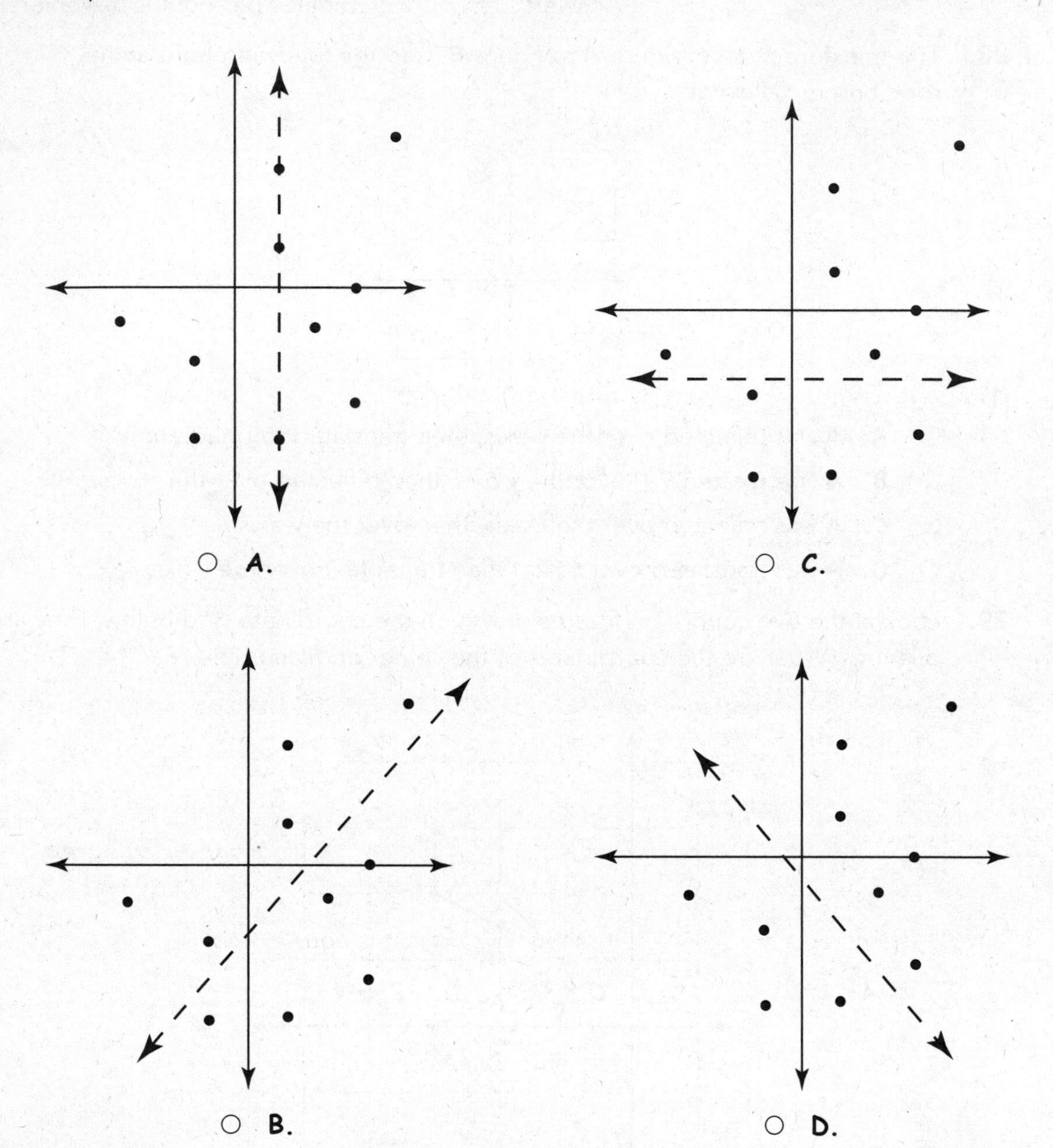

# Practice Test End-of-Year Assessment Session II

## (Calculator Permitted.)

Today you will be taking Session II of the Grade 8 Mathematics End-of-Year Assessment (EOY). For this session you are permitted to use a scientific calculator.

Read the directions carefully in each question. If you do not know the answer to a question, skip it and go on. If time permits, you may return to earlier questions. **Do your best to answer every question!** Read each question carefully. You may circle, highlight, or underline important words. You may also use the Grade 8 Reference sheet for some formulas.

This End-of-Year Assessment is divided into two sessions. Session I, no calculator permitted, and Session II, calculator permitted.

- Some are **multiple-choice** questions with only **one-right answer.** Here you fill in the correct choice. (Fill in one letter only.)

  *Example: What is the y-intercept of this line?* $2y = 5x + 12$

  ○ **A.** 2 ○ **B.** 3 ○ **C.** 5 ○ **D.** 6

  *The correct answer is D. 6*

- Some are **new multiple-choice** type questions with **more than one right answer.** Here you should check all that are correct (all that apply).

  *Example: Which statements show a value that is equivalent to* $2^3$*?*

  ☐ **A.** $4^2$ ☐ **B.** $8^1$ **C.** $\sqrt{16}$ **D.** $-2 \times -4$

  *The correct answers are:*

  ☑ **B.** $8^1$ *and* ☑ **D.** $-2 \times -4$ *since both are equivalent to* $2^3$ *or 8.*

- Other **new multiple-choice** examples are **Yes/No** statements. Here you should check Yes √ or No √ for each statement.

  *Example: Use the equation* $\mathbf{y = 2x + 3}$ *to answer the following.*

  **A.** *This equation represents a curved line.* Yes ------ No ------

  **B.** *This equation has a slope of 2.* Yes ------ No ------

  **C.** *The line* $y = 1/2x + 3$ *is parallel to this line.* Yes ------ No ------

  **D.** *The point (4, 11) falls on this line.* Yes ------ No ------

The correct answers are

**A.** No √ **B.** Yes √ **C.** No √ **D.** Yes √

- **Short-Constructed** questions ask you to write your answer with no multiple-choice options.

  *Example: What is the slope of a horizontal line?* Answer: zero

## SESSION II WITH CALCULATOR (55 MINUTES)

**Name:** ______________________ **Date:** ______________________

**1.** The distance from Earth to a nearby planet is 679,000,000 miles away. Use scientific notation to write the distance an astronaut would travel on a *round-trip* from Earth to this planet.

- ○ **A.** $1.358 \times 10^9$
- ○ **B.** $13.58 \times 10^8$
- ○ **C.** $134.8 \times 10^7$
- ○ **D.** $1,358 \times 10^6$

**2.** Refer to the graph below. Both describe the speed of two different animals. Which animal is moving faster? How do you know?

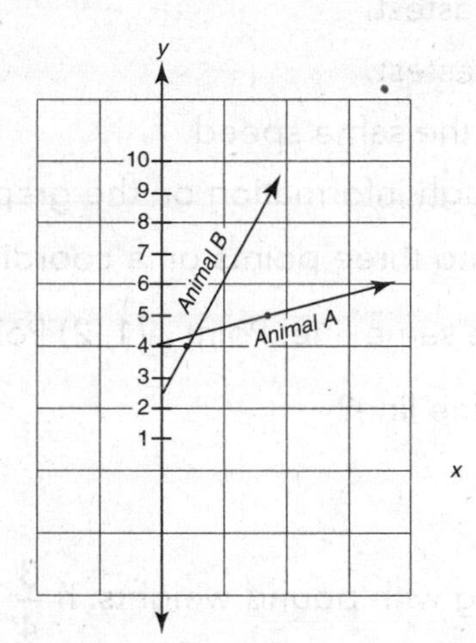


- ○ **A.** Animal A because the $y$-intercept is 4
- ○ **B.** Animal A because it has a slope of $\frac{1}{4}$
- ○ **C.** Animal B because it begins at (0, 0)
- ○ **D.** Animal B because it has $\frac{4}{1}$ as its slope

**3.** The graph and equation shown below describe how fast two different race horses performed in last week's race. Which horse ran the fastest?

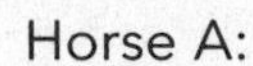

Horse A:

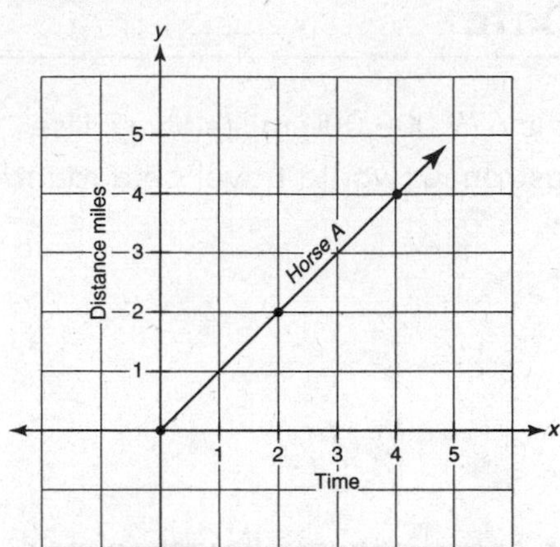


Horse B:

$y = 1/3x + 8$

- O **A.** Horse A ran the fastest.
- O **B.** Horse B ran the fastest.
- O **C.** They both ran at the same speed.
- O **D.** There is not enough information on the graph.

**4.** You are given the following three points on a coordinate grid.

Points $A$ and $B$ are on the same line. Point $A(1, 2)$ Point $B(4, 5)$

Is point $C(3, 4)$ on the same line?

- O **A.** yes
- O **B.** no

**5.** A bar of soap is balancing with pound weights. If $\frac{3}{4}$ of a bar of soap balances with $\frac{1}{2}$ of a pound, how much does the whole bar of soap weigh?

- O **A.** $\frac{3}{4}$ of a pound
- O **B.** $\frac{1}{3}$ of a pound
- O **C.** $\frac{2}{3}$ of a pound
- O **D.** 1 pound

**6.** Which expression says that "the sum of two times a number and 8 is 16?"

- O **A.** $2 + n + 8 = 16$
- O **B.** $2(8) + n = 16$
- O **C.** $2 + 8 = 2n + 16$
- O **D.** $2n + 8 = 16$

**7.** What is the relationship between $x$ and $y$ in the graph below?

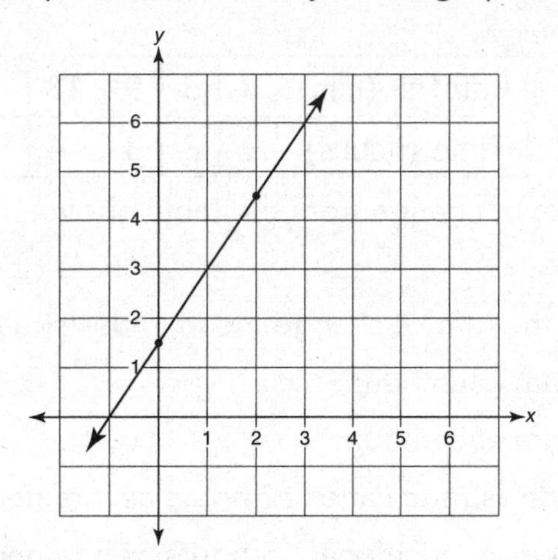


○ **A.** $y = 2x - 4$

○ **B.** $y = \frac{3}{2}x + \frac{3}{2}$

○ **C.** $y = -\frac{1}{2}x + 2$

○ **D.** $y = -4x + 2$

**8.** The florist decided to measure how quickly a certain plant was growing after being fed new plant food. Look at the graph below to answer the following questions:

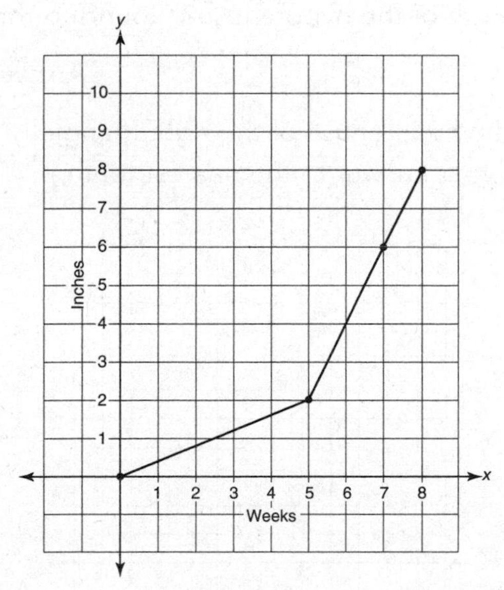


**Part A:** How tall was the plant <u>before</u> it received the new plant food?

Answer: ________________

**Part B:** What is the rate of growth after the plant began receiving the new plant food?

Answer: ________________

**9.** The science group collected data about the recent rainfall and organized the data into the chart below.

| Rainfall (mm) | 3 | 6 | 9 | 12 | 15 |
|---|---|---|---|---|---|
| Time (hours) | 1 | 2 | 3 | 4 | 5 |

**Part A:** Find the rate of change from the table above.

Rate of change is ________________

**Part B:** What does the rate of change mean in this situation?

○ **A.** It rained 6 mm each hour.

○ **B.** It rained 3 mm each hour.

○ **C.** It rained twice as much after 12 hours as it rained in 3 hours.

○ **D.** It rained twice as much after 15 hours as it rained in 9 hours.

**10.** What is the length of the hypotenuse of the right triangle drawn below? (Hint: Use the Pythagorean theorem.)

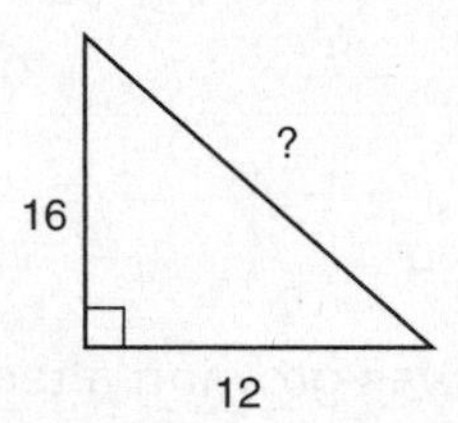


Answer: ________________

**11.** You are given a right triangle. Two sides measure 5 inches, and 6 inches. What is the approximate length of the hypotenuse? (Round to the nearest tenth.)

Answer: ________________ inches

**12.** Find the length of the hypotenuse of the right triangle drawn on the coordinate grid below. (Round your answer to the nearest tenth.)

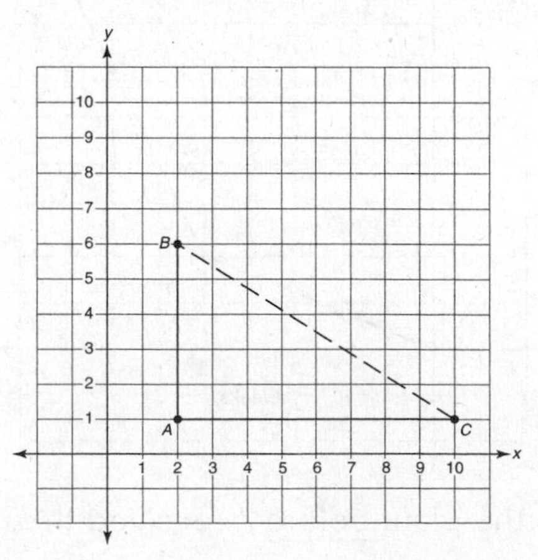

**13.** How does the volume of this cylinder and the volume of this cone compare? (*Hint:* In the diagram, one is measured in inches and the other is measured in feet. Change all to the same measurement unit first.) *You may use your formula sheet if needed.*

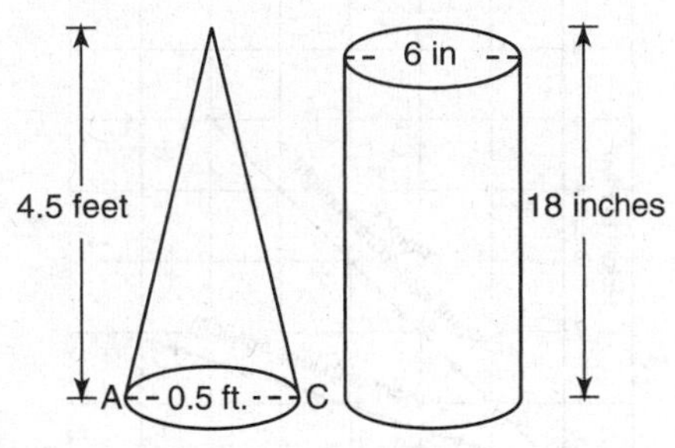


- ○ **A.** They are the same.
- ○ **B.** The volume of the cone is 1/3 the volume of the cylinder.
- ○ **C.** The volume of the cylinder is 2 times the volume of the cone.
- ○ **D.** The volume of the cone is 1/4 the volume of the cylinder.

**14.** If the *diameter* of a cone shaped container measures 5 inches, and the *height* of that cone measures 9 inches, how many cubic inches of liquid will that cone hold? (Round your answer to the nearest hundredth.) *You may use your formula sheet if needed.*

Answer: ________________

**15.** If the diameter of Eliza's beach ball is 18 inches, how much air can it hold? Round your answer to the nearest cubic inch. *(You may use your formula sheet if needed.)*

Answer: ________________

**16.** What is the relationship between the numbers of additional hours of sunlight this plant *X* receives each day and the additional height in the mature plant?

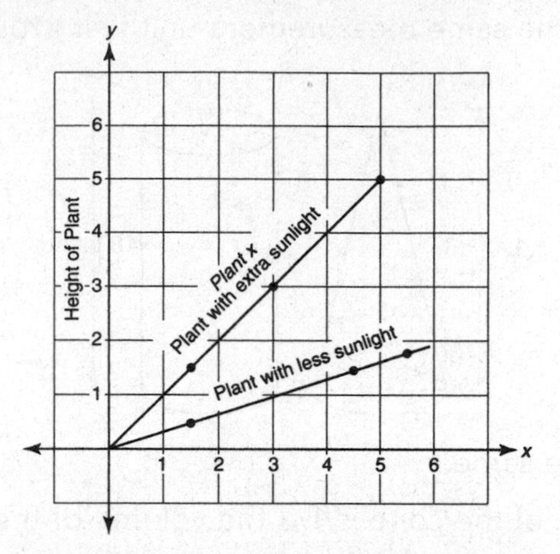


- ○ **A.** You look at the *y*-intercept for this information.
- ○ **B.** You look at where the two lines intercept.
- ○ **C.** You look at the slope of one line compared to the slope of the other.
- ○ **D.** You look at whether the graphs are increasing or decreasing.

**17.** What is the significance of the *y-intercept* in the graph below.

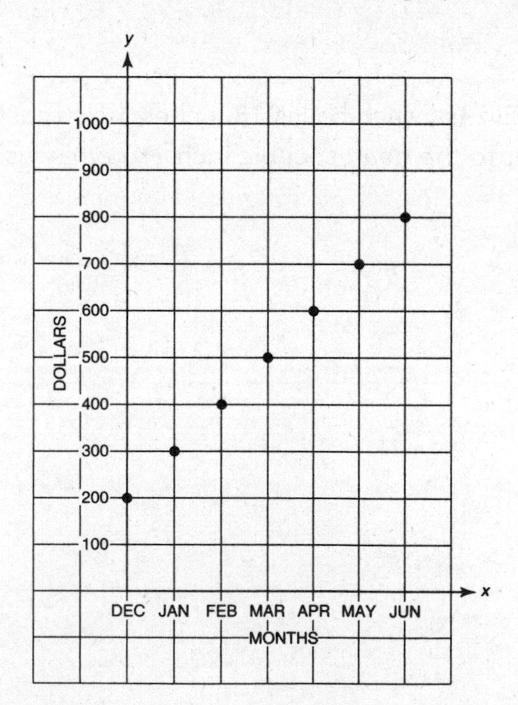


- ○ **A.** Janie saved $200 each month.
- ○ **B.** Janie had $200 in her account before she began saving each month.
- ○ **C.** Janie owed the bank $200 before she began saving each month.
- ○ **D.** $200 is the most she can save in any one month.

**18.** Carmen uses a two-way frequency table to help organize and display data that she collected during her survey. Use her table below to answer the questions.

| Number of Bedrooms | Number of Children | | | | | |
|---|---|---|---|---|---|---|
| | 0 | 1 | 2 | 3 | 4+ | Totals |
| 1 | 2 | 1 | 0 | 0 | 0 | **3** |
| 2 | 1 | 5 | 6 | 3 | 1 | **16** |
| 3 | 0 | 2 | 4 | 2 | 2 | **10** |
| 4+ | 0 | 0 | 2 | 2 | 2 | **6** |
| | **3** | **8** | **12** | **7** | **5** | **35** |

**Part A:** How many people did Carmen survey in all?

Answer: ________________

**Part B:** What seems to be the number of children most people have?

Answer: ________________

**Part C:** What number of bedrooms do most people have?

Answer: ________________

**Part D:** What percentage of 3-bedroom homes ave 2 children?

Answer: ________________

**19.** The town Pizza Palace is planning to do a survey. They want to find out what pizza toppings boys usually prefer on their pizzas. They will distribute their survey to 50 people. Which would be the best group of people to use if they want a fair and useful sample?

○ **A.** Fifty people as they enter the local supermarket.

○ **B.** Fifty students as they walk out of the local high school.

○ **C.** Fifty boys as they walk off the town soccer field after a practice.

○ **D.** The first 50 people who enter the Pizza Palace.

**20.** If the *circumference* of a circle measures $30\pi$, what is the *area* of that circle? (Round your answer to the nearest tenth. Use 3.14 for $\pi$ or use the $\pi$ key on your calculator.)

Answer: ________________

**21.**

**Part A:** What is the *surface area* of a cube with a side length of 4?

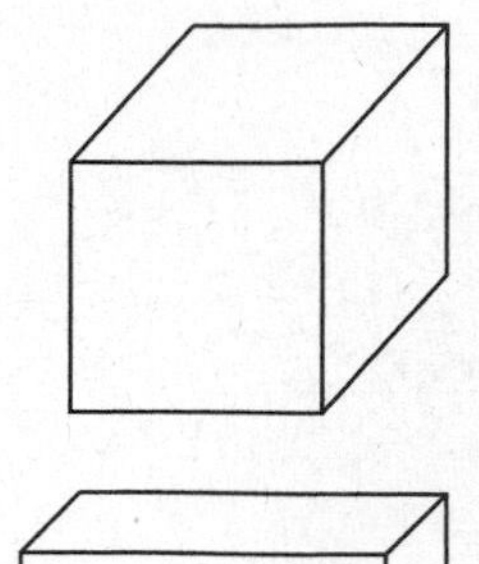

- ○ **A.** 16 square units
- ○ **B.** 64 square units
- ○ **C.** 96 square units
- ○ **D.** 128 square units

**Part B:** How much larger or smaller is the *surface area* of a rectangular box with a front that is 6 units wide and 2 units high, and a side that is 2 units high and 4 units long?

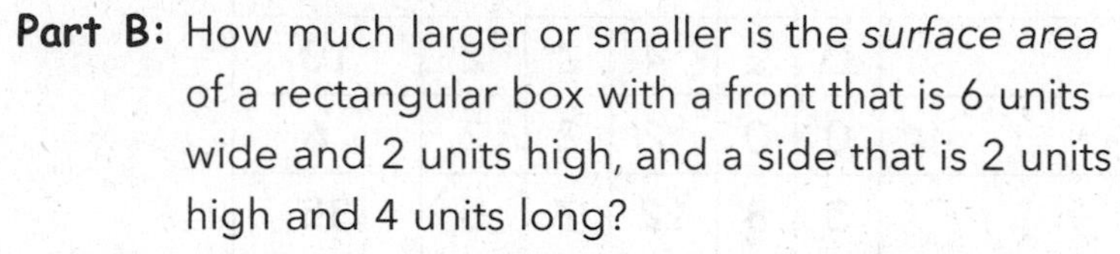

- ○ **A.** The surface area of the cube is 8 units larger.
- ○ **B.** The surface area of the rectangular box is 52 units larger.
- ○ **C.** The surface area of the cube is 52 units larger.
- ○ **D.** They have the same surface area.

**22.** Which solution is the largest integer? (*Figures are not drawn to scale.)*

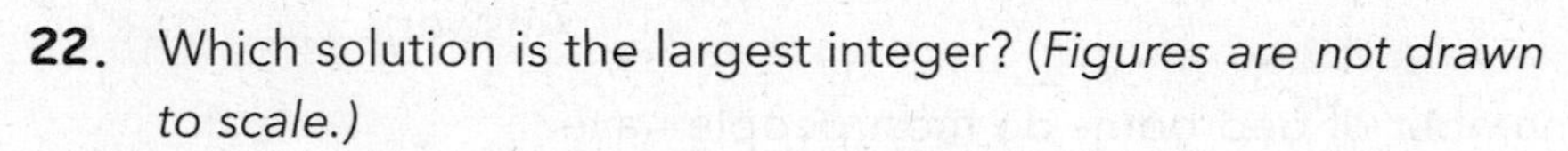

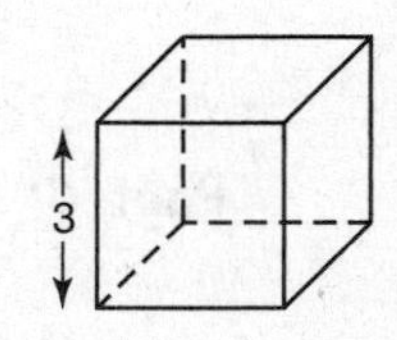


- ○ **A.** The surface area of the cube shown.
- ○ **B.** The volume of this rectangular prism.

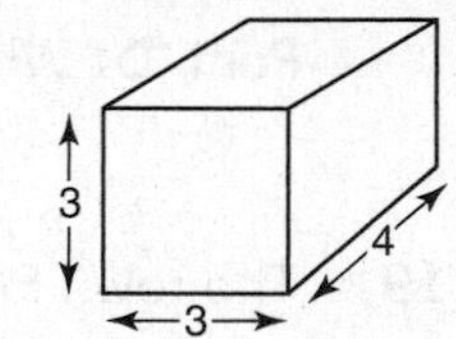


- ○ **C.** The volume of the right cylinder shown with radius = 2 and height = 4.
- ○ **D.** The perimeter of this isosceles triangle.

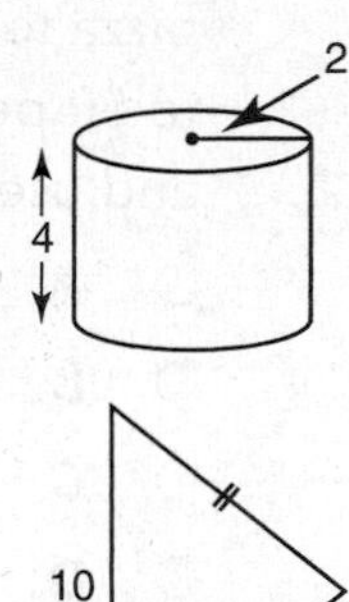


**23.** One of the large department stores at the mall has a jacket on sale for 20% off. The original price of the jacket was \$125. This state also charges a $6\frac{1}{2}$% sales tax on the purchase price. What would be your total cost to buy this jacket?

Answer: ________________

**24.** Charlie's dad is a carpenter. He has a 32" wide door and needs to put a towel rod on the door. The rod is 18" long. If he wants to put the rod exactly in the middle of the door width, how long will the space be from the edge of the door to one end of the towel rod?

- ○ **A.** 14 inches
- ○ **B.** 12 inches
- ○ **C.** 10 inches
- ○ **D.** 7 inches

**25.** Jill is checking her partner Suzie's math homework. She says that Suzie did the following example incorrectly. What error did Suzie make?

$$3(2x - 9) = 33$$
$$6x - 27 = 33$$
$$+27 \quad +27$$
$$6x = 60$$
$$x = 10$$

- ○ **A.** She should have combined the $2x$ and the $-9$ first.
- ○ **B.** She added 27 to 33 instead of subtracting 27 from 33.
- ○ **C.** She incorrectly distributed the 3.
- ○ **D.** Suzie did not make any errors.

**26.** The table below shows the miles John walked each day last week.

| Day | Monday | Tues. | Wed. | Thurs. | Friday | Saturday | Sunday |
|---|---|---|---|---|---|---|---|
| Miles | 2.5 | 0 | 1.5 | 2.0 | 3.0 | 4.2 | 3.5 |

What type of graph would be most useful for estimating the average number of miles he walked in a week?

- ○ **A.** scatter plot
- ○ **B.** circle graph
- ○ **C.** bar graph
- ○ **D.** histogram

**27.** The Brenner family is on vacation in Colorado. They are driving north from the small town of Trinidad to the large city of Denver. The graph below shows the number of miles they drove each hour.

The speed limit changes from 25 miles per hour when they are in a small town to 55 miles per hour on the major highway in less populated areas.

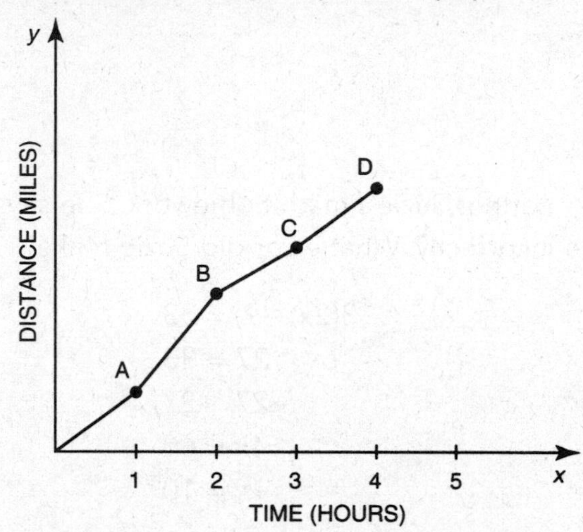


On which part of their trip did they drive the fastest?

- ○ **A.** from when they started to point A
- ○ **B.** from A to B
- ○ **C.** from B to C
- ○ **D.** from C to D

# END-OF-YEAR ASSESSMENT ANSWERS

## Session I No Calculator Permitted (pages 321–328)

**8.NS.1**

1. **B.** $0.8\overline{33}$ $\frac{5}{6}$ means $5 \div 6$; $5 \div 6 = 0.8\overline{33}$

**8.NS.2**

2. **A.** Yes **B.** No **C.** No **D.** No **E.** No
3. **D.** 9 $\sqrt{81} = 9$
4. **Part A:** *F*,5 $\sqrt{26}$ is very close to $\sqrt{25}$ and the $\sqrt{25} = 5$

   **Part B:** *K*,10 $\sqrt{99}$ is almost $\sqrt{100}$ and the $\sqrt{100} = 10$

   **Part C:** *D*,3 $\pi$ is about 3.14 which is approximately 3

   **Part D:** *F*,5 $\sqrt{4} + \sqrt{9} = 2 + 3 = 5$

   **Part E:** *I*,8 $2 \times \sqrt{16} = 2 \times 4 = 8$ exactly

**8.EE.1**

5. **A.** and **E.** There are two equivalent expressions. $\frac{2^{-6}}{2^{-4}} = \frac{2^4}{2^6} = \frac{1}{2^2} = \frac{1}{4}$
6. **B.** 6 $2(6 - 3)^2 - 3(2)^2 = 2\,(3)^2 - 3(4) = 2(9) - 12 = 18 - 12 = 6$

**8.EE.2**

7. **A.** No If $x^2 = 144$ then $x = 12$, because $12 \times 12 = 144$

   **B.** No If $x^3 = 8$ then $x = 2$, because $2^3 = 2 \times 2 \times 2 = 4 \times 2 = 8 = x$

   **C.** No If $x = 3$ then $x^3 \neq x^2 + 3$, because $3^3 \neq 3^2 + 3^1$, $3^3 = 27$ whereas $3^2 + 3^1 = 12$

   **D.** No If $x^3 = 27$, then $x = 3$ (note that $3 \times 3 \times 3 = 9 \times 3 = 27$)

   $x^3 + x^2 = 3^3 + 3^2 = 27 + 9 = 36$

   **E.** Yes If $x^3 = -8$, then $x = -2$, because

   $-2^3 = (-2) \times (-2) \times (-2) = (4) \times (-2) = -8$
8. 9 because $9^2 = 9 \times 9 = 81$
9. **D.** −27 because If $y = -3$ then $y^3 = (-3)(-3)(-3) = (9)(-3) = -27$
10. **A.** 6 $\sqrt[3]{8} + \sqrt[2]{16} = 2 + 4 = 6$

**8.EE.3**

11. ***C.*** 40 times larger $8 \times 10^9 = 8{,}000{,}000{,}000$ and $2 \times 10^8 = 200{,}000{,}000$

$$\frac{8{,}000{,}000{,}000}{200{,}000{,}000} = \frac{80}{2} = \frac{40}{1}$$

12. ***A.*** Nicole is correct.

The United States = $3 \times 10^8 = 300{,}000{,}000$; the world = $7 \times 10^9 = 7{,}000{,}000{,}000$

**70**~~00,000,000~~ ÷ **3**~~00,000,000~~ = $70/3 = 23.\overline{333}$

$23.\overline{333}$ is greater than 20 times more

**8.EE, 7b**

13. $w = -10$

$$\frac{3(w+4)}{3} = \frac{-18}{3} \quad \text{or} \quad 3(w+4) = -18$$

$$3w + 12 = -18$$
$$-12 \quad -12$$
$$w + 4 = -6$$
$$-4 \quad -4$$
$$w = -10$$
$$3w = -30$$
$$w = -10$$

14. $a = -3$

$$2(a-3) + 4a = -24 \quad \text{or} \quad \frac{2(a-3)}{2} + \frac{4a}{2} = \frac{-24}{2}$$

$$2a - 6 + 4a = -24 \qquad a - 3 + 2a = -12$$
$$6a - 6 = -24 \qquad 3a - 3 = -12$$
$$+6 \quad +6 \qquad +3 \quad +3$$
$$6a = -18; \qquad 3a = -9$$
$$a = -3 \qquad a = -3$$

**8.EE.4.1**

15. **D.** 12 $100 \times 5^{-2} + 2^3 = 100 \times \frac{1}{5^2} + 8 = \frac{100}{25} + 8 = 4 + 8 = 12$

**8.EE, 7b**

16. $y = \frac{3}{2}x - 4$

$3x + 6y = 4(3x) - 24$ *distribute the 4*

$3x + 6y = 12x - 24$ *combine like terms*

$-3x \quad -3x$

$\frac{6y}{6} = \frac{9x}{6} \frac{-24}{6}$ *divide each term by 6*

$y = \frac{3}{2}x - 4$

17. **D.** They are parallel and do not intersect.

    Compare $\mathbf{y = 3x + 9}$ to $3y = 9x - 18$

    $$3y = 9x - 18$$
    $$\frac{3y}{3} = \frac{9x}{3} - \frac{18}{3}$$

    *Set each equal to y and then compare them.* $\mathbf{y = 3x - 6}$ and $\mathbf{y = 3x + 9}$

    Since each line has a slope of **3** and different *y*-intercepts, you know the two lines are parallel; parallel lines do not intersect.

**8.EE.8b-1**

18. **D.** Because they do not have the same slope. The slope of one line is $\frac{2}{5}$ and the slope of the other is $\frac{3}{2}$.

**8.EE.8b-2**

19. **B.**

**8.EE.8b-3**

20. **D.** When you set each equation equal to *y* you see both have the same equation.

    $$2y = 3x + 10 \quad \textit{and} \quad 4y = 6x + 20$$

    *divide; set all = to y* $y = \frac{3}{2}x + 5$ *and* $y = \frac{3}{2}x + 5$

**8.F.1.1**

21. **A.** input = Joey Brown's name output = his birthday. For every input there is one and only one output if the relationship is considered a function.

**8.F.1-2**

22. **A.** No **B.** No **C.** No **D.** No **E.** Yes **F.** Yes

**8.F.3.2**

23. No These are non-collinear points. You cannot draw a straight line through all three of them.

24. **B.** is *not* a function.

    Note for B: when $x = 2$ there are two values for $y$.

**8.F.5.1**

25. **B.** decreasing It has a negative slope so it begins higher on the left side and gets lower on the right side.

**6.RP.1**

26. **A.** $\frac{2\,wings}{1\,beak} = \frac{14\,wings}{7\,beaks}$

**7.RP.1**

27. $\frac{1}{2} \div \frac{1}{4} = \frac{\frac{1}{2}}{\frac{1}{4}} = \frac{1}{2} \times \frac{4}{1} = \frac{4}{2} = 2\ miles\ per\ hour$

**8.G.1b**

28. **A.** *Reflect* (flip) **A** over the *y-axis,* then *translate* (slide) it down to quadrant IV. Remember, quadrant IV is the bottom right quadrant.

**8.G.3**

29. **D.** point (1, 1) Triangle *ABC* pivots around point (1, 1).

**8.SP.2**

30. **B.** graph B Notice how there are about the same number of data points above the line of best fit as there are below it.

# END-OF-YEAR ASSESSMENT ANSWERS

## Session II Calculator Permitted (pages 331-340)

**8.EE.4.2**

1. **A.** $1.358 \times 10^9 = 2(6.79 \times 10^8)$ = the scientific notation form for 679,000,000 multiplied by 2 for the *round-trip*.

**8.EE.5.1**

2. **D.** Animal B is moving faster. A steeper slope shows the animal moves faster (the vertical height is distance) for each second it is running (the horizontal movement on the graph relates to time).

3. **A.** Horse A ran the fastest. Looking at the graph for Horse A you see the slope is $\frac{1}{1}$. When you plot Horse B's equation you see the slope is $\frac{1}{3}$.

   Horse A's graph shows a steeper line.

   Horse A: For every 1-unit of time Horse A runs 1-unit of a mile.

   Horse B runs slower. For every 3-units of time Horse B runs 1-unit of a mile.

   Horse B is actually running 3 times slower than Horse A.

**8.EE.8c**

4. **A.** Yes the slope of the segment between points $A$ and $B$ is $\frac{5-2}{4-1} = \frac{3}{3} = 1$;

   the slope of the segment between points $A$ and $C$ $\frac{2-4}{1-3} = \frac{-2}{-2} = 1$

   since the segments of the line have the same slope, the three points are on the same line. They are collinear points.

**8.EE.C.Int.1**

5. **C.** $\frac{2}{3}$ *of a pound* $\quad \frac{3}{4}x = \frac{1}{2} \quad \left(\frac{4}{3}\right)\frac{3}{4}x = \frac{1}{2}\left(\frac{4}{3}\right)$ therefore

   $x = \frac{1}{2}\left(\frac{4}{3}\right)$ *or* $\frac{4}{6} = \frac{2}{3}$ *of a pound*

6. **D.** $2n + 8 = 16$

**8.F.2**

7. **B.** notice that the line intersects the *y*-coordinate at $\frac{3}{2}$ *or* $1\frac{1}{2}$ and it has a slope of $\frac{3(\textit{rise is } 3)}{2(\textit{run is } 2)}$

**8.F.4**

8. **Part A:** The plant was 2 inches tall before it was fed the new plant food

   **Part B:** $\frac{4}{2}$ or $\frac{2}{1}$ The slope of the line tells you the rate of growth. You can tell the slope just by counting the change in *rise* over the change in *run* from any two points on the line segment.

9. **Part A:** The rate of change is $\frac{3}{1}$.

   **Part B:** **B.** It rained 3 mm each hour.

**8.G.7.1**

10. 20 cm $16^2 + 12^2 = x^2$ Remember: $(\text{side})^2 + (\text{side})^2 = (\text{the hypotenuse})^2$

    $256 + 144 = x^2$

    $400 = x^2$

    Now take the square root of both sides; (20)(20) = 400, so $x = 20$

11. 7.8 $5^2 + 6^2 =$ the hypotenuse squared Let $c$ = hypotenuse $a^2 + b^2 = c^2$

    $25 + 36 = c^2$

    $61 = c^2$ Now take the square root of both sides. Since 61 is not a perfect square number, use your calculator to help. $\sqrt{61}$ *is about* 7.8102497

**8.G.8**

12. 9.4 $a^2 + b^2 = c^2$

    $8^2 + 5^2 = c^2$

    $64 + 25 = 89 = c^2$ The square root of 89 is about 9.4339811

**8.G.9**

13. **A.** The volume of the cone is the same as the volume of the cylinder.

    First, change all dimensions to inches so you can compare.

    Vol. cylinder: (Area of Base)(Height) = (Area Circle)(Height)

    **Vol. cylinder** $= \pi r^2 h = [(3.14)(3)(3)(18)] = 508.68$ cubic inches

    **Vol. cone** $= \frac{(\textit{Area Base})(\textit{height})}{3} = \left(\frac{1}{3}\right)(3.14)(3^2)(18) = 503.68$ cubic inches

    Remember to change all units to inches!

14. 58.9 cubic inches

$$\text{Volume of cone} = \frac{(Area\ of\ Base)(height)}{3} = \frac{\pi r^2 h}{3} = \frac{(3.14)(2.5)(2.5)(9)}{3}$$
$$= (3.14)(2.5)(2.5)(3) = 58.9 \text{ cubic inches}$$

**8.G.9**

15. 3,052 cubic inches or *3,054 cubic inches

$$\frac{(Area\ of\ Base)(height)}{3} = \frac{\pi r^2 h}{3} = \frac{(3.14)(2.5)(2.5)(9)}{3}$$ volume of the beach ball

(This information is on your PARCC Reference Sheet.)

$$\frac{4}{3}\pi 9^3 = \frac{4}{3}(3.14)(9)(9)(9) = \frac{4}{3}(3.14)(729) = 3{,}052.08$$

*If you used the π symbol on your calculator your answer would have been more exact. Your answer would have been 3,053.6281 or about 3,054.

**8.SP.3**

16. **C.** By looking at the *slope* of each line you can see how one is rising faster (it has a steeper slope); it is increasing in height faster than the plant with less extra sunlight.

17. **B.** The *y-intercept* in this situation is (0, 200). It means that Janie had $200 in her savings account in December before she started to add $100 each month.

**8.SP.4**

18.

**Part A:** 35 Carmen survey 35 people in all.

**Part B:** 2 12 people said they had 2 children.

**Part C:** 2 16 people said they had 2 bedrooms.

**Part D:** 40% 4 out of 10 = $\frac{2}{5}$ of the 3-bedroom homes had 2 children.

19. **C.**

**7.G.4**

20. 706.9 square units approximately if you used the π symbol on your calculator.

706.5 square units if you used 3.14 to represent π

(Note: Since the formula for circumference is $\pi d$, we know the *diameter* of the circle is 30 and therefore the *radius* of the circle is 15.)

The formula for area of a circle is $\pi r^2$ which here is $\pi 15^2$.

**7.G.4-6**

21.

**Part A: C.** The surface **area** of the cube is 96 square units

The surface area of one face of the cube is $4 \times 4 = 16$.

Since a cube has 6 faces (4 sides + top + bottom) just multiply
$6 \times 16 =$ **96 square units**

**Part B: A.** The surface area of the cube is 8 units larger than the *surface area* of the rectangular solid.

*Surface area* of the rectangular solid is:

Top + bottom $(2 \times 6 \times 4) = 48$
Side + side $(2 \times 2 \times 4) = 16$
Back + front $(2 \times 2 \times 6) = \underline{24}$
**Total = 88 sq. units**

*Surface Area* of Cube – Rectangular Solid = 96 – 88 =
**8 units larger**

22. **A.** The largest integer is the *surface area* of the cube.

The *surface area* of the cube is (6 sides)(3)(3) = (6)(9) = **54**

The *volume* of the rectangular prism is (3)(3)(4) = (9)(4) = **36**

The *volume* of the right cylinder is about $(3.14)(2^2)(4)$ is about **50.24**

or about **50.27** if you used the $\pi$ symbol instead of 3.14.

The *perimeter* of the isosceles triangle is 10 + 5 + 5 = **20**

**7.RP.3**

23. $106.50

$125 × .80 = $100 cost of jacket on sale; + tax = $100 + .065(100) = $106.50

**7.EE.3**

24. **D.** 7 inches 32 – 18 = 14 inches 14 ÷ 2 = 7 inches on each side of the rod.

**6.EE.2**

25. **D.** Suzie did not make any errors.

**6.S.24P/7.S.24P**

26. **A.** Scatter plot You could draw a *line of best fit* to estimate the average number of miles he walked each week.

27. **B.** From A to B The line segment has the steepest slope here.

# Grade 8 Common Core State Standards

**Standards** define what students should understand and be able to do.

**Clusters** summarize groups of related standards. Note that standards for different clusters may sometimes be closely related, because mathematics is a connected subject.

**Domains** are larger groups of related standards. Standards for different domains may sometimes be closely related.

See the sample below:

Grade 8: The Number System 8.NS (**The Domain**)

**Know that there are numbers that are not rational, and approximate them by rational numbers.**

CCSS.MATH.CONTENT.8.NS.A.1 (A **standard** within the domain)

Know that numbers that are not rational are called irrational. Understand informally that every number has a decimal expansion; for rational numbers show that the decimal expansion repeats eventually, and convert a decimal expansion which repeats eventually into a rational number.

CCSS.MATH.CONTENT.8.NS.A.2 (A **standard** within the domain)

Use rational approximations of irrational numbers to compare the size of irrational numbers, locate them approximately on a number line diagram, and estimate the value of expressions (e.g., $\sqrt{2}$). *For example, by truncating the decimal expansion of $\sqrt{2}$, show that $\sqrt{2}$ is between 1 and 2, then between 1.4 and 1.5, and explain how to continue on to get better approximations.*

Standards 8.NS.A.1 and 8.NS.A.2 make up a **Cluster**.

On the following pages you will find detailed descriptions of the CCSS Grade 8 Math standards. For additional information go to *http://www.corestandards.org*.

## 8.EE Expressions and Equations (Work with Radicals and Integer Exponents)

**A.** EE.A.1 Know and apply the properties of integer exponents to generate equivalent numerical expressions. For example, $3^2 \times 3^{-5} = 3^{-3} = 1/3^3 = 1/27$.

EE.A.2 Use square root and cube root symbols to represent solutions to equations of the form $x^2 = p$ and $x^3 = p$, where $p$ is a positive rational number. Evaluate square roots of small perfect squares and cube roots of small perfect cubes. Know that $\sqrt{2}$ is irrational.

EE.A.3 Use numbers expressed in the form of a single digit times an integer power of 10 to estimate very large or very small quantities, and to express how many times as much one is than the other. *For example, estimate the population of the United States as 3 times $10^8$ and the population of the world as 7 times $10^9$, and determine that the world population is more than 20 times larger.*

EE.A.4 Perform operations with numbers expressed in scientific notation, including problems where both decimal and scientific notation are used. Use scientific notation and choose units of appropriate size for measurements of very large or very small quantities (e.g., use millimeters per year for seafloor spreading). Interpret scientific notation that has been generated by technology.

**B. Understand the connections between proportional relationships, lines, and linear equations.**

EE.B.5 Graph proportional relationships, interpreting the unit rate as the slope of the graph. Compare two different proportional relationships represented in different ways. For example, compare a distance-time graph to a distance-time equation to determine which of two moving objects has greater speed.

EE.B.6 Use similar triangles to explain why the slope $m$ is the same between any two distinct points on a non-vertical line in the coordinate plane; derive the equation $y = mx$ for a line through the origin and the equation $y = mx + b$ for a line intercepting the vertical axis at $b$.

**C. Analyze and solve linear equations and pairs of simultaneous linear equations.**

EE.C.7 Solve linear equations in one variable.

EE.C.7a Give examples of linear equations in one variable with one solution, infinitely many solutions, or no solutions. Show which of these possibilities is the case by successively transforming the given equation into simpler forms, until an equivalent equation of the form $x = a$, $a = a$, or $a = b$ results (where $a$ and $b$ are different numbers).

EE.C.7b Solve linear equations with rational number coefficients, including equations whose solutions require expanding expressions using the distributive property and collecting like terms.

EE.C.8 Analyze and solve pairs of simultaneous linear equations.

EE.C.8a Understand that solutions to a system of two linear equations in two variables correspond to points of intersection of their graphs, because points of intersection satisfy both equations simultaneously.

EE.C.8b Solve systems of two linear equations in two variables algebraically, and estimate solutions by graphing the equations. Solve simple cases by inspection. *For example, $3x + 2y = 5$ and $3x + 2y = 6$ have no solution because $3x + 2y$ cannot simultaneously be 5 and 6.*

EE.C.8c Solve real-world and mathematical problems leading to two linear equations in two variables. *For example, given coordinates for two pairs of points, determine whether the line through the first pair of points intersects the line through the second pair.*

## 8.F Functions

### A. Define, evaluate, and compare functions.

F.A.1 Understand that a function is a rule that assigns to each input exactly one output. The graph of a function is the set of ordered pairs consisting of an input and the corresponding output.[1]

F.A.2 Compare properties of two functions each represented in a different way (algebraically, graphically, numerically in tables, or by verbal descriptions). *For example, given a linear function represented by a table of values and a linear function represented by an algebraic expression, determine which function has the greater rate of change.*

F.A.3 Interpret the equation $y = mx + b$ as defining a linear function, whose graph is a straight line; give examples of functions that are not linear. *For example, the function $A = s^2$ giving the area of a square as a function of its side length is not linear because its graph contains the points (1, 1), (2, 4), and (3, 9), which are not on a straight line.*

### B. Use functions to model relationships between quantities.

F.B.4 Construct a function to model a linear relationship between two quantities. Determine the rate of change and initial value of the function from a description of a relationship or from two $(x, y)$ values, including reading these from a table or from a graph. Interpret the rate of change and initial value of a linear function in terms of the situation it models, and in terms of its graph or a table of values.

F.B.5 Describe qualitatively the functional relationship between two quantities by analyzing a graph (e.g., where the function is increasing or decreasing, linear or nonlinear). Sketch a graph that exhibits the qualitative features of a function that has been described verbally.

## 8.G Geometry

**A. Understand congruence and similarity using physical models, transparencies, or geometry software.**

G.A.1 Verify experimentally the properties of rotations, reflections, and translations:

G.A.1a Lines are taken to lines, and line segments to line segments of the same length.

G.A.1b Angles are taken to angles of the same measure.

G.A.1c Parallel lines are taken to parallel lines.

G.A.2 Understand that a two-dimensional figure is congruent to another if the second can be obtained from the first by a sequence of rotations, reflections, and translations; given two congruent figures, describe a sequence that exhibits the congruence between them.

G.A.3 Describe the effect of dilations, translations, rotations, and reflections on two-dimensional figures using coordinates.

G.A.4 Understand that a two-dimensional figure is similar to another if the second can be obtained from the first by a sequence of rotations, reflections, translations, and dilations; given two similar two-dimensional figures, describe a sequence that exhibits the similarity between them.

G.A.5 Use informal arguments to establish facts about the angle sum and exterior angle of triangles, about the angles created when parallel lines are cut by a transversal, and the angle-angle criterion for similarity of triangles. *For example, arrange three copies of the same triangle so that the sum of the three angles appears to form a line, and give an argument in terms of transversals why this is so.*

**B. Understand and apply the Pythagorean Theorem.**

G.B.6 Explain a proof of the Pythagorean Theorem and its converse.

G.B.7 Apply the Pythagorean Theorem to determine unknown side lengths in right triangles in real-world and mathematical problems in two and three dimensions.

G.B.8 Apply the Pythagorean Theorem to find the distance between two points in a coordinate system.

**C. Solve real-world and mathematical problems involving volume of cylinders, cones, and spheres.**

G.C.9 Know the formulas for the volumes of cones, cylinders, and spheres and use them to solve real-world and mathematical problems.

## 8.SP Statistics and Probability

**A. Investigate patterns of association in bivariate data.**

SP.A.1 Construct and interpret scatter plots for bivariate measurement data to investigate patterns of association between two quantities. Describe patterns, such as clustering, outliers, positive or negative association, linear association, and nonlinear association.

SP.A.2 Know that straight lines are widely used to model relationships between two quantitative variables. For scatter plots that suggest a linear association, informally fit a straight line, and informally assess the model fit by judging the closeness of the data points to the line.

SP.A.3 Use the equation of a linear model to solve problems in the context of bivariate measurement data, interpreting the slope and intercept. *For example, in a linear model for a biology experiment, interpret a slope of 1.5 cm/hr as meaning that an additional hour of sunlight each day is associated with an additional 1.5 cm in mature plant height.*

SP.A.4 Understand that patterns of association can also be seen in bivariate categorical data by displaying frequencies and relative frequencies in a two-way table. Construct and interpret a two-way table summarizing data on two categorical variables collected from the same subjects. Use relative frequencies calculated for rows or columns to describe possible association between the two variables. *For example, collect data from students in your class on whether or not they have a curfew on school nights and whether or not they have assigned chores at home. Is there evidence that those who have a curfew also tend to have chores?*

# Mathematical Practices

APPENDIX B

These practices are the same for grades K-12.

The standards for Mathematical Practice describe the varieties of expertise that mathematics educators at all levels should seek to develop in their students.

1. Make sense of the problems and persevere in solving them.
2. Reason abstractly and quantitatively.
3. Construct viable arguments and critique the reasoning of others.
4. Model with mathematics.
5. Use appropriate tools strategically.
6. Attend to precision.
7. Look for and make use of structure.
8. Look for and express regularity in repeated reasoning.

# Additional Online Resources

Core Content Standards: *http://www.corestandards.org*

NJ Model Curriculum: Mathematics

*http://www.state.nj.us/education/modelcurriculum/math*

## PARCC Blueprints

The Blueprints list the standards that will be included on the PBA assessments and on the EOY assessments. They include Evidence Statements and Clarifications; the MP (math practices) emphasized and whether students will be able to use a calculator or not. Scroll down to select grade-level and click on PBA or EOY.

*http://www.parcconline.org/assessment-blueprints-test-specs*

## PARCC Accessibility Features and Accommodations Manual

*http://www.parcconline.org/parcc-accessibility-features-and-accommodations-manual*

PARCC Assessments: *http://www.parcconline.org*

Online Practice PARCC Sample Assessments: *http://practice.parcc.testnav.com/#*

PARCC Testing Schedule: *http://www.parcconline.org/new-jersey*

(Note: At this time, PARCC has just released online interactive Practice Assessments for all grade levels for the PBA in Mathematics. For more information, go to http://www.parcconline.org/math-pba-here

# PARCC Grade 8 Mathematics Assessment Reference Sheet

## Conversions

1 inch = 2.54 centimeters
1 meter = 39.37 inches
1 mile = 5,280 feet
1 mile = 1,760 yards
1 mile = 1.069 kilometers

1 kilometer = 0.62 miles
1 pound = 16 ounces
1 pound = .454 kilograms
1 kilogram = 2.2 pounds
1 ton = 2,000 pounds

1 cup = 8 fluid ounces
1 pint = 2 cups
1 quart = 2 pints
1 gallon = 4 quarts
1 gallon = 3.785 liters
1 liter = 0.264 gallon
1 liter = 1,000 cubic centimeters

## Formulas

| | |
|---|---|
| Triangle | $A = \frac{1}{2}bh$ |
| Parallelogram | $A = bh$ |
| Circle | $A = \pi r^2$ |
| Circle | $C = \pi d$ or $C = 2\pi r$ |
| General Prism | $V = Bh$ |
| Cylinder | $V = \pi r^2 h$ |
| Sphere | $V = \frac{4}{3}\pi r^3$ |
| Cone | $V = \frac{1}{3}\pi r^2 h$ |
| Pythagorean Theorem | $a^2 + b^2 = c^2$. |

# Put New Jersey Students on the Path to Success

## NEW JERSEY STATE ASSESSMENT TESTS

Preparing students for success is so important. That's why these newly revised study guides are the perfect tools to help students get ready for the PARCC (The Partnership for Assessment of Readiness for College and Careers) tests in the State of New Jersey. Through a series of high quality, computer-based K–12 assessments in Mathematics and English Language Arts, teachers, schools, and parents can rest assured that students are on track for success. Each title features:

- Two full-length practice tests with answers and explanations
- In-depth review through engaging lessons, hints, and tips for each PARCC test
- Content reflective of classroom lessons and aligned to Common Core Curriculum
- An explanation and overview of the PARCC assessment

Give students the help they need to achieve their very best—now and in the future.

**New Jersey Grade 3 Math Test**
*Thomas Walsh, M.A., M.Ed., Ed.D. and Dan Nale, M.Ed., MBA*
**ISBN 978-1-4380-0562-1, $12.99, *Can$14.99***

**New Jersey Grade 4 ELA/Literacy Test**
*Kelli Eppley, M.S.*
**ISBN 978-1-4380-0561-4, $12.99, *Can$14.99***

**New Jersey Grade 5 ELA/Literacy Test**
*Mark Riccardi and Kimberly Perillo*
**ISBN 978-1-4380-0560-7, $12.99, *Can$14.99***

Reflective of the NEW PARCC ASSESSMENT

**Each book: Paperback, approx. 216–288 pp., 7 13/16" x 10"**

Available at your local book store
or visit **www.barronseduc.com**

**Barron's Educational Series, Inc.**
250 Wireless Blvd.
Hauppauge, NY 11788
Order toll-free: 1-800-645-3476

**In Canada:** Georgetown Book Warehouse
34 Armstrong Ave.
Georgetown, Ont. L7G 4R9
Canadian orders: 1-800-247-7160

Prices subject to change without notice.

(#156) R1/15